Namibia's Fisheries:
Ecological, Economic and Social Aspects

NAMIBIA'S FISHERIES:
ECOLOGICAL, ECONOMIC AND SOCIAL ASPECTS

Edited by

Ussif Rashid Sumaila, David Boyer,
Morten D. Skogen and Stein Ivar Steinshamn

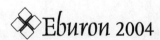 2004

ISBN 90 5972 017 2

Eburon Academic Publishers
P.O. Box 2867
2601 CW Delft
The Netherlands
phone: +31 (0)15 - 2131484 / fax: +31 (0)15 - 2146888
info@eburon.nl / www.eburon.nl

Cover design: Hermkens Design, Amsterdam

© 2004. All rights reserved. No part of this
publication may be reproduced, stored in a
retrieval system, or transmitted, in any form
or by any means, electronic, mechanical, photo-
copying, recording, or otherwise, without the
prior permission in writing from the proprietor(s).

Contents

Acknowledgements IX

Foreword XI
Dr. Abraham Iyambo
Hon. Minister of Fisheries and Marine Resources of Namibia

Namibia's fisheries: Introduction and overview 1
Ussif Rashid Sumaila, David Boyer, Morten D. Skogen and
Stein Ivar Steinshamn

PART 1   ECOLOGICAL ASPECTS

1. A direct estimate of the Namibian upwelling flux 11
   Morten D. Skogen

2. The effects of internal and external control on the northern
   Benguela ecosystem 29
   Johanna J. Heymans

3. Biodiversity of the Namibian Exclusive Economic Zone:
   A brief review with emphasis on online databases 53
   Maria Lourdes D. Palomares and Daniel Pauly

4. On Namibia's marine fish diversity 75
   Gabriella Bianchi, Elizabeth Lundsør and Hashali Hamukuaya

5. Reconstruction and interpretation of marine fisheries catches from
   Namibian waters, 1950 to 2000 99
   Nico E. Willemse and Daniel Pauly

6. Management regulations of Namibian angling fish species 113
   Johannes Andries Holtzhausen and Carola Heidrun Kirchner

7. Aggregation dynamics and behaviour of the Cape horse mackerel
   (*Trachurus trachurus capensis*) in the northern Benguela –
   implications for acoustic abundance estimation 135
   Bjørn Erik Axelsen, Jens-Otto Krakstad and Graça Bauleth-D'Almeida

PART 2   ECONOMIC ASPECTS

8. A brief overview of current bioeconomic studies of Namibian
   fisheries 165
   Ussif Rashid Sumaila and Stein Ivar Steinshamn

9. Economic value of fish stocks and the national wealth of Namibia....... 187
   Glenn-Marie Lange

10. Benefits and costs of the Namibianisation policy 203
    *Claire W. Armstrong, Ussif Rashid Sumaila, Anna Erastus
    and Orton Msiska*

11. Economic valuation of the recreational shore fishery:
    a comparison of techniques 215
    *Jonathan Barnes, Fredrik Zeybrandt, Carola Kirchner, Alison Sakko
    and James MacGregor*

12. The Namibian - South African hake fishery:
    Costs of non-cooperative management 231
    Claire W. Armstrong and Ussif Rashid Sumaila

13. A stochastic feedback model for optimal management of Namibian
    sardine 245
    Stein Ivar Steinshamn, Arne-Christian Lund and Leif K. Sandal

PART 3   SOCIAL AND INSTITUTIONAL ASPECTS

14. Institutional and industrial perspectives on fisheries management
    in Namibia 267
    Bendigt Maria Olsen

15. Against all odds: taking control of the Namibian fisheries 289
    Per Erik Bergh and Sandy Davies

16. Marine fisheries management in Namibia: Has it worked? 319
    Paul Nichols

17. Co-management: Namibia's experience with two large-scale
    industrial fisheries - sardine and orange roughy 333
    David Boyer and Burger Oelofsen

APPENDICES

Glossary of Acronyms 357

Contributors 361

Acknowledgements

I am grateful to Bjørn Hersoug for encouraging me to put together a team of editors and authors to do an edited book on Namibian fisheries. The idea was to start with participants in the research programme on Developing Country Fisheries of the Research Council of Norway (RCN), and then invite researchers and managers outside this group who are interested in Namibian fisheries to join the effort. I must also thank my co-editors for agreeing to join me, even though they knew that this was going to consume a lot of their time. The authors of the various chapters also deserve special thanks for helping to make the book a reality. The contributions of referees to scholarly work are invaluable – I thank the various persons who helped improve the quality of the book by reviewing the chapters contained therein. Other contributors to the effort in terms of proofreading, editing, word processing work, etc. are Gundula Weingartner, Chuck Hollingworth, Heather Keith, Vicky and William Cheung – I thank you all very much for your inputs. Mr. Maarten Fraanje and his team at Eburon (the Publishers), were patient with me during the course of this work, and I thank them for their patience. Some of my inputs to this book were carried out when I was at the Chr. Michelsen Institute, Bergen, Norway, and during my academic visits to the University of Namibia, St Francis Xavier University, the University of Tromsø and the University of West Indies at Cave Hill, Barbados. I thank colleagues at these institutions for their support during these visits. I wish to thank the research programme on Developing Country Fisheries not only for supporting some of the research reported in this book, but also for funding the publication of the book. Finally, I thank the Sea Around Us Project of the Fisheries Centre at the University of British Columbia, the Pew Charitable Trusts and Oceana for partially supporting my contribution to this book.

Ussif Rashid Sumaila
August 2004, Vancouver, Canada

Foreword

It is my pleasure and privilege to introduce this book. As its title suggests this volume provides much material on the ecological, economical and social aspects of Namibia's marine fisheries. It will certainly be of interest to a wide audience, from the general reader who is interested in how Namibia's fisheries sector has developed, to fisheries managers, researchers, students and investors. The book complements the growing body of documentation on Namibia's fisheries, but focuses on subject areas that have not been particularly well researched to date. It is thus both timely and to a large extent fills a gap.

The importance of Namibia's marine capture fisheries and fish processing sectors cannot be overemphasised. Largely as a result of upwelling of the nutrient-rich Benguela Current, the waters off Namibia's coast are highly productive. Prior to Namibian Independence in 1990, uncontrolled fishing on a massive scale by foreign fleets greatly reduced the abundance of all the major fish stocks. This period was followed by a dramatic recovery of the resources following Independence in 1990 and the implementation of a resource management system that incorporates a highly effective, cost-efficient system of monitoring, control and surveillance. Namibia's early post-Independence policy to put in place a system of research, control and conservation measures for the fisheries has yielded benefits, as described in the text. Today, we target more than 20 commercial species. In 2003 our fishing fleet of 335 vessels, 80% of which were Namibian flag, landed over 600,000 tonnes. For us, fishing and processing is a major contributor to the national economy as well as providing formal employment for around 14,000 Namibians.

In order to have optimal biological and economical sustainability of the resources, I am an ardent supporter of well-versed conservation and management measures based on best available scientific evidence and the principle of precautionary approach to inform management. With this approach, I am convinced that we can reap the sustainable benefits from our marine fish resources in perpetuity.

The Benguela Current creates a highly variable marine environment, characterised by ever fluctuating conditions that impact in various ways and intensities on our fish stocks. Some of these impacts are good – high primary productivity due to upwelling of nutrient water being one example. Others are not so good. The book evaluates these factors and provides some indications as to how our complex ecosystem functions. It also introduces the much-lauded ecosystem approach to fisheries management (EAF) - a significant and novel evolution of traditional 'single species' management. Although the implementation of EAF will be difficult, I am pleased that Namibia has made a start in this direction, despite being a developing country with limited financial, human and material resources.

Marine biodiversity and apparent regime shifts that have occurred during the past 50 years are described. Anomalous events such as sulphur eruptions, intrusions of warm water from the tropics, low dissolved oxygen plumes are naturally occurring features of the Benguela Current and can have, at times, negative impacts. These events are not yet fully understood nor can they be predicted with any reasonable level of confidence. The implications of these for fishery management are elaborated in the text.

The Namibianisation policy has opened the doors to hundreds of previously disadvantaged citizens who today constitute the majority of fishing rights holders. Rapid entry of newcomers to the industry has brought with it challenges upon the incumbents in terms of mobilising and managing resources, in a new business environment. Readers will find informative analyses and review of this policy. The Book describes the "dark" history of the fisheries starting from when commercial resources were in pristine states and then plundered by foreign fleet before the Independence of Namibia in 1990. It further elucidates the challenges faced by the government in rebuilding the depleted stocks. This is the first book of its kind where bio-economic analyses are incorporated to inform management. The relatively new concept of 'environmental accounting' is applied in the Namibian context, and valuable insights given on the impacts of macro economic policies (taxes, interest rates) on Namibia's marine fisheries sector.

The book also explores the possibilities for co-management of fisheries, the role of stakeholders and highlights the need for regional collaboration with our neighbours if we are to successfully research, manage and sustainably harvest trans-boundary stocks.

I was informed by the lead editor, Dr Ussif Rashid Sumaila, that each manuscript underwent a peer review process, and that final acceptance was the prerogative of the editors. It behoves me to point out, however, that it has taken me some time to write this Foreword. The reason for this, is that when I first read through the chapters, I was obliged to point out to Dr Sumaila that I had a few, let me say 'reservations' about some of the statements made and conclusions drawn by some of the authors. Sumaila, working with the relevant authors, did his best to incorporate some of my most compelling comments. In the end we agreed to make it clear that my writing this Foreword does not imply that I agree totally with all that is said in the book. In fact there are views expressed herein that I strongly disagree with – I challenge the reader to try and work out which parts of which chapters perturb me! But being a strong believer in academic freedom, and due to the fact that I believe that the book contains material that will stimulate debate and potentially provide some very useful insights for fisheries management in Namibia, I am very happy indeed to introduce it.

Dr Abraham Iyambo
Minister for Fisheries and Marine Resources
Republic of Namibia

NAMIBIA'S FISHERIES: INTRODUCTION AND OVERVIEW

Ussif Rashid Sumaila, David Boyer, Morten D. Skogen and Stein Ivar Steinshamn

Namibia's fisheries have received and continue to receive a lot of attention from fisheries scholars. This is partly because these are some of the few relatively well-managed fisheries in the world. This book seeks to provide analysis and discussions of various aspects of Namibia's fisheries, even though to cover all aspects of the fisheries would be a daunting task beyond the scope of a single book. Therefore, the book does not attempt this, rather it presents specific examples to demonstrate some of the achievements and the challenges that Namibian fisheries have faced and continue to face. Many of these challenges are generic to other fisheries worldwide, and hence it is hoped that others may learn both from the successes and failures illustrated in this book.

Why produce a book about Namibia's fisheries? Namibia's fisheries are in many respects similar to the many large-scale modern fisheries to be found in most regions of the world. As has often been stated in fisheries and scientific circles (and recently in the public domain), many of these fisheries are over-capitalised and the fish stocks fully- or over-exploited (see for example, Pauly *et al.*, 2002, Myers and Worm, 2003). Namibia's fisheries provide a relatively more encouraging story.

BACKGROUND TO NAMIBIAN FISHERIES

Marine fishing in Namibia is a relatively young industry. Colonisation by Europeans occurred rather late by African standards, in the late 19th and early 20th century, largely due to an arid and hostile natural landscape that appeared to have little worth colonising. Prior to this, few indigenous people visited the desolate coastline and although there are fascinating accounts of how these small bands of nomads subsisted on shellfish and shallow water fish, their harvests can be measured in kilograms rather than tonnes. European and North American sailors had plundered Namibia's rich seal and guano resources from several small offshore islands, and whales in the open waters. But again, these were rather isolated events that, despite the devastating impact on the birds and mammals harvested, had a limited impact on marine ecosystems.

Even after the colonists arrived, fishing largely remained a subsistence activity. The coastline offers little shelter, and the riches that were to be harvested later in the century remained undiscovered. Between the two World Wars a small snoek fishery was attempted, apparently with some spectacular results. Not least of these were some large catches of sardine; a portent of the riches hidden below the waves. The 2nd World War put an end to these activities, but on the resumption of peace a new type of fisher arrived along the coast of south western Africa employing all the technological advances of the day. Large mechanised vessels, hydraulic winches hauling synthetic nets and acoustic fish-finding equipment all entered the fishery in the 1950s and 1960s to enable fishers to harvest ever greater catches. By the early 1950s sardine catches had risen to around a quarter of a million tonnes, and by the late 1960s this fishery became one of the largest the world had ever seen, or is likely to. Officially reported landings peaking at 1.4 million tonnes, but with discards and illegal landings included, catches probably exceeded 2 million tonnes. This led to one of the most spectacular crashes witnessed within fisheries as catches declined to just 300 000 tonnes three years later and despite a recovery in the mid-1970s, to a catch of just 12 000 tonnes in 1980.

Throughout the 1960s, as the fish stocks in their own waters were becoming depleted, a number of deepwater fleets from European and the Eastern Bloc countries sailed to Namibian waters, targeting first the abundant horse mackerel stock and then hake. It is estimated that by independence in 1990, 20 million tonnes of these valuable fish had been caught in Namibian waters by foreign fleets, with hardly any benefit accruing to Namibians and the Namibian nation (Bonfil et al., 1998; Sumaila and Vasconcellos, 2000).

Independence was a watershed for fisheries in Namibia, as indeed it was for all facets of life in Namibia. Finally, the control of this once valuable resource was vested in Namibians, and a vigorous policy to enable the recovery of the resources, and in particular the prosperity of the industry, to previous levels was implemented. The current Namibian fisheries management objective is similar to that of most industrial fishing nations: to utilise the living marine resources on a sustainable basis for the benefit of the nation, and to manage these fisheries based on scientific information and principles. However, unlike many nations, Namibia has a fisheries management system that incorporates many of the accepted best-practices as outlined in the major international fisheries conventions. Aspects of this are described in many of the Chapters that follow (for example, Nichols, Holtzhausen and Kirchner).

The Namibian fishing sector operates under circumstances that are somewhat unusual and, in combination, increase the likelihood of successful sustainable utilisation. Namibian fisheries management started with an almost clean slate in 1990 at the country's independence. The authorities were able to implement a completely new fisheries management system, with few of the historical, cultural, social and political encumbrances that new policies so often hold. The Namibian constitution provides for sustainable utilisation of natural resources, and that these resources should be managed according to the best-available scientific advice for the benefits of both current and future generations. As the fisheries sector is one of the most important economic sectors in Namibia, politicians placed great importance on 'getting it right'. The leaders of the country were politically very strong and therefore able to take unpopular but necessary decisions, and plan beyond the timeframe of the next elections. In addition, the industry is relatively simple to manage with few large companies and no artisanal fishers.

The Benguela upwelling system supports some of the highest productivity in the world, and this tends to lead to a relatively simple ecosystem with each species having relatively simple interactions with other components of the system (Boyer and Hampton, 2001). Hence, the biological complexities that often seem to defy management in many marine systems are somewhat simplified and allow for enhanced understanding.

Environmental conditions of Namibia's two maritime borders (the Lüderitz Upwelling Cell in the south and Angola-Benguela Front in the north) form natural barriers limiting the migration of many fish stocks, especially pelagic species. Thus, many of Namibia's fish stocks are either not shared or only shared to a limited extent. Once again, so many of the political complications introduced through stocks being managed by several nations are, to a large extent, avoided in Namibia. Similarly, most fish stocks occur within 100 nautical miles or so of the coast, hence being far from the limits of the EEZ and thus are not straddling stocks.

Namibia only has two harbours and therefore landings are relatively easy to monitor. Compare this with the nations of the North Sea, where not only does each of the eight bordering nations have numerous landing places, but vessels are able to offload in other countries, far away from the watchful eyes of the controlling authorities.

In addition, Namibia's geographic neighbours, Angola and South Africa, have been supportive and co-operative, providing assistance in the apprehension of illegal fishers, sharing data, etc. This is further evidenced by the various international and regional agreements that have been reached between these and other countries (SADC Fisheries Protocol, SEAFO, the BENEFIT Research Programme and BCLME Programme).

As a result of the above factors, many of the confounding factors that cloud the successes and failures of management regimes are clarified, and while in many other nations, failure to successfully manage a fishery can be blamed on many different players, often each blaming the other without accepting responsibility, in Namibia this is less easy to do.

Due to good management practices and fortunate circumstances, Namibia has developed a fisheries management system that, according to our perceptions of fisheries management, includes many of the recognised 'best practices' (Sainsbury and Sumaila, 2003). Namibia's fisheries are largely found in the northern Benguela ecosystem; a system that has many characteristics that are conducive to successful management and control of fisheries (see Part 1 of book). This, added to some fortuitous economic, social and political circumstances have enabled a management system to be implemented that has resulted in fisheries becoming one of the economic success stories of this developing country. But Namibia's fisheries management has not been entirely successful. For example, the sardine stocks are as depleted as ever, despite an extremely conservative management policy, while the newly developed orange roughy fishery blossomed and collapsed in just four short years, also in the face of apparently responsible and conservative management. The control of the fishery is to a large extent now vested in Namibian hands, but distribution of fisheries benefits is still problematic, even though the Namibianization programme of the government is helping to deal with the problem.

BOOK CONTENT

The book does not attempt to provide a comprehensive account of all aspects of Namibian fisheries. In fact, this book is meant to be complimentary to an earlier volume, namely, 'A Decade of Namibian Fisheries Science', Volume 23 (2001) of the *South African Journal of Marine Science*. This Volume covered in more detail the physical and biological processes that underlie the productivity of the Namibian marine ecosystem. The contribution of the current book lies in its attempt to bring together the economic, social and biological aspects of the fisheries of Namibia in one volume.

The book opens with a Foreword written by the Honourable Minister of Fisheries and Marine Resources of Namibia. While many such books are usually prefaced by the leading politician in the field in a country, Namibia is, once again, unique here. Minister Iyambo has not only led the development of fisheries policy and implemented many of the management policies described within this book, he is also a fisheries scientist and academic by training who cares and supports academic work even when some of the findings of the research are critical of his Ministry. This attitude of the Minister is truly a gift to Namibia, as it allows new ideas that benefit the country to flourish.

The main body of the book is organized into three Parts. Part 1 describes some of the critical ecological aspects; from the physical functioning of the system through to the biodiversity, and includes two chapters dealing with more traditional fisheries biology and the application of this science to fisheries management.

The second Part deals with economic aspects of Namibia's fisheries, but, unlike traditional economic treatises, these chapters emphasise the need to consider economics holistically, in terms of the biological, social and political context. The economic value of Namibian fish stocks to the nation is described, together with a detailed economic study of the Namibian linefishery. The benefits and costs of the government's policy to enable Namibians to fully participate in the fishery are also studied. The study of the economic benefits of Namibia and South Africa cooperating in managing the shared hake stocks of the Benguela marine ecosystem was the subject of one of the chapters.

The final section of the book addresses some social and institutional aspects of Namibian fisheries. Some chapters describe Namibia's attempts at involving the industry in management decisions, while other chapters describe some of the innovative management, control and surveillance structures and policies that have been implemented to enable Namibia to claim back and maintain control of its fisheries.

THE FUTURE

Fisheries, globally, have in recent years gained a reputation for mismanagement. The environmental, economic and social disasters that have accompanied this mismanagement have been highlighted in the academic literature, and all too graphically in the world's media. In contrast, Namibian fisheries have often been lauded as the almost unique success story amongst all the doom and gloom. Commentators often point out that if Namibia, a developing African country, can manage their fish stocks rationally and sustainably, and with increasing benefits to their own people, why can't more developed countries do the same.

It is tempting to suggest that Namibian managers should continue with their current policies. However, as described within this book, and elsewhere, while many aspects of Namibian fisheries have been hugely successful, others have not. Clearly some policies need to be reviewed to address remaining shortcomings.

External factors are also changing. Climate change may affect the biological functioning of the Benguela marine ecosystem, and declining fish stocks elsewhere will encourage foreign vessels to attempt to enter Namibian fisheries, both legally and illegally. An increasing population within Namibia, together with debilitating rates of AIDS/HIV infection, will pose further challenges that will have to be met by the managers of Namibia's economic sectors, including the fisheries sector. Also, it appears that there is still some scope for increasing the potential of Namibia's fisheries by undertaking some restoration efforts. All of these potential challenges and opportunities need close attention from policy makers and managers. It will require continued innovation in the design and implementation of fisheries polices that will ensure that fisheries benefits continue to flow to both current and future generations of Namibians.

To this end, this book provides insights that would be useful to fisheries managers in their effort to continue developing innovative approaches for the effective and efficient management of Namibia's fisheries.

The need for ecosystem-based management of Namibia's marine fisheries was expressed variously in the ecological section of the book (Axelsen et al., Heymans, Skogen, Willemse and Pauly). The importance of marine biodiversity in maintaining ecosystem structure and functioning is not yet understood, and simple diversity indicators that may be useful for fisheries management are not yet available. While theoretical work progresses, it is recommended that basic data of fish diversity are collected as part of standard monitoring resources surveys (Bianchi et al.; Palomares and Pauly).

The need for more economic studies of Namibian fisheries to support management was discussed in the chapter by Sumaila and Steinshamn. Lange stressed the need to develop fisheries accounts that estimate the economic value of stocks, the economic loss incurred through over-exploitation and depletion of stocks, and the share of resource rent recovered by quota fees. The need for transboundary management of the resources of the Benguela marine ecosystems was highlighted in Armstrong and Sumaila. While the economic importance of the recreational angling fishery was the subject of the chapter by Barnes et al. The authors argue that there is a consumer surplus in the fishery that can be captured. According to the findings in Steinshamn et al. drastic measures, which are beginning to be implemented, are needed to help bring back large quantities of the sardine stock. Finally, the Namibianization policy implemented after independence seems to have opened up the fisheries sector to previously disadvantaged Namibians, and therefore has justified its costs. To make sure that this policy continues to meet its objectives, it needs to be assessed and revised from time to time (Armstrong et al.).

Finally but not the least, the social and institutional section of the book contains a number of useful insights for fisheries management. From the chapter by Olsen, for instance, we learn that the institutional arrangements for consultation between the fishing industry and the Ministry of Fisheries and Marine Resources need to be improved. And, according to the work in Bergh and Davies, annual compliance targets need to be set, monitored and analysed, and tools to promote voluntary compliance among fishers need to be adopted to improve monitoring, control and surveillance of Namibian fisheries. Boyer and Oelofsen make the point that incorporating co-management into management process can be beneficial, but such a development needs to be implemented with caution.

REFERENCES

Armstrong, C.W., U.R. Sumaila, A. Erastus, and O. Msiska (2004): Benefits and costs of the Nambianisation policy. In: *Namibia's Fisheries: Ecological, Economic and Social Aspects* (U.R. Sumaila, D. Boyer, M. Skogen, and S.I. Steinshamn, eds.), pp. 203-214. Eburon, Delft.

Armstrong, C.W. and U.R. Sumaila (2004): The Namibian-South African hake fishery: Costs of non-cooperative management. In: *Namibia's Fisheries: Ecological, Economic and Social Aspects* (U.R. Sumaila, D. Boyer, M. Skogen and S.I. Steinshamn, eds.), pp. 231-244. Eburon, Delft.

Axelsen, B.E., J.O. Krakstad, and G. Bauleth-D'Almeida (2004): Aggregation dynamics and behaviour of the Cape horse mackerel in the Namibian Benguela – implications for acoustic abundance estimation. In: *Namibia's Fisheries: Ecological, Economic and Social Aspects* (U.R. Sumaila, D. Boyer, M. Skogen and S.I. Steinshamn, eds.), pp. 135-164. Eburon, Delft.

Barnes, J., F. Zeybrandt, C. Kirchner, A. Sakko, and J. MacGregor (2004): Economic valuation of the recreational shore fishery: a comparison of techniques. In: *Namibia's Fisheries: Ecological, Economic and Social Aspects* (U.R. Sumaila, D. Boyer, M. Skogen and S.I. Steinshamn, eds.), pp. 215-230. Eburon, Delft.

Bergh, P.E. and S. Davies (2004): Against all odds: Taking control of the Namibian Fisheries. In: *Namibia's Fisheries: Ecological, Economic and Social Aspects* (U.R. Sumaila, D. Boyer, M. Skogen and S.I. Steinshamn, eds.), pp. 289-318. Eburon, Delft.

Bianchi, G., E. Lundsør, and H. Hamukuaya (2004): On Namibia's marine fish diversity. In: *Namibia's Fisheries: Ecological, Economic and Social Aspects* (U.R. Sumaila, D. Boyer, M. Skogen and S.I. Steinshamn, eds.), pp. 75-98. Eburon, Delft.

Bonfil, R. (1998): Case study: distant water fleets off Namibia. In: Distant water fleets: an ecological, economical and social assessment (R. Bonfil, G. Munro, U.R. Sumaila, H. Valtysson, M. Wright, T. Pitcher, D. Preikshot, N. Haggan and D. Pauly. Eds.), Fisheries Centre Research Report 6(6). University of British Columbia. Vancouver, Canada. 77 pp.

Boyer, D.C. and I. Hampton (2001): An overview of the living marine resources of Namibia. In: *A decade of Namibian Fisheries Science. South African Journal of Marine Science* 23: 5-35.

Boyer, D. and B. Oelofsen (2004): Co-management: Namibia's experience with two large-scale industrial fisheries - sardine and orange roughy. In: *Namibia's Fisheries: Ecological, Economic and Social Aspects* (U.R. Sumaila, D. Boyer, M. Skogen and S.I. Steinshamn, eds.), pp. 333-356. Eburon, Delft.

Heymans, J.J. (2004): The effects of internal and external control on the northern Benguela ecosystem. In: *Namibia's Fisheries: Ecological, Economic and Social Aspects* (U.R. Sumaila, D. Boyer, M. Skogen and S.I. Steinshamn, eds.), pp. 29-52. Eburon, Delft.

Holtzhausen, J.A. and C.H. Kirchner, (2004): Management regulations for Namibian angling fish species. In: *Namibia's Fisheries: Ecological, Economic and Social Aspects* (U.R. Sumaila, D. Boyer, M. Skogen and S.I. Steinshamn, eds.), pp. 113-134. Eburon, Delft.

Lange, G.M. (2004): Economic value of fish stocks and the national wealth of Namibia. In: *Namibia's Fisheries: Ecological, Economic and Social Aspects* (U.R. Sumaila, D. Boyer, M. Skogen and S.I. Steinshamn, eds.), pp. 165-186. Eburon, Delft.

Myers, R. A. & B. Worm (2003): Rapid worldwide depletion of predatory fish communities. *Nature*, 423, 280-283.

Nichols, P. (2004): Marine fisheries management in Namibia: Has it worked? In: *Namibia's Fisheries: Ecological, Economic and Social Aspects* (U.R. Sumaila, D. Boyer, M. Skogen and S.I. Steinshamn, eds.), pp. 319-332. Eburon, Delft.

Olsen, B.M. (2004): Institutional and industrial perspectives on fisheries management in Namibia. In: *Namibia's Fisheries: Ecological, Economic and Social Aspects* (U.R. Sumaila, D. Boyer, M. Skogen and S.I. Steinshamn, eds.), pp. 267-288. Eburon, Delft.

Palomares, M.L.D. and D. Pauly (2004): Biodiversity of the Namibian Exclusive Zone: A brief review with emphasis on online databases. In: *Namibia's Fisheries: Ecological, Economic and Social Aspects* (U.R. Sumaila, D. Boyer, M. Skogen and S.I. Steinshamn, eds.), pp. 53-74. Eburon, Delft.

Pauly, D., V. Christensen, S. Guenette, T.J. Pitcher, U.R. Sumaila, C.J. Walters, R. Watson and D. Zeller (2002): Towards sustainability in world fisheries. *Nature*, 418, 689-695.

Payne, A.I.L., S.C. Pillar and R.J.M. Crawford (eds.) (2001): *A Decade of Namibian fisheries science. South African Journal of Marine Science* 23. 466 pp.

Sainsbury, K. and U.R. Sumaila (2003): Incorporating ecosystem objectives into management of sustainable marine fisheries, including 'best practice' reference points and use of Marine Protected Areas. In: Sinclair, M. and G. Valdimarson (eds.): *Responsible Fisheries in the Marine Ecosystem*. CAB International, pp. 343-361.

Skogen, M.D. (2004): A direct estimate of the Namibian upwelling flux. In: *Namibia's Fisheries: Ecological, Economic and Social Aspects* (U.R. Sumaila, D. Boyer, M. Skogen and S.I. Steinshamn, eds.), pp. 11-28. Eburon, Delft.

Steinshamn, S.I, A-C. Lund, and L. Sandal (2004): A stochastic feedback model for optimal management of Namibian sardine. In: *Namibia's Fisheries: Ecological, Economic and Social Aspects* (U.R. Sumaila, D. Boyer, M. Skogen and S.I. Steinshamn, eds.), pp. 245-266. Eburon, Delft.

Sumaila, U.R. and M. Vasconcello (2000): Simulation of ecological and economic impacts of distant water fleets on Namibian fisheries. *Ecological Economics*, 32, 457-464.

Sumaila, U.R. and S.I. Steinshamn (2004): A brief overview of current bioeconomic studies of Namibian fisheries. In: *Namibia's Fisheries: Ecological, Economic and Social Aspects* (U.R. Sumaila, D. Boyer, M. Skogen and S.I. Steinshamn, eds.), pp. 165-186. Eburon, Delft.

Willemse, N.E. and D. Pauly (2004) Reconstruction and interpretation of marine fisheries catches from Namibian waters, 1950 to 2000. In: *Namibia's Fisheries: Ecological, Economic and Social Aspects* (U.R. Sumaila, D. Boyer, M. Skogen and S.I. Steinshamn, eds.), pp. 99-112. Eburon, Delft.

1   A DIRECT ESTIMATE OF THE NAMIBIAN UPWELLING FLUX

*Morten D. Skogen**

Abstract

The coastal upwelling regions associated with the four Eastern Boundary Currents of the world are important because of their high productivity, and with respect to primary productivity the Benguela system is assumed to be the most productive of these. The high biological productivity is made possible by the presence of cold, nutrient-rich water in the surface layer as a result of wind-forced coastal upwelling. To quantify this upwelling off Namibia, a biophysical model is used to examine the intra- and interannual variability, spatial extent and intensity of the upwelling outside Namibia over a 14 year period (1983 to 1997). Based on the modelled vertical velocity field, the annual (all years) mean vertical volume flux off Namibia is estimated to be 2.2 Sverdrup.

INTRODUCTION

The Benguela region, which is generally considered to be the part of the southeast Atlantic lying between 14°S and 37°S, is one of four major Eastern Boundary Currents (EBC) regions of the world. An overview of the principal characteristics of the Benguela system can be found in Shannon and Nelson (1996). The coastal upwelling regions associated with the four EBCs (the Canary, Benguela, California and Peru-Humboldt Currents) are important because of their high productivity. Although their area is only 0.1% of the world ocean, they account for 5% of global primary production and 17% of global fish catch (Pauly and Christensen, 1995). This high biological productivity is made possible by the presence of cold, nutrient-rich water in the illuminated surface layer as a result of equatorward wind-forced coastal upwelling. By using satellite-based chlorophyll images (SeaWiFS), Carr (2002) suggests the Benguela system to be the most productive EBC with respect to

---

* This work was supported by the Norwegian Science Foundation under grant 119622/730.

potential productivity both for the areal mean (gC/m²/day) and total primary production (gC/year), while the Peru-Humboldt system is by far the most productive one with respect to landings of small pelagics such as anchovies, sardines and herrings (FAO statistics database).

Upwelling is a dominating feature in the whole Benguela system between Cape Frio (18°S) and Cape Point (34°S) (Boyd, 1987). The upwelling also extends along much of the south coast as far as 25°E (Schumann et al., 1982). The wind and pressure field, topographic features and orientation of the coast result in the formation of a number of upwelling cells. These upwelling cells are normally located near regions of cyclonic wind stress curl, and are in most cases in regions where there is a change in orientation of coastline. Lutjeharms and Meeuwis (1987) distinguish eight different cells (Kunene, Namibia (northern Namibia), Walvis Bay (central Namibia), Lüderitz, Namaqua, Columbine, Peninsula and the Agulhas cells), whereas Shannon and Nelson (1996) have included three more along the south coast (Plettenberg, St. Francis and Recife). The typical seaward extent of the influence of the upwelling, excluding major filaments, is between 150 and 250 km for the cells on the west coast, and with respect to seaward extent, sea surface temperature and the frequency of upwelling the Lüderitz cell has been identified as the most intense one (Lutjeharms and Meeuwis, 1987). This cell effectively divides the Benguela into northern and southern sections. The cells south of Lüderitz are less intense, but of fairly equal intensity, whereas the cell intensity to the north decreases equatorward.

The strong upwelling, which displays intra-, interannual and decadal variability, gives a high primary productivity, and supports large commercial fisheries. Variations in fish recruitment, growth and distribution is largely dependent upon variations in the environmental conditions (Lasker, 1985; Cochrane & Hutchings, 1995), and one of the main tasks in marine science is to understand these relationships. A complicating and limiting factor in such studies is data availability. The number of data points from field observations is usually limited, and their time and spatial coverage are sparse. One approach has therefore been the use of simple proxies or indices to explain observed variations.

Owing to the lack of direct indices of upwelling strength and phytoplankton production, various kinds of indirect upwelling indices have been developed as proxies. The use of Sea Surface Temperature (SST) images (Cole and McGlade, 1998; Cole, 1999), sea level rise (Waldron et al., 1997a, b) and modelled primary production (Skogen, 2004) are such examples. Given that upwelling intensity is proportional to the rate surface water is carried offshore, Ekman transport has also often been used as an upwelling index (e.g. Bakun and Parrish, 1990; Johnson and Nelson, 1999).

The present work is motivated by the fact that upwelling indices already to some extent have been used to explain biological variability in the Benguela upwelling system. The key question is how to quantify upwelling, and in the present work it is proposed to use a wind and density driven coupled physical, chemical and biological ocean model (Skogen, 1999). The reason for this is that a full three-dimensional numerical model with realistic forcing is an inexpensive and flexible tool that can calculate variables and relationships with high resolution in both space and time. The focus has been on the northern Benguela (Namibia), but the method is general and can be extended to other areas of interest. After a short model description, several upwelling indices are proposed and compared. The temporal and spatial distribution of the upwelling is then investigated before a gross estimate of the total amount of upwelled water outside Namibia is given. It should be noted that there is large presence of mesoscale structures in the Namibian upwelling. The first internal baroclinic Rossby radius is likely to be of the order of, or smaller, than, 25 km, thus these structures are not fully accounted for by the present modelling approach (20 km horizontal resolution).

Material and methods
The NORWegian ECOlogical Model system (NORWECOM) is a coupled physical, chemical and biological model applied to study primary production and dispersion of particles such as fish larvae or pollution. The model is fully described in Skogen (1993) (see also Aksnes et al., 1995; Skogen et al., 1995). The model has been validated by comparison with field data in the North Sea/Skagerrak and Norwegian Sea (North East Atlantic) in Svendsen et al. (1995, 1996), Berntsen et al. (1996), and Skogen et al. (1997). The Benguela implementation has been validated in Skogen (1999).

The circulation model is based on the three-dimensional, primitive equation, time-dependent, wind and density driven Princeton Ocean Model (POM) (Blumberg and Mellor, 1987). The prognostic variables of this model are: three components of the velocity field, temperature, salinity, turbulent kinetic energy, a turbulent macroscale and the water level. The governing equations of the model are the horizontal momentum equations, the hydrostatic approximation, the continuity equation, conservation equations for temperature and salinity and a turbulence closure model for calculating the two turbulence variables (Mellor and Yamada, 1982). The equations and boundary conditions are approximated by finite difference techniques in an Arakawa C-grid (Mesinger and Arakawa, 1976).

For the present study, the model is used with a horizontal resolution of 20x20 km on a grid covering the African coast and the open ocean from about 12-46°S and 4-30°E (Figure 1). The bottom topography is interpolated

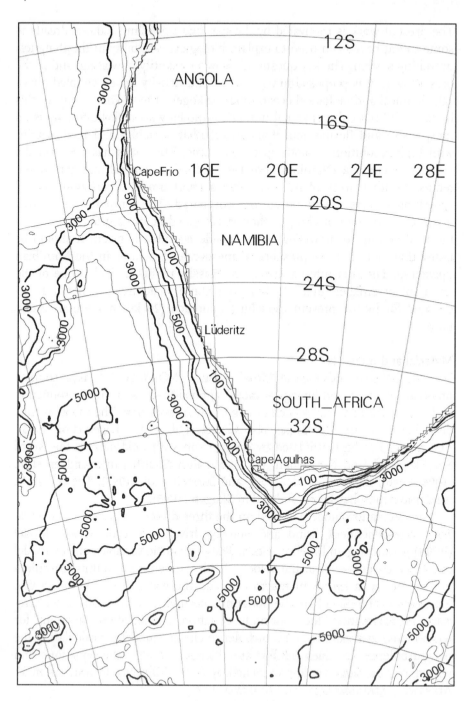

Figure 1. Model domain with bottom topography

and extrapolated from data from the South African Naval Hydrographic Office, and from the GEBCO Digital Atlas (IOC, IHO and BODC, 1994). In the vertical dimension, 18 bottom-following sigma layers are used.

Forcing variables are 12-hourly hindcast atmospheric pressure fields and wind stress fields, interpolated from the NCEP/NCAR Reanalysis Project (Kalnay et al., 1996). Initial values (monthly fields) for salinity and temperature are interpolated from the 1° x 1° NOAA atlas (Levitus et al., 1994; Levitus and Boyer, 1994), whereas the initial fields for velocities and water elevation are derived from the density fields using the thermal wind equation and assuming zero net flux in each water column. Interpolations between these monthly fields are also used at all open boundaries. To absorb inconsistencies between the forced boundary conditions and the model results, a 7 grid-cell "Flow Relaxation Scheme" (FRS) zone (Martinsen & Engedahl, 1987) is used around these open boundaries. In the zone, each prognostic variable, φ, is simply updated by the translation $\varphi = (1-\beta)\, \varphi_{int} + \beta\, \varphi_{ext}$, where φ_{int} contains the time-integrated, unrelaxed values calculated in the entire model domain, i.e. also in the areas covered by the FRS-zone, and φ_{ext} is the specified external solution in the zone. φ is the new value and β a relaxation parameter that varies from 0, at the end of the zone facing the interior model domain, to 1 at the outer end of the zone.

To account for a lack of appropriate data on the surface heat fluxes, a "relaxation toward climatology" method is used for the surface layer (Cox and Bryan, 1984). For this purpose, monthly climatological fine-scale satellite-derived SST fields from the CORSA project at the Joint Research Centre in Ispra, Italy (*http://me-www.jrc.it/CORSA*) are used. During calm wind conditions, the surface temperature field adjusts to the climatological values after about 10 days (Oey and Chen, 1992). In deep water (below 500 meters), a weak relaxation to climatological salinity and temperature fields is performed (Sarmiento and Bryan, 1982) so that the model does not drift too far away from known values. The time factor in this relaxation increases with depth. The net evaporation precipitation flux is set to zero. At present, freshwater runoff from rivers and tidal forcing are not included in the Benguela implementation of the model.

The biological model is coupled (one way coupling physics to biology) to the physical model through the subsurface light field, the hydrography and the horizontal and vertical movement of water masses. The prognostic variables are: inorganic nitrogen, phosphorus and silicate, two different types of phytoplankton (diatoms and flagellates), detritus (dead organic matter), light in the water column and the turbidity. Phytoplankton production is affected by nutrients, surface irradiance, temperature and the attenuation/turbidity of the water. The attenuation coefficient is a function of the chlorophyll concentration only, so selfshading is included. The incident irradiation is modelled using a formulation based on Skartveit and Olseth (1986, 1987). Data for global radiation are taken from the Surface Radiation Budget (Darnell et al., 1996). Phytoplankton respiration is a function of temperature alone, while mortality is a constant fraction, and is assumed to account also for zooplankton predation, which, in this context, is included as a forcing function. For both dead and living algae, sinking rates are applied. Nitrogen and phosphorus are regenerated from the dead algae (detritus) at a constant rate.

Initial values for the nutrient fields (annual means) are interpolated from the 1° x 1° NOAA atlas (Conkright et al., 1994). These annual fields are also used at the open boundaries. A small initial density of diatoms and flagellates (2.75 mgN/m^3) is set in the model.

Results and Discussion

The model has been run for 14 different years (1983/84 to 1996/97). After a 2-month spin-up, starting on May 1, the model integration was performed one by one year (July 1 to June 30). Before each new simulation, the model was reinitialised to avoid longterm drifts. Data for all prognostic variables were stored as monthly means in the whole model domain in all sigma layers. In addition, the depth-integrated production (gC/m^2/month) was calculated.

Upwelling indices

The use of a three-dimensional model to quantify upwelling is advantageous for several reasons. First it is inexpensive compared with field data. Second, a model can give a full three-dimensional picture in the entire model domain with a high resolution in both space and time instead of only spot measurements at sea from field data. This means that upwelling can be estimated in any geographical domain. Last, the model includes subsurface information, whereas SST and ocean colour from satellite only have surface information and cannot penetrate clouds. The largest problem with a model is that the estimates are in a virtual world, and will depend on errors and uncertainties in the forcing fields and process formulations. Therefore model validation (Dee, 1995) is a very important task, and the model should go through validation exercises (Skogen, 1999) where it is compared with relevant data.

There are several candidates for an upwelling index. As the Ekman layer moves away from the coast, water from below upwells to replace it. Instead of trying to compute Ekman transport directly, an alternative task is to compute the volume flux (m^3/s) of upwelled water using the modelled vertical velocities through a surface at a fixed depth.

One common method of quantifying upwelling is the use of satellite SST (e.g. Cole and McGlade, 1998; Cole, 1999). This method can easily be implemented in the model, simply by using the modelled temperature field. Over much of the area between Cape Frio (18°S) and Cape Point (34°S), there exists a well-developed longshore system of thermal fronts demarcating the seaward extent of the upwelled water. South of Lüderitz (26.5°S) there is usually a single well-defined front, coinciding approximately with the shelf break. Further north, the surface manifestation of the front is more diffuse, and multiple fronts are evident on occasions (Shannon and Nelson, 1996). Therefore, in addition to the use of actual SST, the difference between the temperature offshore and the temperature on the coast inside the upwelling can be used as a proxy for the upwelling (e.g. Nykjaer and van Camp, 1994). Both these approaches have been used in the model, with a default distance between coastal and offshore locations of 100 km (5 grid cells).

Finally the model has shown (Skogen, 1999) that it is possible to identify several local coastal maxima in the diatom production. In fact, except for the Agulhas cell, all known major upwelling cells can be identified from a locally increased diatom production. Upwelling is transporting extra nutrients to the surface layer, so primary production is having an integrative function and acts like an upwelling index.

To summarize, four proposed proxies for calculating the modelled upwelling intensity will be examined: vertical volume flux through a surface (W), sea surface temperature (T), difference between offshore and coastal SST (D) and diatom primary production (P). The indices have been calculated from monthly means of these quantities. In addition, to adjust for the seasonal cycle, the same set of indices were also calculated from the normalized monthly anomalies calculated as the ratio between the individual monthly and mean (14 years) climatological monthly values. For example,

$$PA_{may} = P_{may} / \overline{P_{may}},$$

where $\overline{P_{may}}$ is the mean May production, will be the anomaly diatom production in May (PA=1 if P equals the long-term mean).

Intercomparison of upwelling indices

Even if the proposed modelled indices are all assumed to quantify upwelling, differences between them are likely to appear. To get an idea about how the different indices compare at a certain location, an intercomparison of the modelled indices has been done for an area (Figure 2) just north of Lüderitz (24.5-26.5°S). This small strip along the coast (13600 km²) has been selected since it includes the local maximum modelled diatom production.

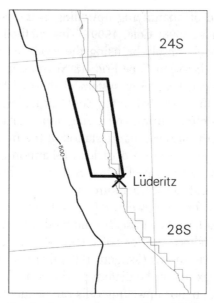

Figure 2. Area for the intercomparison of modelled upwelling indices outside Lüderitz.

Annual indices for the 14 years (83/84 to 96/97), computed as the mean of the monthly indices for that respective year (July to June), were estimated using all proposed proxies for upwelling. One of these time series, anomaly diatom production, is shown in Figure 3. This figure suggests an increasing trend through the 14 year period, with interannual anomalies between ±15% off the mean, due to an increase in the NCEP/NCAR wind stress in the period. Time series for all eight indices have been compared by computing the correlation between them. The results are given in Table 1.

Table 1. Modelled upwelling indices intercomparison. Correlation numbers between the different indices outside Lüderitz. P is the index based on diatom production, PA is the anomaly production index, W and WA are the vertical velocity indices at 40 meters, T and TA the SST indices, and D and DA the indices from the difference between offshore and coastal SST.

	P	PA	W	WA	T	TA	D	DA
P	1.00	0.99	0.98	0.91	-0.98	-0.97	-0.10	0.31
PA		1.00	0.97	0.93	-0.98	-0.97	0.01	0.41
W			1.00	0.95	-0.96	-0.96	-0.06	0.36
WA				1.00	-0.89	-0.89	0.03	0.41
T					1.00	1.00	-0.08	-0.47
TA						1.00	-0.08	-0.47
D							1.00	0.89
DA								1.00

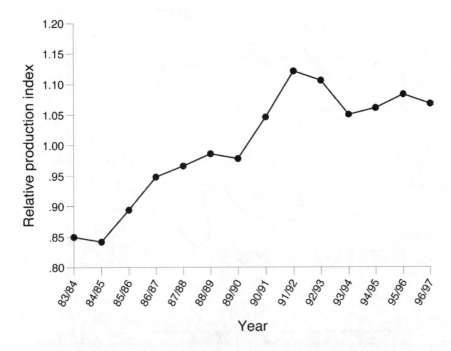

Figure 3. Annual upwelling indices from the anomaly diatom production at Lüderitz.

There is an almost linear relationship between modelled diatom production, vertical velocity and SST, but only a weak correspondence between these and the difference between offshore and coastal SST. However, the latter is sensitive to the distance between points where the temperature difference is calculated from, and by doubling the distance to 200 km, a correlation of r ≈ 0.65 with the other indices is found.

A similar exercise (not shown) has been done in the southern Benguela in an area outside Saldanha Bay. All Saldanha Bay correlations are significant lower. The offshore and coastal SST difference indices are the most typical ones, while the vertical velocity is very different from the others.

The main explanation for the areal differences, with lower correlations between the indices at Saldanha Bay in the southern Benguela than outside Lüderitz in the northern Benguela, can be explained from the wind fields in the two areas (see Figure 4). The upwelling and offshore Ekman transport is mainly wind driven, and there is a substantial correlation between wind stress and both upwelling frequency and offshore penetration of upwelled water (Lutjeharms and Meeuwis, 1987). At Lüderitz the wind direction is almost constant throughout the year, leading to perfect conditions and an almost permanent upwelling, whereas off Saldanha Bay there is a clear sea-

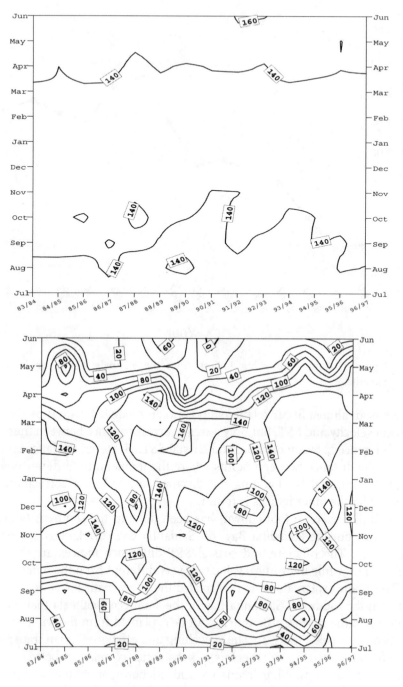

Figure 4. Monthly mean wind direction computed from 12-hourly winds from the NCEP/NCAR Reanalysis Project (Kalnay *et al.*, 1996) outside Lüderitz (top) and Cape Point (bottom).

sonal cycle with lower upwelling intensity during the austral winter. A wind stability parameter (WSP) was calculated to represent this variability in the wind direction for different locations in the southern and northern Benguela. This parameter is based on the ratio of the absolute value of the vector sum of the average monthly wind vectors and the scalar sum of the individual average monthly wind vector:

$$WSP = \frac{\left|\sum \vec{v}\right|}{\sum |\vec{v}|} \times 100\% \qquad (1)$$

This ratio showed that there were great differences in stability between the southern and the northern Benguela. At Lüderitz the stability was 99.1%, at Palgrave Point it is estimated to 96.8%, while it was only 64.3% at Cape Point. This low variability in wind direction in the northern Benguela creates a stable circulation system (Shannon, 1985) with relatively small fluctuations in the upwelling.

Upwelling cells

In Figure 5 the mean (all years) modelled annual diatom production outside Namibia is shown. From the figure the main upwelling cells in the northern Benguela can easily be identified. The southernmost upwelling area stretches from Lüderitz to Walvis Bay with different intensities, and overlaps with the known Walvis Bay (central Namibia) and Lüderitz cells. A smaller cell is seen just north of Palgrave Point at 20°S, just south of the cell identified by Lutjeharms & Meeuwis (1987) with its centre at 19°S (northern Namibia), while the last major upwelling area goes from about 19°S up to the Angolan border at Kunene River. The figure suggest the northernmost (Kunene) cell as the largest and most intense one, not in agreement with literature that identifies the Lüderitz cell as the principal one. This can be explained from the underlying nutrient fields used by the model. The modelled vertical velocity identifies Lüderitz as the most intense one, but that figure (not shown) is more scattered and does not show the cells as clearly as the modelled production field.

To get an idea of the 3-D structure of a cell, the mean (all years) annual cross-sectional velocity field along 25° 20'S inside the Lüderitz cell is shown in Figure 6. The figure suggests an average upward transport of about 13 m/d at 40 m depth at the coast, and a very thin offshore-moving Ekman layer of approximately 20 m depth. The seaward extent of the upwelling, associated with the onshore-moving subsurface water, is about 200 km at this location, in agreement with other estimates (e.g. Lutjeharms & Meeuwis, 1987).

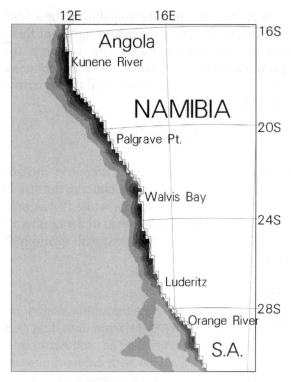

Figure 5. Mean (all years) modelled annual diatom production in the northern Benguela.

Variability and intensity

Even if the wind at Lüderitz is almost constant throughout the year, there is a large variation in the upwelling intensity due to changes in the wind stress. The stress is highest in the austral summer and lowest in winter. Focusing on the point of average (all year) maximum vertical velocity within the Lüderitz upwelling cell at 40 m depth (14° 55'E, 26° 10'S), this results in an order of magnitude difference between winter and summer intensity in the upwelling regime in the maximum point. A maximum upwelling in summer and a minimum during winter is in agreement with Shannon (1985). The modelled monthly mean vertical velocity (metre/day) in this point is shown in Figure 7. The annual trend seen in Figure 3 is also clearly seen in this figure, and is due to an increase in the wind stress through this 14-year period. This is in contradiction to the results by J.-P. Roux (unpubl. data), who based his analysis on a time series of daily pseudo-wind stress measurements of upwelling favourable (southerly) wind at Dias Point Light House suggesting that the Lüderitz upwelling cell has shifted to a reduced upwelling regime after 1988.

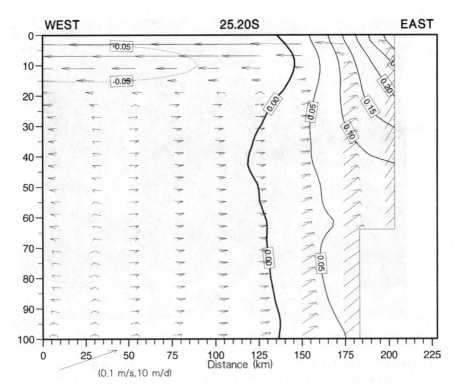

Figure 6. Mean (all years) modelled cross-sectional velocity field along 25° 20'S inside the Lüderitz upwelling cell.

From such velocities it is possible to give an estimate of the vertical volume flux, and thereby the total amount of upwelled water, through a given surface for the whole Namibian shelf area. Counting all points of upward transport from Kunene River to Orange River and from the coast to 200 km offshore, the annual (all years) mean vertical transport through the 40 m depth surface is estimated to 2.2 Sverdrup (1 Sv. = 10^6 m^3/s). The Lüderitz upwelling cell contributes to about 40% of this. The numbers for December and June are 2.6 (42%) and 1.6 (26%) Sv. respectively. These numbers are of the same order of magnitude as a previous estimate based on the entrainment of a cool filament around a warm core ring. It was estimated that this process could represent an average volume flux of upwelled surface water from the Benguela shelf over a period of 2-3 months of 1.5 Sv. (Lutjeharms et al., 1991; Duncombe Rae et al., 1992a,b).

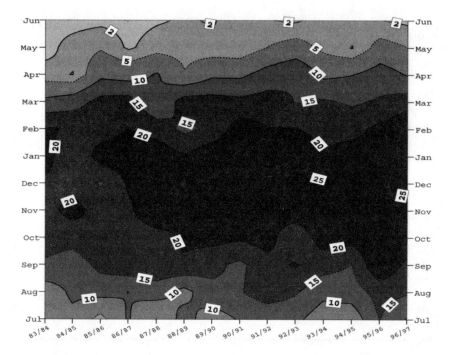

Figure 7. Modelled monthly mean vertical velocity (m /day) at the point of maximum upwelling intensity within the Lüderitz upwelling cell at 40 m depth.

CONCLUDING REMARKS

In this chapter a biophysical numerical model (Skogen, 1999) is used to examine the interannual variability, spatial extent and intensity of upwelling off Namibia over a 14-year period (1983 to 1997). The model agrees with other works on the positioning of the upwelling cells, their seaward extent and on the order of magnitude of the volume flux of upwelled water. The present work shows a large variability in upwelling intensity through the year with a maximum in December and a minimum in June, and a positive trend toward stronger upwelling through the period due to an increase in the wind stress.

To use a numerical model to estimate upwelling strength has the advantage of being an inexpensive and flexible tool in the sense that estimates can be given hindcast with high resolution in space and time in any area of interest. As such, the present model represents a useful tool for gaining new insight in the complex dynamics between physics and biology in nature. However, limitations have to be taken into account when interpreting the results. Open boundaries are calculated from climatological seasonal fields, and the in/out flows assume geostrophy. Clearly the horizontal resolution is a limiting factor. A 20 km grid spacing cannot account for small energetic structures related to the instability of the upwelling front, or for the development of coastally trapped waves. Moreover, satellite images for water colour (as an index for chlorophyll) clearly display patches of biological activity, in addition to the large-scale structures as portrayed in the present work. Also, the use of reanalysis atmospheric fields on a coarse grid (1.875° resolution) are unlikely to sample ideally the cross-shore variability of the alongshore wind stress component, and thus bias the estimate of the true wind effect in the coastal areas. Finally, the model does not incorporate real surface heat fluxes, only a relaxation toward monthly climatological sea surface temperature fields.

References

Aksnes, D.L., Ulvestad, K.B., Balino, B., Berntsen, J., Egge, J., and Svendsen, E. (1995): Ecological modelling in coastal waters: towards predictive physical-chemical-biological simulation models. *Ophelia* 41: 5-36.

Bakun, A. and Parrish, R.H. (1990): Comparative studies of coastal pelagic fish reproductive habitats: the Brazilian sardine *Sardinella aurita*. *Journal de Conseil internationa pourl l'Exploration de la Mer* 46: 269-283.

Berntsen, J., Svendsen, E. and Ostrowski, M. (1996): Validation and Sensitivity study of a Sigma-coordinate Ocean Model using the SKAGEX dataset. *ICES C.M.* 1996/C:5, 28 pp.

Blumberg, A.F., and Mellor, G.L. (1987): A description of a three-dimensional coastal ocean circulation model. In: N. Heaps (ed.), *Three-Dimensional Coastal Ocean Models*, Vol. 4, 1-16. American Geophysical Union.

Boyd, A. (1987): The oceanography of the Namibian shelf. Ph.D. thesis, Univ. of Cape Town, South Africa, 190 pp.

Carr, M.-E. (2002): Estimation of potential productivity in Eastern Boundary Currents using remote sensing. *Deep-Sea Research* II 49: 59-80.

Cochrane, K.L. and Hutchings, L. (1995): A structured approach to using biological and environmental parameters to forecast anchovy recruitment. *Fisheries Oceanography* 4(2): 102-127.

Cole, J. (1999): Environmental conditions, satellite imagery, and clupeoid recruitment in the northern Benguela Upwelling system. *Fisheries Oceanography* 8(1): 25-38.

Cole, J. and McGlade, J. (1998): Temporal and spatial patterning of sea surface temperature in the northern Benguela upwelling system: Possible environmental indicators of clupeoid production. *South African Journal of marine Science* 19: 143-157.

Conkright, M.E., Levitus, S. and Boyer, T. (1994): *World Ocean Atlas 1994*, Volume 1: Nutrients. Tech. rep. NOAA Atlas, NESDIS 1. National Oceanic and Atmospheric Administration, Washington, DC. 150 pp.

Cox, M.D., & Bryan, K. (1984): A Numerical model of the ventilated thermocline. *Journal of Physical Oceanography* 14: 674-687.

Darnell, W.L., Staylor, W.F., Ritchey, N.A., Gupta, S.K., & Wilber, A.C. (1996): Surface Radiation Budget:

long-term global dataset of shortwave and longwave fluxes. http://www.agu.org/eos_elec/95206e.html.

Dee, D.P. (1995): A pragmatic approach to model validation. In: D.R. Lynch and A.M. Davies (eds.): *Quantitative Skill Assessment for Coastal Ocean Models*. American Geophysical Union. ISBN 0-87590-261-8, DC, USA.

Duncombe Rae, C.E., Shillington, F.A, Agenbag, J.J., Taunton-Clark, J. and Grüundlingh, M.L. (1992a): An Agulhas ring in the South Atlantic Ocean and its interaction with the Benguela upwelling frontal system. *Deep Sea Research* 139(11/12): 2009-2027.

Duncombe Rae, C.E., Boyd, A.J. and Crawford, R.J.M. (1992b): Predation of anchovy by an Agulhas ring: a possible contributory cause for the very poor year-class of 1989. *South African Journal of marine Science* 12: 167-173.

IOC, IHO and BODC (1994): The general bathymetric chart of the oceans GEBCO digital atlas. http://www.bodc.ac.uk/projects/gebco.

Johnson, A.S. and Nelson, G. (1999): Ekman estimates of upwelling at Cape Columbine based on measurements of longshore winds from a 35 year time series. *South African Journal of marine Science* 21: 433-436.

Kalnay, E., Kanamitsu, M., Kistler, R., Collins, W., Deaven, D., Gandin, L., Iredell, M., Saha, S., White, G., Woollen, J., Zhu, Y., Leetmaa, A., Reynold, R., Chelliah, M., Ebisuzaki, W., Higgins, W., Janowiak, J., Mo, K.C., Roplewski, C., Wang, J., Jenne, R. and Joseph, D. (1996): The NCEP/NCAR 40-Year Reanalysis Project. Bulletin of the American Meteorological Society 123 pp.

Lasker, R. (1985): What limits clupeiod production. *Canadian Journal of Fisheries and Aquatic Science* 42(Supl.1): 31-38.

Levitus, S. and Boyer, T. (1994): *World Ocean Atlas 1994*, Volume 4: Temperature. Tech. rep. NOAA Atlas, NESDIS 4. National Oceanic and Atmospheric Administration, Washington, DC. 117 pp.

Levitus, S., Burgett, R. and Boyer, T.P. (1994): *World Ocean Atlas 1994*, Volume 3: Salinity. Tech. rep. NOAA Atlas, NESDIS 3. National Oceanic and Atmospheric Administration, Washington, DC. 99 pp.

Lutjeharms, J.R.E. and Meeuwis, J.M. (1987): The extent and variability of the South-East Atlantic upwelling. *South African Journal of marine Science* 5: 51-62.

Lutjeharms, J.R.E., Shillington, F.A. and Duncombe Rae, C.M. (1991): Observations of extreme upwelling filaments in the Southeast Atlantic Ocean. *Science* 253: 774-776.

Martinsen, E.A. and Engedahl, H. (1987): Implementation and testing of a lateral boundary scheme as an open boundary condition in a barotropic ocean model. *Coastal Engineering* 11: 603-627.

Mellor, G.L. and Yamada, T. (1982): Development of a turbulence closure model for geophysical fluid problems. *Reviews of geophysics and space physics* 20: 851-875.

Mesinger, F. and Arakawa, A. (1976): Numerical methods used in atmospheric models. GARP Publication Series 17: 64 pp.

Nykjaer, L. and van Camp, L. (1994): Seasonal and interannual variability of coastal upwelling along northwest Africa and Portugal from 1981 to 1991. *Journal of Geophysical Research* 99(C7): 1419-1420.

Oey, L.-Y. and Chen, P. (1992): A model simulation of circulation in the Northeast Atlantic Shelves and Seas.

Journal of Geophysical Research 97(C12): 20087-20115.

Pauly, D. and Christensen, V. (1995): Primary production required to sustain global fisheries. *Nature* 374: 255-257.

Sarmiento, J.L. and Bryan, K. (1982): An Ocean transport model for the North Atlantic. *Journal of Geophysical Research* 87(C1); 394-408.

Schumann, E.H., Perrins, L-A. and Hunter, I.T. (1982): Upwelling along the south coast of the Cape Province, South Africa. *South African Journal of Science* 78(6): 238-242.

Shannon, L.V. (1985): The Benguela ecosystem 1. Evolution of the Benguela physical features and processes. *Oceanography and Marine Biology: an Annual Review* 23: 105-182.

Shannon, L.V. and Nelson, G. (1996): The Benguela: Large scale features and processes and system variability. In: G. Wefer, W.H. Berger, G. Siedler and D.J. Webb (eds.): *The South Atlantic: Present and Past Circulation*. Springer-Verlag, Berlin Heidelberg.

Skartveit, A. and Olseth, J.A. (1986): Modelling slope irradiance at high lattitudes. *Solar Energy* 36(4): 333-344.

Skartveit, A. and Olseth, J.A. (1987): A model for the diffuse fraction of hourly global radiation. *Solar Energy* 37: 271-274.

Skogen, M.D. (1993): A User's guide to NORWECOM, the NORWegian ECOlogical Model system. Tech. rep. 6. Institute of Marine Research, Division of Marine Environment, Pb.1870, N-5024 Bergen. 23 pp.

Skogen, M.D. (1999): A biophysical model applied to the Benguela upwelling system. *South African Journal of Marine Science* 21: 235-249.

Skogen, M.D. (2004): Clupeoid larval growth and plankton production in the Benguela upwelling system. *Fisheries Oceanography*, in press.

Skogen, M.D., Svendsen, E., Berntsen, J., Aksnes, D. and Ulvestad, K.B. (1995): Modelling the primary production in the North Sea using a coupled 3 dimensional Physical Chemical Biological Ocean model. *Estuarine, Coastal and Shelf Science* 41: 545-565.

Skogen, M.D., Svendsen, E. and Ostrowski, M. (1997): Quantifying volume transports during SKAGEX with the Norwegian Ecological Model system. *Continental Shelf Research* 17(15): 1817-1837.

Svendsen, E., Fossum, P., Skogen, M.D., Eriksrod, G., Bjorke, H., Nedraas, K. and Johannessen, A. (1995): Variability of the drift patterns of Spring Spawned herring larvae and the transport of water along the Norwegian shelf. *ICES C.M.* 1995/Q:25. 29 pp.

Svendsen, E., Berntsen, J., Skogen, M.D., Ådlandsvik, B. and Martinsen, E. (1996): Model simulation of the Skagerrak circulation and hydrography during SKAGEX. *Journal of Marine Systems* 8(3-4): 219-236.

Waldron, H.N., Brundrit, G.B. and Probyn, T.A. (1997a): Anchovy biomass is linked to annual potential new production in the southern Benguela: support for the "optimal environmental window" hypothesis. *South African Journal of marine Science* 18: 107-112.

Waldron, H.N., Probyn, T.A. and Brundrit, G.B. (1997b): Preliminary annual estimates of reginal nitrate supply in the southern Benguela using coastal sea level fluctuations as a proxy for upwelling. *South African Journal of Marine Science* 18: 93-105.

2 THE EFFECTS OF INTERNAL AND EXTERNAL CONTROL ON THE NORTHERN BENGUELA ECOSYSTEM

*Johanna J. Heymans**

Abstract

Internal and external control of the northern Benguela ecosystem over the past 30 years was examined using the steady-state Ecopath model of the ecosystem, created for the 1970s and simulating to the present. The effect of an external factor such as fishing was combined with the internal "wasp-waist" control of the ecosystem by forage species such as anchovy and sardine (i.e. top-down control on their prey and bottom-up control on their predators). The effect of including these external and internal control factors increased the fit of the model to the observed data by 50%. Six primary production forcing anomalies were subsequently predicted by Ecosim, three with "wasp-waist" forcing included and three without. The three scenarios with internal control increased the fit of the model to 65%. When no internal control was included, the goodness of fit only increased marginally. The primary production forcing functions predicted by Ecosim in these six scenarios were then correlated to environmental variables such as summer and winter sea surface temperature as well as pseudo-wind stress anomalies. Significant correlations were obtained for summer sea surface temperature and pseudo-wind stress anomalies when "wasp-waist" control was invoked.

INTRODUCTION

Upwelling ecosystems, by their very nature, are affected by external forces such as environmental variation and fishing. The northern Benguela ecosystem is no exception. The effects of these external forces on the northern Benguela ecosystem have been described by various authors, *viz.* Boyer *et al.*

---

* The author would like to thank Chris Bartholomae, Astrid Jarre, Claude Roy, Lynne Shannon, Nico Willemse, and the Ministry of Fisheries and Marine Resources for the data and models provided for this work. Thank you also to Villy Christensen, Robyn Forrest, Lynne Shannon and to two anonymous reviewers for their assistance and comments.

(2001); Boyer and Hampton (2001); Cole (1999); Cole and Villacastin (2000); Kreiner et al. (2000) and Reid et al. (2000).

The environmental forces driving upwelling in the northern Benguela ecosystem are persistent southerly winds, which impact sea surface temperature (SST), nutrient supply and primary productivity (Shannon et al., 1990). Primary production, in turn, drives fish production and Jarre-Teichmann and Christensen (1998) suggest that 15-30% of total primary production is needed to sustain major pelagic fish communities in upwelling areas. Thus, these external forces also impact the important forage species in the northern Benguela ecosystem.

Kreiner et al. (2000) showed that wind anomalies are a good predictor of sardine spawner biomass and recruitment, while Reid et al. (2000) suggested that sardine play a role in controlling primary productivity and during periods of low sardine abundance, the result could be eutrophication due to under-utilisation of phytoplankton (Boyer et al., 2001). In addition, Cole (1999) found that sea surface temperature usually acts as a good proxy for recruitment success in both anchovy and sardine, although for one year in his study (1987), both anchovy and sardine recruitment was high, probably due to the good oceanographic conditions for onshore retention of eggs and high nutrient enrichment (Cole, 1999).

Internal forces, such as predator-prey interactions, competition, etc., also affect ecosystems. The internal control of predators and prey in an ecosystem could be top-down (i.e. predators control their prey), bottom-up (i.e. prey control the abundance of predators) or a mixture of those. Cury et al. (2000) suggest that in the Benguela ecosystem there is limited evidence of top-down control of forage fish by predator populations, with more substantial bottom-up control of predators by forage fish. They postulate that one might expect to observe strong interactions between primary production and pelagic fish, mediated through zooplankton, by means of "wasp-waist" control. Under "wasp-waist" control, zooplankton is highly vulnerable to predation by small pelagic fish such as anchovy, sardine, etc. and therefore, competition for zooplankton prey is reduced when fishing is increased (Cury et al., 2000). Thus, in the northern Benguela ecosystem, both zooplankton and top predators are controlled by small pelagic fish such as anchovy, sardine, etc., thus both the top and the bottom of the ecosystem are being controlled by the middle trophic level, and removing large quantities of species from the system, as was done in the 1970s and 1980s, would have severe effects on the ecosystem.

External control of the Benguela ecosystem includes both fishing and environmental forcing. The effects of fishing and internal forcing have been studied in the southern Benguela system by Cury et al. (2000) and Shannon et al. (2000), while the effects of fishing and internal control in the northern

Benguela ecosystem have been attempted by J.-P. Roux (unpublished work). However, the combined effect of the internal forcing of "wasp-waist" control and *both* anthropogenic and natural external forcing (i.e. fishing and environmental variation) on the northern Benguela ecosystem has not been attempted. This paper therefore examines the effects of environmental forcing functions, fish catches and "wasp-waist" control of forage fish on the northern Benguela ecosystem to see if the combined effect of these factors could explain the variation in the ecosystem over time.

THE ECOSYSTEM

The northern Benguela ecosystem (Figure 1) is bound to the north by the Angolan front at approximately 15°S (Shannon *et al.*, 1987; Lutjeharms and Meeuwis, 1987) and to the south by the Lüderitz upwelling cell, with the official limit set close to the South African-Namibian border at 29°S (Brown *et al.*, 1991). There are four main upwelling cells in the northern Benguela: the Kunene cell, Namibia cell (or Cape Frio cell, Figure 1), Walvis Bay cell and Lüderitz cell (Lutjeharms and Meeuwis, 1987). The temperatures of all upwelled waters south of the Angola Front vary between 10 and 19°C.

Upwelling in the northern Benguela is wind driven, with the climatic forcing being produced mainly by the South-East Atlantic High Pressure anticyclone that causes persistent equatorward winds north of 32°S (Estrada and Marrase, 1987). The seaward extent of the major wind-induced upwelling cells ranges from nearly 150 km at the Namibia cell (19°S) to close to 300 km at the Lüderitz cell (27°S) (Lutjeharms and Meeuwis, 1987). The wind stress for all upwelling cells is from the south-east and the climatic Ekman drift is suitable for year-long upwelling along the whole coast (Lutjeharms and Meeuwis, 1987).

There are five major species of commercial importance in the northern Benguela ecosystem: Cape anchovy (*Engraulis capensis*), sardine (*Sardinops sagax*), horse mackerel (*Trachurus capensis*) and hake (*Merluccius capensis* and *M. paradoxus*). The two species of hake are usually reported in catch statistics as one group. The total cumulative catch of anchovy, sardine, horse mackerel and hake since the Second World War is approximately 4, 15, 12, and 12 million tonnes respectively (Boyer *et al.*, 2000).

Anchovy was important in the ecosystem and the fishery until recently, but since the mid 1990s very little has been caught, and surveys in the late 1990s indicated that their biomass is very low (Boyer and Hampton, 2001). Shackleton (1987) has examined the interaction between anchovy and sardine (the two main forage species) over the previous 100 years by looking at the fossil fish scales in a core off Walvis Bay. She found that sardine and anchovy seemed to decline over two 20-year periods, with sardine recovering ahead of anchovy. Subsequent to the recovery, the two stocks alternated in dominance, with the combined stock size remaining constant over those periods (Shackleton, 1987). Overall, the scale analysis showed that the stocks decreased by a factor of 8 over the century, and that the community was always sardine dominated (Shackleton, 1987).

The estimates of sardine biomass and catch show a clear reduction since the start of the fishery. Excessive fishing and recruitment failure seems to have been the reason for the stock collapse in the 1960s and 1970s (Cram, 1981 in Fossen et al., 2001). Fishing mortality increased since the mid 1970s and was an important cause of mortality until 1990 (Thomas, 1986 in Fossen et al., 2001). The fishing mortality of northern Benguela sardine was much

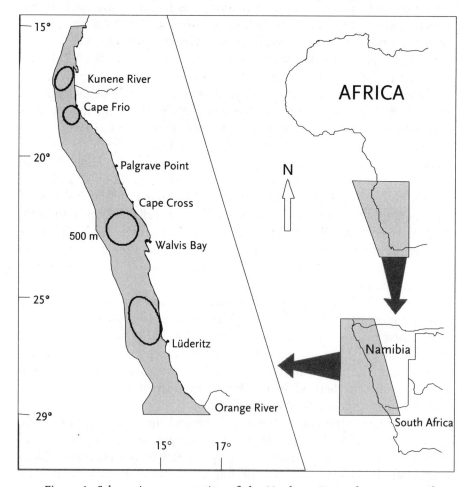

Figure 1. Schematic representation of the Northern Benguela ecosystem (from Heymans, 1997) showing the 500 m depth contour and upwelling cells (redrawn from Lutjeharms and Meeuwis, 1987).

lower in the 1990s (Fossen et al., 2001), but by the end of that decade the sardine stock had still not recovered. According to Fossen et al. (2001) the fishing pressure during the past four decades has been at least one of the reasons for the increased total mortality. High fishing mortality contributed to the stock collapse and low abundance during the 1970s and 1980s, but current mortality rates seem to be controlled mainly by other factors, such as a change in growth rate and/or higher predation mortality (Fossen et al., 2001). The higher predation mortality is probably due to their low biomass causing higher vulnerability to predators (Fossen et al., 2001).

Hakes are the most commercially important demersal fish in the northern Benguela system (Boyer et al., 2000). Two species of hake are caught along the Namibian coast – deep-water hake (*Merluccius paradoxus*) and shallow-water Cape hake (*M. capensis*). In Namibia, shallow-water hake is more abundant than the deep-water species, but the biomass of deep-water hake has increased since 1990 (Burmeister, 2001). Cape hake is mainly distributed off central and northern Namibia, while deep-water hake extends from the southern Benguela (Boyer et al., 2000). There are indications that adult Cape hake feed on juvenile deep-water hake (Boyer et al., 2000).

Horse mackerel is the only mesopelagic species that is currently fished in the northern Benguela (Boyer et al., 2000). Cape horse mackerel (*Trachurus capensis*) have a wide distribution, from Lüderitz to Angola and from the surf zone to beyond the shelf break, but their highest concentrations are between 17°S and 22°S (Boyer et al., 2000). Cunene horse mackerel (*Trachurus trechae*), in contrast, occurs in northern Namibia and southern Angola and overlaps with Cape horse mackerel at the southern end of their range (Boyer et al., 2000).

METHODOLOGY

Heymans et al. (2004) compared the three Ecopath models of the northern Benguela ecosystem constructed by Jarre-Teichmann (1998); Jarre-Teichmann and Christensen (1998); Shannon and Jarre-Teichmann (1999) and Heymans and Baird (2000). The earliest of these models (1971-1977), constructed by Jarre-Teichmann (1998); Jarre-Teichmann and Christensen (1998) and adapted in Heymans et al. (2004), was subjected to time series fitting here, using Ecosim in the Ecopath with Ecosim (*EwE* version 5 Beta) suite. The Ecopath with Ecosim methodology, functions, equations, etc. are well explained in Christensen et al. (2000); Christensen and Walters (2004); Walters et al. (1997) and Walters et al. (2000), however, a brief description is given here.

Ecopath is a mass balance approach that creates a static model of the ecosystem and uses two master equations to parameterise the model; one for the energy balance of each group (Eq. 1) and the other to depict the production term (Eq. 2 and 3) (Christensen and Walters, 2004). The energy balance for each group is calculated *sensu* Winberg (1956), who defined consumption as the sum of somatic and gonadal growth, metabolic costs and waste product (Christensen and Walters, 2004). Consumption is calculated with the formula:

$$\text{Consumption} = \text{production} + \text{respiration} + \text{unassimilated food} \quad (1)$$

The total production rate P_i for each group i is given by:

$$P_i = Y_i + M2_i \cdot B_i + E_i + BA_i + M0_i \cdot B_i \quad (2)$$

where Y_i is the total fishery catch rate of i, $M2_i$ is the instantaneous predation rate for group i, E_i the net migration rate (emigration • immigration), BA_i is the biomass accumulation rate for i, while $M0_i$ is the 'other mortality' rate for i. P_i is calculated as the product of B_i, the biomass of i and $(P/B)_i$, the production/biomass ratio for i. The $(P/B)_i$ rate under most conditions corresponds to the total mortality rate, Z (Allen, 1971), commonly estimated as part of fishery stock assessments (Christensen and Walters, 2004). This can be re-written as:

$$P_i = \frac{Y_i + E_i + BA_i + \sum_j Q_j \cdot DC_{ji}}{EE_i} \quad (3)$$

where EE_i, the ecotrophic efficiency of group i, describes the proportion of the production that is utilised in the system as described, Q_j is the total consumption rate for group j, and DC_{ji} is the fraction of predator j's diet contributed by prey i. Q_j is calculated as the product of B_j, the biomass of group j and $(Q/B)_j$, the consumption/biomass ratio for group j (Christensen and Walters, 2004).

Ecosim uses a set of differential equations to calculate the changes in biomass of each group over time, by using the harvest rates as well as external forcing functions imposed on the ecosystem (Christensen et al., 2000).

The equations are derived from the Ecopath master (Eq. 2), and take the form:

$$\frac{dB_i}{dt} = g_i \sum_j Q_{ji} - \sum_j Q_{ij} + I_i - (M0_i + F_i + e_i) \cdot B_i \quad (4)$$

where dB_i/dt represents the growth rate during the time interval dt of group i in terms of its biomass, B_i, g_i is the net growth efficiency, $M0_i$ the non-predation ('other') natural mortality rate, F_i is fishing mortality rate, e_i is emigration rate, I_i is immigration rate (assumed constant over time, and hence independent of events in the ecosystem modelled), and $e_i \cdot B_i - I_i$ is the net migration rate of Eq. 2 (Christensen and Walters, 2004). The two summations estimate consumption rates, the first calculating the total consumption by group i, and the second the predation by all predators on the same group i. The consumption rates, Q_{ji}, are calculated based on the 'foraging arena' concept, where B_i's are divided into vulnerable and invulnerable components *sensu* Walters *et al.* (1997). The transfer rate (v_{ij}) between the vulnerable and invulnerable components determines if control is top down (i.e. Lotka–Volterra), bottom up (i.e. donor-driven), or of an intermediate type (Christensen and Walters, 2004). Top-down versus bottom-up control is a continuum in the model, where low v's implies bottom-up and high v's top-down control. Consumption at each timestep is calculated by:

$$Q_{ij} = \frac{a_{ij} \cdot v_{ij} \cdot B_i \cdot B_j \cdot T_i \cdot T_j \cdot S_{ij} \cdot M_{ij}/D_j}{v_{ij} + v_{ij} \cdot T_i \cdot M_{ij} + a_{ij} \cdot M_{ij} \cdot B_j \cdot S_{ij} \cdot T_j/D_j} \quad (5)$$

where a_{ij} is the rate of effective search for i by j, T_i represents prey relative feeding time, T_j the predator relative feeding time, S_{ij} the user-defined seasonal or long term forcing effects, M_{ij} the mediation forcing effects, and D_j represents effects of handling time as a limit to consumption rate (see Walters *et al.*, 1997, 2000 and Christensen and Walters, 2004 for further information on these algorithms).

Time series data

Catch estimates for all species in the northern Benguela ecosystem were obtained from Willemse (2002) and Willemse and Pauly (*this volume*). The biomass estimates for each group were obtained from various sources and are explained below. The biomass estimates of anchovy were obtained from the Namibian Ministry of Fisheries and Marine Resources (see Heymans, 1997) for the time period 1990-1995. No estimates of biomass were available for anchovy prior to 1990 and subsequent to 1995 no surveys for anchovy were done, due to the low biomass (Boyer *et al.*, 2001). The catch and biomass estimates for anchovy are shown in Figure 2.

Biomass estimates of sardine from 1990 to 2000 were obtained from Fig. 4 in Boyer *et al.* (2001). Biomass estimates for 1971 to 1989 were obtained from Thomas (1986), and correlate well with the biomass trajectory given in Fossen *et al.* (2001). See Figure 3 for the catch and biomass estimates of sardine in the northern Benguela ecosystem.

Biomass estimates of hake were not readily available prior to 1990, but Butterworth and Geromont (2001) gives CPUE for ICSEAF Divisions 1.3+1.4 and 1.5 for 1965-1988, and from General Linear Modelling for 1991-1996 (Figure 4). The summer and winter biomass estimates made by the Research Vessel Dr. Fridjof Nansen for 1990 to 1997 are also given by Butterworth and Geromont (2001) (see Figure 4), and Burmeister (2001) gives depth-stratified biomass estimates for both species of hake from 1990 to 1999 (Figure 5). The

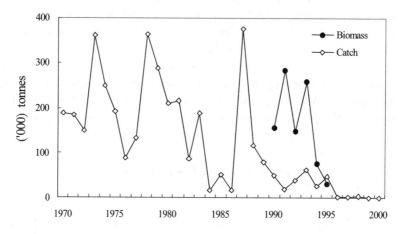

Figure 2. Catch and biomass estimates (tonnes) of anchovy from 1970 to 2000 in the northern Benguela ecosystem. Biomass estimates obtained from the Ministry of Fisheries and Marine Resources and catches from Willemse and Pauly (*this volume*).

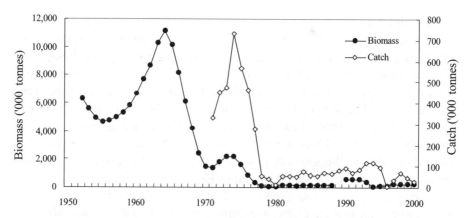

Figure 3. Catch and biomass estimates of sardine from 1951 to 2000 in the northern Benguela ecosystem. Biomass estimates obtained from Willemse and Pauly (*this volume*) and Boyer *et al.* (2001) and catches from Willemse and Pauly (*this volume*).

biomass estimates obtained from Burmeister (2001) and the CPUE for ICSEAF Divisions 1.3+1.4 were used to compare the biomass estimates for hake (Butterworth and Geromont, 2001). The two species of hake are not specified in the landings and therefore they were combined into one stock. See Figure 5 for catch and biomass of hake.

Time series data for horse mackerel were not as easy to obtain as for hake and sardine, even though horse mackerel has a large biomass at present in the system. Biomass estimates were not available prior to 1989, but estimates from 1989 to 1998 were obtained from acoustic surveys (Vaske and Klingel-

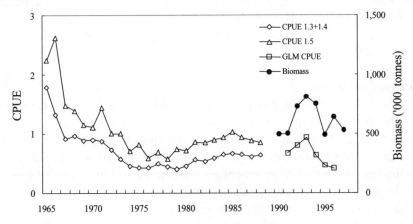

Figure 4. CPUE estimates and biomass estimates of hake stocks in the northern Benguela ecosystem for 1965 to 1997 obtained from Butterworth and Geromont (2001).

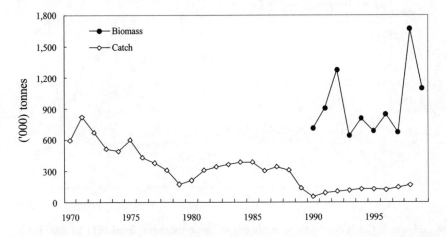

Figure 5. Catch and biomass estimates of hake from 1971 to 2000 in the northern Benguela ecosystem. Biomass estimates obtained from Burmeister (2001) and catches from Willemse and Pauly (*this volume*).

hoeffer (2001) and Nico Willemse (*pers. comm.*). See Figure 6 for catch and biomass of horse mackerel.

Time series fitting
The time series data described above, for catch and biomass of the four most important groups (anchovy, sardine, hake and horse mackerel) were used to calculate harvest rates (catch/biomass). These harvest rates were used to drive the 1970s model to the present time period by changing the biomass of the different groups with the changes in harvest rate. Ecosim estimates a goodness-of-fit measure to estimate the variance between the predicted and observed parameters. The goodness-of-fit measure is a weighted sum of squares (SS) deviation of log biomasses from log predicted biomasses (Christensen and Walters, 2004).

Vulnerability changes
In Ecosim it is assumed that prey are split into two groups in accordance with the "foraging arena theory": those that come out to feed and are therefore vulnerable to predation, and those that hide from predators but also don't get to feed (Christensen and Walters, 2004). Predation rates are therefore limited by the flow of prey between vulnerable and invulnerable groups, and the more time a prey spends feeding, the higher is its vulnerability to predators (Walters *et al.*, 2000). This vulnerability is also dependent on the biomass of the predator with respect to its unfished biomass (Christensen

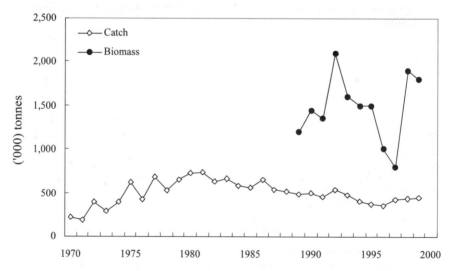

Figure 6. Catch and biomass estimates of horse mackerel from 1971 to 2000 in the northern Benguela ecosystem. Biomass estimates obtained from the Namibian Ministry of Fisheries and Marine Resources and catches from Willemse and Pauly (*this volume*).

and Walters, 2004). The vulnerabilities are scaled between 0 and 1, with 0 defining bottom-up flow control, 1 delineating top-down flow control and 0.3 serving as the default for mixed control (Christensen and Walters, 2004). It is consequently possible to test the two extremes of predator control, i.e. "top down" or "bottom up", as well as the intermediate "wasp-waist" control by forage species (Christensen et al., 2000). As a result of the work done by Shannon et al. (2000) and Cury et al. (2000) in the southern Benguela ecosystem, the vulnerabilities in this model were adapted to include the "wasp-waist" control of forage fish, viz. sardine and anchovy. "Wasp-waist" control by forage fish occurs when these fish exert top-down control on their prey and bottom-up control on their predators (Cury et al., 2000; Shannon et al., 2000). "Wasp-waist" control of sardine and anchovy on zooplankton and other prey were included by increasing the vulnerability of the prey to sardine and anchovy from the default of 0.3 to 0.6, and by decreasing vulnerability of sardine and anchovy to its predators to 0.1.

Estimating primary production anomalies

The effects of the external forcing of environmental variables were inspected by first letting Ecosim derive forcing functions, or primary production anomalies, that would correlate best with the empirical time series data, and reduce the sum of squares of the output time series. Time series fitting was done by using a splicing value of five and by only fitting values from 1971 to 1999, excluding 2000. Six scenarios were tested, the first three with "wasp-waist" control and the final three without:

- "wasp-waist" control and forcing function on phytoplankton only;
- "wasp-waist" control and forcing function on phytoplankton and macrophytes;
- "wasp-waist" control and forcing function on phytoplankton and macrophytes, but with the weighting on hake CPUE reduced to 50%;
- no "wasp-waist" control, but with a forcing function on phytoplankton only;
- no "wasp-waist" control, but with a forcing function on phytoplankton and macrophytes;
- no "wasp-waist" control, but with a forcing function on phytoplankton and macrophytes, and with the weighting on hake CPUE reduced to 50%.

Correlations with environmental variables

Average summer and winter sea surface temperature data for the area 24-26°S, 9-11°E were obtained for 1971-1995. The data were extracted by Claude Roy[1] from the COADS release laboratory dataset (Woodruff et al., 1987) by

---

[1] Email: claude.roy@ird.fr

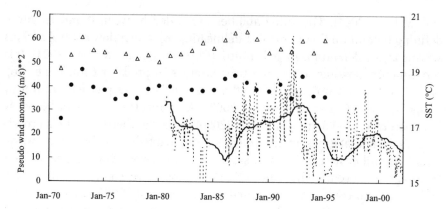

Figure 7. Environmental variables used for correlating to the primary production anomalies estimated by Ecosim. Triangles and dots are average summer and winter SST from 1971 to 1995 respectively. The dashed lines are the daily mean pseudo wind stress as the north-south wind speed in m.s.$_{-1}$ and the solid black line is the three-year average trend-line of pseudo wind stress. Data obtained from Claude Roy (SST) and Ministry of Fisheries and Marine Resources (pseudo-wind stress anomalies).

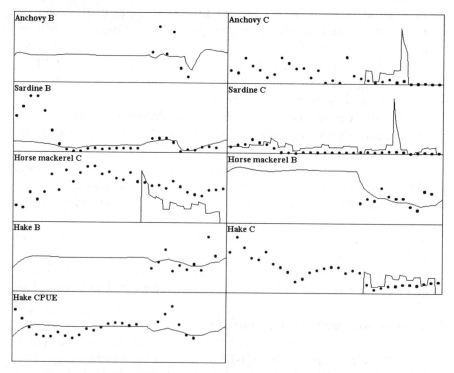

Figure 8. First run of the 1971-1977 model with time series data incorporated (SS = 216). The x-axis represents the time series modelled (1971-2000) and the y-axis shows the relative change in the measured parameter, with C being catch, B being biomass and CPUE being catch per unit effort.

means of an updated version of the CODE software (Mendelssohn and Roy, 1996), and used by Roy in Shannon and Crawford (2003). Daily mean pseudo wind stress data (as the north-south wind speed in m.s^{-1}) for 1980-2002 were obtained from the Namibian Ministry of Fisheries and Marine Resources in Swakopmund (Chris Barthalomae[2], *pers. comm.*). These data are shown in Figure 7.

The three-year average summer and winter sea surface temperature, as well as the three-year average pseudo wind stress, were then correlated to the three-year average forcing functions obtained for the six scenarios.

RESULTS

Time series fitting

The first run of the model excluded any internal forcing and only included the effects of the catch rate of the four groups discussed above. In this run, the sum of squares deviation was 216 (Table 1) and the results from this run are shown in Figure 8. In the second run of the model, "wasp-waist" control by sardine was included, which decreased the sum of squares from 216 to 146 (Table 1). In addition, including "wasp-waist" control of anchovy (0.6 for prey and 0.1 for predators) decreased the sums of squares marginally to 145.

Environmental forcing

EwE was used to derive forcing functions, or primary production anomalies, which would correlate best with the input data as a time series, and reduce the sums of squares of the output time series. Six scenarios were tested, three with "wasp-waist" control and three without. The resulting 3-year average of the six forcing functions are shown in Figure 9.

The sum of squares was reduced to 56 in Scenarios 1 and 2, and to 53 in Scenario 3. However, without "wasp-waist" control the sum of squares was not as low, being 133, 135 and 132 respectively for scenarios 4, 5 and 6 (Table 1). Thus, scenario 3, the estimated production anomaly on both phytoplankton and macrophytes with a reduced weighting on the CPUE of hake, reduced the sum of squares the most.

Ecosim calculates the probability distribution for the F statistic $SS_{reduced}/SS_{base}$ by using the null hypothesis that there are no real productivity anomalies (Christensen *et al.*, 2000), i.e. that the emergent primary production anomalies obtained in scenarios 1-6 can be explained by chance alone. This is done by using a Monte Carlo simulation procedure to account for autocorrelation in the model residuals (Christensen *et al.*, 2000). The

---

[2] Email: cbartholomae@mfmr.gov.na

probability that the emergent primary production anomaly explained environmental effects by chance alone was lowest in scenarios 1, 2 and 3 (P < 0.001), and highest in scenario 5 (P = 0.002), with scenarios 4 and 6 being intermediate (P = 0.001). However, Christensen et al. (2000) warns that even if there is a statistically significant reduction in sum of squares by using the search procedure, the estimated relative primary production values could still be misleading, and one can only postulate that *"assuming that primary production was in fact variable and that this did cause changes in relative abundance throughout the foodweb, then our best estimate of the historical pattern of variation is the one obtained by the fitting procedure"* (Christensen et al., 2000).

Comparison of predicted primary production anomaly and environmental variables

Two types of environmental factors that have been hypothesised to drive the northern Benguela ecosystem and that are tested here are SST (Cole, 1999) and wind anomalies (Kreiner et al., 2000). The correlation between the three-year average summer and winter sea surface temperatures, three-year average pseudo wind stress and the three-year averages of the forcing functions

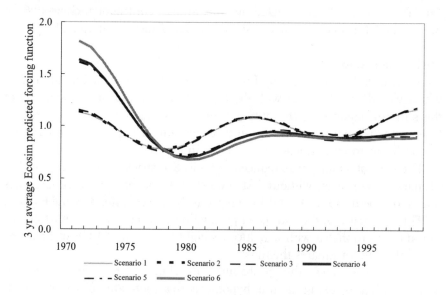

Figure 9. Three-year running average primary production anomaly predicted as a forcing function by Ecosim when using the time series data described in the text. Scenario 1 is a forcing function estimated only for phytoplankton, scenario 2 a forcing function for phytoplankton and macrophytes, and scenario 3 a forcing function for phytoplankton and macrophytes, but the weighting of hake CPUE was reduced to 50% in this scenario. Scenarios 4-6 are the same as scenarios 1-3 but the vulnerability settings were kept at 0.3, i.e. no "wasp-waist" control was imposed.

obtained from the six scenarios are shown in Figures 10-12. Correlation coefficients and significance are given in Table 1.

From the correlation coefficients in Table 1 and Figures 10-12 it was evident that the forcing functions estimated by Ecopath did not correlate well with winter SST, while there was a significant positive correlation with summer SST and a significant negative correlation with wind stress in scenarios 1-3. Scenarios 4, 5 and 6 (namely those that excluded "wasp-waist" control) did not show significant correlations with any of the environmental variables. The scenario that best correlated to summer SST was scenario 2, while scenario 1 correlated best to wind stress. The correlation with wind stress was significant at P = 0.01 while the correlation to summer SST was only significant at P = 0.05.

Finally, plots of the best fit of the model to the data, for the predicted biomass and catch estimates using wasp-waist control and assuming a forcing function on both phytoplankton and macrophytes (i.e. scenario 2) are given in Figure 13.

Table 1. Sum of squares deviation of log biomasses from log predicted biomasses, as well as correlation coefficients (r) and significance of correlations between environmental variables (average three-year averages of summer SST, winter SST and wind stress) and the six scenarios. Scenario 1 is a forcing function estimated only for phytoplankton, scenario 2 a forcing function for phytoplankton and macrophytes, and scenario 3 a forcing function for phytoplankton and macrophytes, but the weighting of hake CPUE was reduced to 50% in this scenario. Scenarios 4-6 are the same as scenarios 1-3 but the vulnerability settings were kept at 0.3, i.e. no "wasp-waist" control was imposed.

	Sum of Squares	Summer SST	Winter SST	Pseudo-wind stress
First run	216	-	-	-
"Wasp waist"	146	-	-	-
Scenario 1	56	0.470*	0.388	-0.215*†
Scenario 2	56	0.478*	0.393	-0.212*†
Scenario 3	53	0.408*	0.343	-0.209*†
Scenario 4	133	-0.111	0.009	0.028
Scenario 5	135	-0.103	0.018	0.124
Scenario 6	132	-0.178	-0.040	0.081
Degrees of Freedom	-	23	23	195
$r_{0.05\,2}$	-	0.396	0.396	0.142
$r_{0.01\,2}$	-	0.505	0.505	0.185

* = significant (P = 0.05) correlation
† = significant (P = 0.01) correlation

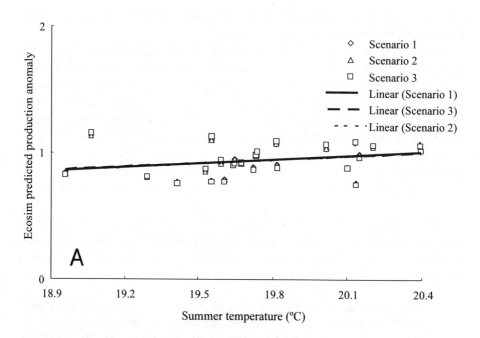

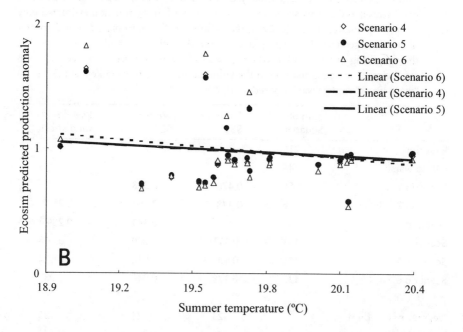

Figure 10. Correlations between three-year average summer sea surface temperature (°C) and the three-year average Ecosim predicted production anomalies for Scenarios 1-3 (Figure 10A) and 4-6 (Figure 10B).

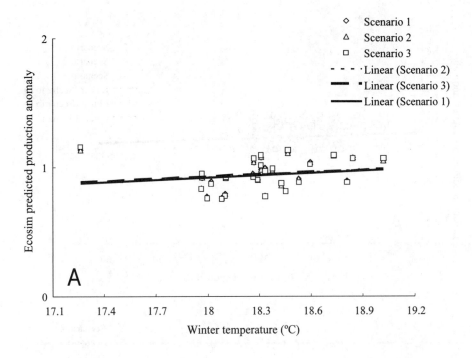

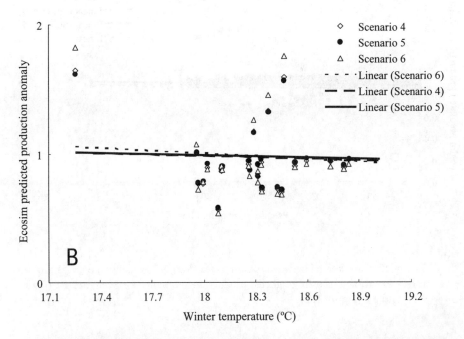

Figure 11. Correlations between the three-year average winter sea surface temperature (°C) and the three-year average Ecosim predicted production anomalies for Scenarios 1-3 (Figure 11A) and 4-6 (Figure 11B).

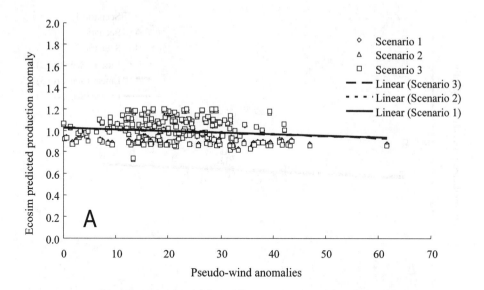

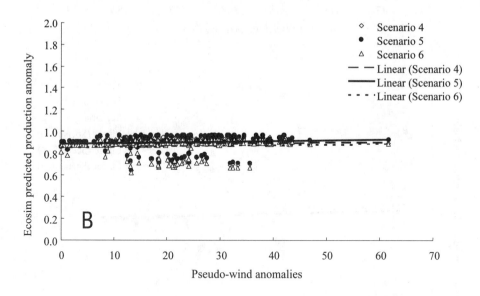

Figure 12. Correlations between the three-year average pseudo wind stress and the three-year average Ecosim predicted production anomalies for Scenarios 1-3 (Figure 12A) and 4-6 (Figure 12B).

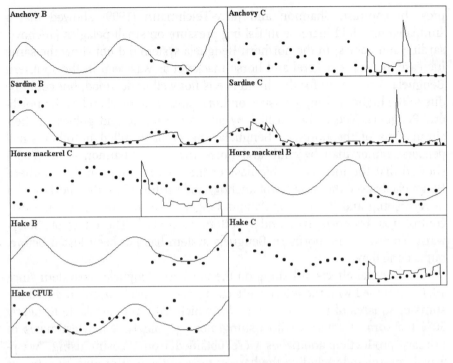

Figure 13. Best run of the 1971-1977 model with time series data incorporated (SS = 56). The x-axis represents the time series modelled (1971-2000) and the y-axis shows the relative change in the measured parameter, with C being catch, B being biomass and CPUE being catch per unit effort.

Discussion and Conclusions

The world's fisheries have been massively exploited since the industrialisation of the fishing fleets (Watson and Pauly, 2001). This exploitation has had a devastating effect on the important predatory fish species (Myers and Worm, 2003) and the food web as a whole (Christensen et al., 2003; Jackson et al. 2001; Pauly and Palomares, 2001). Similarly, in the southern Benguela ecosystem, fishing over the past 50 years has caused the previously important forage fish species such as anchovy and sardine to collapse (Cury et al., 2000) and fishing over the past 40 years in the northern Benguela ecosystem caused the collapse of the sardine and anchovy populations while the catches of hake and horse mackerel have declined (Boyer and Hampton, 2001).

According to Shannon et al. (2000) and Cury et al. (2000), modelling the heavy fishing pressure on small pelagic fish in the southern Benguela ecosystem caused stocks of anchovy and sardine to collapse and favoured groups such as chub mackerel and horse mackerel, which compete for zooplankton prey. In contrast, Shannon and Jarre-Teichmann (1999) showed that a simulated four-fold increase in fishing pressure on small pelagics (anchovy, sardine and gobies) in the northern Benguela system did not show the same increase in chub mackerel and horse mackerel as was seen in the southern Benguela. The reason for the difference is not well understood, but could be due to the higher fishing pressure on horse and chub mackerel in the northern Benguela. The niche vacated by anchovy, sardine and gobies in their simulations of the northern Benguela was probably filled by other small pelagics, rather than by chub and horse mackerel. Shannon et al. (2000) showed that the increase in biomass of these competitors in the southern Benguela delayed the recovery of anchovy and sardine since the alleviation of heavy fishing and it took populations longer to recover when "wasp-waist" control was assumed than under bottom-up control. The effect of "wasp-waist" control in the northern Benguela system has not been tested before, but is done here.

The external effects of fishing on the northern Benguela ecosystem since 1970, combined with the internal effects of "wasp-waist" control, reduced the sums of squares of the four groups for which data were available to nearly 50% the sum of squares with fishing alone (Table 1). When estimates of primary production anomalies were obtained from Ecosim using "wasp-waist" control and including the harvest rates of the four groups, the sum of squares was reduced by 65% (Table 1), while the primary production anomalies without "wasp-waist" control only reduced the sum of squares marginally.

The primary production anomalies correlated significantly with summer sea surface temperature (Figures 10 and 11) and pseudo-wind anomalies (Figure 12), the two environmental variables postulated to influence the ecosystem by Cole (1999); Kreiner et al. (2000) and Reid et al. (2000). These anomalies correlated best with summer sea surface temperature when including a forcing function on both phytoplankton and macrophytes, while the pseudo-wind stress anomalies correlated best with a forcing function on phytoplankton only. The positive correlation between summer sea surface temperature and the forcing functions that included "wasp-waist" control were significant at $P = 0.05$, while the negative correlation between pseudo wind stress anomalies and the forcing functions that included "wasp-waist" control were significant to $P = 0.01$ (Table 1).

The environmental variable that, as a result, seems to have the most significant effect is wind stress, which Kreiner et al. (2000) suggested to be a good predictor of sardine spawner biomass and recruitment. The effect of wind on upwelling in general is well known and has been studied extensively by Shannon et al. (1990); Andrews and Hutchings (1980) and Nelson and

Hutchings (1983), while Stenevik *et al.* (2001) and Kreiner *et al.* (2000) showed that wind has a significant effect on sardine in particular.

On the other hand, summer sea surface temperature also seems to play an important role in the ecosystem. Sea surface temperature, however, is also related to wind stress, as upwelling is driven by wind and by definition affects sea surface temperature. Thus, it would seem that using wind stress anomalies for future predictions would be more sensible.

Finally, including the internal forcing of forage fish as "wasp-waist" control as well as that of external forcing of pseudo-wind stress and/or summer sea surface temperature explains more of the variability in the ecosystem than any of these variables alone. The resulting estimated biomasses fits best to the biomass time series available (Figure 13) and indicates that both internal and external forces are at work in the northern Benguela ecosystem.

References

Allen, R.R. (1971): Relation between production and biomass. *Journal of the Fisheries Research Board of Canada* 28: 1573–1581.

Andrews, W.R.H., and Hutchings, L. (1980): Upwelling in the southern Benguela current. *Proceedings in Oceanography* 9: 1-81.

Boyer, D.C., Boyer, H.J., Fossen, I., and Kreiner, A. (2001): Changes in abundance of the northern Benguela sardine stock during the decade 1990-2000, with comments on the relative importance of fishing and the environment. *South African Journal of marine Science* (A Decade of Namibian Fisheries Science) 23: 67-84.

Boyer, D.C., Cole, J., and Bartholomae, C. (2000): Southwestern Africa: Northern Benguela Current Region. *Marine Pollution Bulletin* 41(1-6): 123-140.

Boyer, D.C., and Hampton, I. (2001): An overview of the living marine resources of Namibia. *South African Journal of Marine Science* (A Decade of Namibian Fisheries Science) 23: 5-35.

Brown, P.C., Painting, S.J., and Cochrane, K.L. (1991): Estimates of phytoplankton and bacterial biomass and production in the northern and southern Benguela ecosystems. *South African Journal of Marine Science* 11: 537-564.

Burmeister, L.-M. (2001): Depth-stratified density estimates and distribution of the Cape hake *Merluccius capensis* and *M. paradoxus* off Namibia deduced from survey data, 1990-1999. *South African Journal of Marine Science* (A Decade of Namibian Fisheries Science) 23: 347-356.

Butterworth, D.S., and Geromont, H.F. (2001): Evaluation of a class of possible simple interim management procedures for the Namibian hake fishery. *South African Journal of Marine Science* (A Decade of Namibian Fisheries Science) 23: 357-374.

Christensen, V., Guénette, S., Heymans, J. J., Walters, C., Watson, R., Zeller, D., and Pauly, D. (2003): Hundred-year decline of North Atlantic predatory fishes. *Fish and Fisheries* 4: 1-24.

Christensen, V. and Walters, C.J. (2004):

Ecopath with Ecosim: methods, capabilities and limitations. *Ecological Modelling* 172: 109-139.

Christensen, V., Walters, C., and Pauly, D. (2000): *Ecopath with Ecosim: A User's guide.* Fisheries Centre, University of British Columbia and ICLARM, Vancouver, BC and Penang, Malaysia, 131 pp.

Cole, J. (1999): Environmental conditions, satellite imagery, and clupeoid recruitment in the northern Benguela upwelling system. *Fisheries Oceanography* 8(1): 25-38.

Cole, J., and Villacastin, C. (2000): Sea surface temperature in the northern Benguela upwelling system, and implications for fisheries research. *International Journal of Remote Sensing* 21(8): 1597-1617.

Cram, D.L. (1981): Hidden elements in the development and implementation of marine resource conservation policy: the case of the South West African/Namibian fisheries. In: *Resource Management and Environmental Uncertainty: Lessons from Coastal Upwelling Fisheries.* M.H. Glantz and J.D. Thompson (Eds.). New York, Wiley: 137-155.

Cury, P., Bakun, A., Crawford, R.J.M., Jarre, A., Quiñones, A., Shannon, L.J., and Verheye, H. M. (2000): Small pelagics in upwelling systems: patterns of interaction and structural changes in "wasp-waist" ecosystems. *ICES Journal of Marine Science* 57: 603-618.

Estrada, M., and Marrase, C. (1987): Phytoplankton biomass and productivity off the Namibian coast. *South African Journal of Marine Science* 5: 347-356.

Fossen, I., Boyer, D.C., and Plarre, H. (2001): Changes in some key biological parameters of the northern Benguela sardine stock. *South African Journal of Marine Science* (A Decade of Namibian Fisheries Science) 23: 111-121.

Heymans, J.J. (1997): Network Analysis of the Carbon Flow Model of the northern Benguela Ecosystem, Namibia. Ph.D. Thesis. Zoology Department, University of Port Elizabeth, 206 pp.

Heymans, J.J. and Baird, D. (2000): Network analysis of the northern Benguela ecosystem by means of NETWRK and Ecopath. *Ecological Modelling* 131: 97-119.

Heymans, J.J., Shannon, L.J. and Jarre, A. (2004): Changes in the northern Benguela ecosystem over three decades: 1970s, 1980s and 1990s. *Ecological modelling* 172(2-4): 175-195.

Jackson, J.B.C., Kirby M.X., Berger, W.H., Bjorndal, K.A., Botsford, L.W., Bourque, B.J., Bradbury, R.H., Cooke, R., Erlandson, J., Estes, J.A., Hughes, T.P., Kidwell, S., Lange, C.B., Lenihan, H.S., Pandolfi, J.M., Peterson, C.H., Steneck, R.S., Tegner, M.J., Warner, R.R. (2001): "Historical Overfishing and the Recent Collapse of Coastal Ecosystems." *Science* 293: 629-638.

Jarre-Teichmann, A. (1998): The potential role of mass-balance models for the management of upwelling ecosystems. *Ecological Applications* 8 (1) Sup.: 93-103.

Jarre-Teichmann, A. and Christensen, V. (1998): Comparative modelling of trophic flows in four large upwelling ecosystems: global vs. local effects. In: Durant, M.H., Cury, R., Mendelssohn, R., Roy, C., Bakun, A. and Pauly, D. (Eds.). *From Local to Global Changes in Upwelling Systems*, pp. 423-443. ORSTROM, Paris.

Kreiner, A., Mouton, D., Daskalov, G., Clement, A. and Wiggert, J. (2000): Namibian sardine fisheries and its interaction with environmental conditions. Presentation given at the

Workshop on Interannual Climate Variability and Pelagic Fisheries, Nouméa, Nouvelle Calédonie, 6th - 24th November 2000.

Lutjeharms, J.R.E., and Meeuwis, J.M. (1987): The extent and variability of South-East Atlantic upwelling. *South African Journal of Marine Science* 5: 51-62.

Mendelssohn, R. and Roy, C. (1996): Comprehensive Ocean Data Extraction Users Guide. U.S. Dep. Comm. NOAA Tech. Memo. NOAA-TM-NMFS.SWFSC-228, La Jolla, California. 67 pp.

Myers, R.A., and Worm, B. (2003): Rapid worldwide depletion of predatory fish communities. *Nature* 423: 280-283.

Nelson, G., and Hutchings, L. (1983): The Benguela upwelling area. *Proceedings in Oceanography* 12(3): 333-356.

Pauly, D., and Palomares, M.L.D. (2001): Fishing down marine food web: it is far more pervasive than we thought. In: Conference on Sustainability of Fisheries Rosenstiel School of Marine Sciences, University of Miami, 26-28 November, 2001.

Reid, P.C., Battle, E.J.V., Batten, S.D. and Brander, K.M. (2000): Impacts of fisheries on plankton community structure. *ICES Journal of Marine Science* 57: 495-502.

Shackleton, L.Y. (1987): A comparative study of fossil fish scales from three upwelling regions. *South African Journal of Marine Science* 5: 79-84.

Shannon, L.J. and Crawford, R.J.M. (2003): Report of the SPACC/BENEFIT/IDYLE working group on major turning points in the Benguela ecosystem during the latter half of the 20th century Cape Town, February 2001.

Shannon, L.J., and Jarre-Teichmann, A. (1999): Comparing models of trophic flows in the Northern and Southern Benguela Upwelling systems during the 1980s. In: *Ecosystem Considerations in Fisheries Management*, pp. 527-541, Anchorage, Alaska, Alaska Sea Grant College Program. AK-SG-99-01.

Shannon, L.J., Cury, P. and Jarre, A. (2000): Modelling effects of fishing in the Southern Benguela ecosystem. *ICES Journal of Marine Science* 57: 720-722.

Shannon, L.V., Agenbag, J.J. and Buys, M.E.L. (1987): Large- and mesoscale features of the Angola-Benguela front. *South African Journal of Marine Science* 5: 11-34.

Shannon, L.V., Lutjeharms, J.R.E., and Nelson, G. (1990): Causative mechanisms for intra-annual and interannual variability in the marine environment around Southern Africa. *South African Journal of Science* 86: 356-373.

Stenevik, E.K., Sundby, S. and Cloete, R. (2001): Influence of buoyancy and vertical distribution of sardine *Sardinops sagax* eggs and larvae on their transport in the northern Benguela ecosystem. *South African Journal of Marine Science* (A Decade of Namibian Fisheries Science) 23: 85-97.

Thomas, R.M. (1986): The Namibian pilchard: the 1985 season, assessment for 1952-1985 and recommendations for 1986. *Collection of scientific Papers of the International Commission of South East Atlantic Fisheries* 13(2): 243-269.

Vaske, B. and Klingelhoeffer, E. (2001): Review of stock assessment for Cape horse mackerel off Namibia. Paper prepared for the Horse mackerel Workshop, 26-29 March 2001, MFMR, Swakopmund, Namibia.

Walters, C., Christensen, V., and Pauly, D. (1997): Structuring dynamic models of exploited ecosystems from trophic mass-balance assessments. *Re-*

views in *Fish Biology and Fisheries* 7: 139-172.

Walters, C., Pauly, D., Christensen, V., and Kitchell, J. F. (2000): Representing Density Dependent Consequences of Life History Strategies in Aquatic Ecosystems: EcoSim II. *Ecosystems* 3: 70-83.

Watson, R., and Pauly, D. (2001): Systematic distortions in world fisheries catch trends. *Nature* 414: 534-536.

Willemse, N.E. (2002): Major trends in the marine fisheries catches off Namibia, 1950-2000. Department of Biology, Norwegian College of Fishery Science, University of Tromsø, Norway, 65 pp.

Willemse, N. and Pauly, D. (2004): Reconstruction and interpretation of marine fisheries catches from Namibian waters, 1950 to 2000. In *Namibia's fisheries: Ecological, economic and social aspects*, Sumaila, U.R., Boyer, D., Skogen, M. and Steinshamn, S.I. (eds.), pp. 99-112. Eburon, Delft.

Winberg, G.G. (1956): Rate of metabolism and food requirements of fishes. *Translations of the Fisheries Research Board of Canada* 194: 1–253.

Woodruff, S.D., Slutz, R.J., Jenne, R.L. and Steurer, P.M. (1987): A Comprehensive Ocean Atmosphere Data-Set. *Bulletin of the American Meteorological Society* 68: 1239-1250.

3 BIODIVERSITY OF THE NAMIBIAN EXCLUSIVE ECONOMIC ZONE: A BRIEF REVIEW WITH EMPHASIS ON ONLINE DATABASES

Maria Lourdes D. Palomares and Daniel Pauly[*]

Abstract

A summary of available information on Namibian marine biodiversity is presented including the species' scientific and common names in English and Afrikaans, when available. This also includes a listing, when reported, of preferred habitat, status, national and international protective measures by species as well as biological information specific to Namibian locales. This review is based on data currently available from various sources through the Internet, i.e. searchable databases such as FishBase, a global database on fishes of the world, which was updated to include recently published information, including that presented in other chapters of this volume. An updated version of this national database will be maintained and kept available online (at *www.seaaroundus.org*) illustrating the type of 'minimum databases' that we believe maritime countries should create and maintain.

INTRODUCTION

The living marine resources of Namibia are relatively well-known, having been listed and commented upon in the field guide published by Bianchi *et al.* (1999), who relied on an extensive literature. The productivity of these resources, however, is due to their being imbedded in a faunistically or floristically much richer ecosystem. In this contribution, therefore, we briefly review the status of that biodiversity in terms of functional groups, i.e. groups of species with similar functions within the ecosystem (Figure 1). This contribution can thus be seen as an ecosystemic extension of the guide

---

[*] We would like to thank Ms. Catriona Day for providing the list of cephalopods from CephBase, Ms. Eeilis Nic Dhonncha for providing the list of algae from AlgaeBase, Ms. Vasiliki Karpouzi for providing material on seabirds and Ms. Johanna Heymans for her useful comments and for providing material on the Benguela ecosystem.

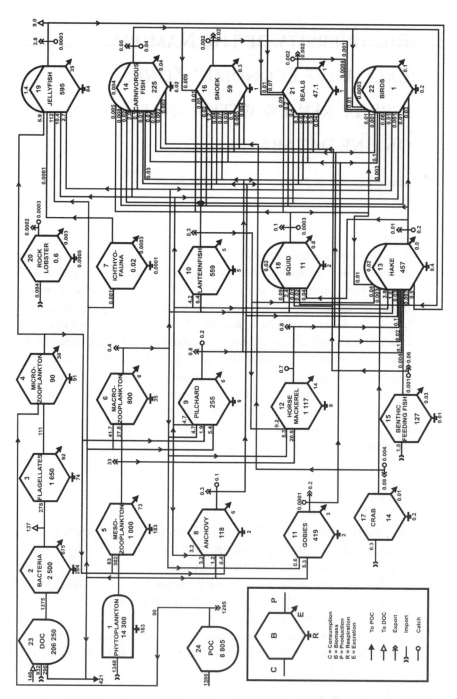

Figure 1. Flow diagram of the northern Benguela upwelling ecosystem adapted from Figure 2 of Heymans and Baird (2000). The species groups identified in this ecosystem model were used to structure the description of the biodiversity of the Namibian Exclusive Economic Zone.

to commercially important marine organisms of Bianchi *et al.* (1999) and an update of the same, as we will comment on the change of status of some of the marine resource species covered by these authors a decade ago.

The ecosystem structure we used follows roughly that of a food web model constructed and documented by Heymans and Baird (2000) and consisting of 24 functional groups (Figure 1). For each of the functional groups, we present the following:

- number of species;
- habitat requirements and other key biological information;
- IUCN status of component species, when available;
- treaties and/or protection measures relevant to these species;
- available sources of additional information on these species, with emphasis on online databases.

Our list is tentative, incomplete, and most likely biased towards fish and crustaceans, which are commercially important and thus better studied (Table 1). However, this list may serve as an example of what we believe is the minimum database each country should create and maintain to document its marine biodiversity.

BRIEF BACKGROUND OF THE NAMIBIAN MARINE ECOSYSTEM

The Namibian continental shelf is narrow, being at its widest off the mouth of Orange River and Walvis Bay, and extends to 90 km offshore between Cape Cross and Conception Bay, with a shelf area of about 110,000 km^2 (Sakko, 1998). The Namibian oceanic zone includes the SE Atlantic abyssal plains, i.e. the Angola and Cape basins, separated by the Walvis Ridge off northern Namibia. Sakko (1998) identifies four major oceanic regions: the epipelagic zone (0-200 m); the mesopelagic (200-1000 m); the bathypelagic (1000-4000 m); and the zone from 4000 m to the sea floor or the abyssopelagic which is generally poorly known; and two major shore habitats: the sandy littoral and the rocky intertidal habitats.

The Namibian EEZ is part of the Benguela system, the eastern boundary current characterized by the equatorward drift in the southeast Atlantic Ocean (Shannon and Pillar, 1986). This large marine ecosystem is divided into two highly productive sub-ecosystems, the northern and southern Benguela systems, separated by a permanent upwelling centre near Lüderitz at 27°S (Heymans and Baird, 2000). The oceanography of the northern limit of the Benguela, off Moçamedes, is similar to the equivalent systems in the South Pacific (the Peru or Humboldt current), characteristically with cool

Table 1. Number of species by habitat and by functional group assembled mainly from data in Sakko (1998) and supplemented with data from Bianchi et al. (1999) and other published sources.

Habitat	Functional group	Number of species	Endemics	Remarks	Reference
Sandy littoral	Macrofaunal invertebrates	<30	None	33.3% r\Restricted to Benguela system.	Fig. 3.3 in Sakko (1998)
Rocky intertidal	Benthic invertebrates	<200	1 Endemic to Namibia (*Discinisca tenuis*)	40% Gastropods and prosobranchs; 11.5% bivalves; 5% crustaceans; 4% polyplacophorans; 0.5% cephalopods; 15% restricted to Benguela system.	Bustamante et al. (1993); Fig. 3.4 and Table 3.1 in Sakko (1998);
0-200 m	Seaweeds	205			Lawson et al. (1990)
0-200 m	Bony fishes	410		Representing 13 orders of which 14.6% belong to Perciformes; 22.2% found at <30 m; 14.3% demersals.	Sakko (1998)
0-200 m	Cartilaginous fishes	83		Representing 10 orders; 36.1% found at <30 m; 21.7% sharks; 14.5% skates and rays.	Sakko (1998)
0-200 m	Zooplankton	>267		91% Copepods, 3.75% chaetognaths, 5.25% planktonic crustaceans.	Carola (1994)
0-200 m	Cephalopods	6		Data from R/V *Dr. Fridtjof Nansen*.	Sakko (1998)
0-200 m	Ichthyoplankton	>100		Eggs and larvae of commercially exploited fish species, e.g. *Austroglossus microlepis*, *Engraulis capensis*, *Hygophum macrochir*, *Lampanyctodes hectoris*, *Maurolicus muelleri*, *Merluccius* spp., *Parablennius pilicornis*, *Sardinops ocellatus*,	Karaseva and Shiganov (1993); Olivar and Barange (1989); Sakko (1998)

Depth	Group	Count	Notes	Reference
0-200 m	Phytoplankton	340	52.9% Diatoms, 47.1% dinoflagellates; 1.2% restricted to Benguela system.	Sakko (1998)
200-1000 m	Bony fishes	500	Mostly demersal species.	Sakko (1998)
1000-4000 m	Bony fishes	57	1 Endemic to Namibia (*Dicrolene pallidus*)	12.3% Exclusively bathypelagic.	Bianchi *et al.* (1999); Nielsen (1990); Sakko (1998)
1000-4000 m	Cartilaginous fishes	21		Sakko (1998)
0-4000 m	Nekton		*Scomberesox saurus scombroides, Sufflogobius bibarbatus, Symbolophorus* sp., *Trachurus capensis*. 10 Species of bony fishes pelagic in shelf waters, 14 species pelagic in shallow coastal waters, 13 species pelagic in habitats ranging from neritic to oceanic, 21 species occur in water column above sea bed in shallow coastal waters, 16 benthopelagic in shallow to deep ocean, no cartilaginous fish specifically neritic, 12 in pelagic as well as deep ocean.	Bianchi *et al.* (1999)
0-4000 m	Sea turtles	5		Bianchi *et al.* (1999)
	Seabirds	62	32.2% Rare or occasional visitors; 19.4% breed in Namibian waters.	Bianchi *et al.* (1999)
	Mammals	31	1 Endemic to Benguela system (*Cephalorhynchus heavisidii*)	74.2% Dolphins and toothed whales; 25.8% baleen whales.	Jefferson *et al.* (1993)

surface waters, while the southern limit of the Benguela, off Cape Town, is characterized by the warm waters of the Agulhas retroflection zone (Shannon, 1989). The fundamental differences in the two parts of the system make the often assumed occurrence of species in the north based on their occurrence in the south a risky proposition.

Table 1 presents a summary of the number of species so far listed occurring in these environments from the various reviews and reports consulted and reflecting the low species endemicity reported by Sakko (1998) of this marine system as a whole. Note that some of the references used here (and also in the online database available at *www.seaaroundus.org*) include species occurring slightly to the south and occasionally to the north of the Namibian border. In such cases, it is assumed that the species also occur within Namibia.

Kruger (1980) observed that the phytoplankton composition of Namibian waters is similar to that of the Mediterranean Sea (73% of Namibian species are also found in Mediterranean waters) and the Southwest Indian Ocean (72% overlap). Shannon and Pillar (1986) noted that the maximum zooplankton abundance lies in a belt parallel to the Namibian coastline but further offshore than the belt of maximum phytoplankton abundance. The high planktonic abundance, notably in the Benguela upwelling system, does not, however, reflect a high level of species diversity.

Macpherson and Gordoa (1996) reported that the benthic (100-800 m) fish assemblages off the coast of Namibia were located in areas covering both the active upwelling centres on the shelf and zones of lower productivity on the lower slope. This study also reports a high abundance of small individuals with low community diversity in the main upwelling centre, predominated by bony fishes and mid trophic level predators (on krill and pelagic crustaceans). Low productivity zones, on the other hand, tend to be dominated by cartilaginous fishes.

PROTECTION OF MARINE BIODIVERSITY IN NAMIBIA

The country has signed and ratified several international treaties and conventions, aimed at or indirectly related to the conservation of marine and coastal resources, *viz.*:

- the Vienna Convention for the Protection of the Ozone Layer;
- the Ramsar Convention on Wetlands;
- the Basel Convention on the Control of Transboundary Movements of Hazardous Wastes and their Disposal;
- the United Nations Framework Convention on Climate Change;

- the Convention on Biological Diversity;
- the Convention on International Trade in Endangered Species of Wild Fauna and Flora (CITES);
- the Migratory Bird Treaty;
- the International Whaling Commission;
- the UN Convention on the Law of the Sea;
- the Subcommittee of Forestry, Fisheries and Wildlife of the South African Development Community;
- the Lomé Convention; and
- the South East Atlantic Fisheries Organization (SEAFO).
- Namibia signed as well several trade agreements impacting on the sustainability of exploited resources within the region, viz.:
- the World Trade Organization;
- the South African Development Community (SADC);
- the Southern Africa Customs Union (SACU);
- the Common Market for Eastern and Southern Africa (COMESA); and
- the Generalized System of Preferences (GSP).

Namibia protects all seabirds occurring in its waters (Bianchi et al., 1999) and the relevant legislation is being updated, along with legislation protecting other groups (McGann et al., 2002). Regular coastal bird surveys along Sandwich Harbour, Walvis Bay and Lüderitz Bay are being conducted as part of a 25-year study by the Directorate of Environmental Affairs, Ministry of Environment and Tourism, Namibian National Biodiversity Programme (see www.dea.gov.na/programmes/biodiversity/birds.htm).

Monitoring surveys have been conducted since the 1960s for plankton, fish (sardines, horse mackerels, hakes, orange roughy, etc.), seals and other important groups. An example is the Namibian lobster research programme launched in the 1960s, which incorporates regular environmental and lobster surveys on the main fishing grounds for *Jasus lalandii* stocks (Grobler and Noli-Peard, 1997). Results obtained by this research programme indicate a recent improvement in the recruitment and abundance of this stock.

Since Independence in March 1990, the Namibian government has been incorporating a number of provisions into its legislation concerning, among others, ownership of marine resources, fisheries in the Exclusive Economic Zone, forfeiture of vessels used in committing fishery offences and international cooperation in enforcing fishery laws (Devine, 1993). Also, the Namibian government exercised its authority to control the access to marine resources by its domestic fishers (FAO, 2000).

The legislation protecting wildlife and natural resources cited by Sakko (1998; see also http://www.mfmr.gov.na/policy/policies.htm) include:

- the Parks and Wildlife Act;
- the Environmental Management Act;
- the Conservation of Biotic Diversity and Habitat Protection Policy;
- the Marine Traffic Act;
- the Merchant Shipping Act;
- the Prevention of Pollution of the Sea by Oil Act;
- the Territorial Sea and Exclusive Economic Zone Act;
- the Sea Fisheries Act of 1992; and
- the New Marine Resources Act of 2000.

MATERIALS AND METHODS

We describe in the following, the data sources we tapped and the method we used to assemble the biodiversity lists presented further below.

The list of commercial marine resources of Namibia by Bianchi *et al.* (1999) was used as a starting point. Branch *et al.* (1994) and Sakko (1998) supplied a considerable part of the marine invertebrate list. Note that Bianchi *et al.* (1999) supplied not only information on the commercial importance of some species, but also on whether some species are potentially important or likely to be affected by commercial fishing (see online database at *www.seaaroundus.org* for more details). The list of marine mammals was improved with additional information from Jefferson *et al.* (1993). Birdlife International (2001; see *www.birdlife.net*) supplied almost all information on seabirds. Information on fish groups was obtained from FishBase (Internet version April 2003; see *www.fishbase.org*). CephBase (*www.cephabase.org*) was used to supplement Bianchi *et al.* (1999), notably on common names, feeding and predator information for cephalopods (noted only for localities near or in Namibian waters). AlgaeBase (see *www.algaebase.org*) was used to supplement the list provided by Bianchi *et al.* (1999) of important algae present or potential use in Namibian marine waters.

The taxonomic list was expanded to include, when applicable, the names of the Order, Suborder, Infraorder, Super Family and Family to which the species belong. These higher hierarchy names were obtained largely through the ITIS biological name search site (see *http://sis.agr.gc.ca/pls/itisca/taxaget?p_ifx=cbif*) accessible through the Canadian Biodiversity Information Facility (CBIF) (see *http://www.cbif.gc.ca/home_e.php*).

The list of threatened species was obtained from the Internet version of the IUCN (1994; see *www.redlist.org*); the list of internationally protected species was obtained from CITES (see *www.unep-wcmc.org/index.html? http://www.unep-wcmc.org/CITES/redirect.htm~main*).

Results and Discussion

Group-specific results
The following describe in some detail results obtained for each of the groups for which information is available. Note that marine turtles, microflagellates, bacteria, macroalgae and phytoplankton are not discussed in detail. The detailed list of species is available as an online searchable database at *www.seaaroundus.org*.

Birds. - Bianchi *et al.* (1999) reported that a total of 62 species of seabirds, including 20 rare visitors, have been recorded in Namibia. However, Bianchi *et al.* (1999) provides detailed information for only 19 commercially important species of seabirds, including 7 (guano producers) of the 12 seabird species which breed along the Namibian coast, the rest being pelagic seabirds most often encountered by fishers at sea. Branch *et al.* (1994) lists 30 coastal birds including the following shorebirds: greater flamingo (*Phoenicopterus ruber*), grey heron (*Ardea cinerea*), greenbacked heron (*Burorides striatus*), little egret (*Egretta garzetta*), curlew sandpiper (*Calidris ferruginea*), turnstone (*Arenaria interpres*), sanderling (*Calidris alba*), grey plover (*Pluvalis squatarola*), whitefronted plover (*Charadrius marginatus*), cape wagtail (*Motacilla capensis*), African black oystercatcher (*Haematopus moquini*), Caspian tern (*Hydroprogne caspia*), common tern (*Sterna hirundo*), Antarctic tern (*Sterna vittata*), Arctic tern (*Sterna paradisaea*), sandwich tern (*Sterna sandvicensis*), swift tern (*Sterna bergii*) and damara tern (*Sterna balaenarum*). Of the 25 endangered bird species listed in the Birdlife International species database for Namibia, 5 are seabirds and 4 are shorebirds.

The online database mentioned above lists 58 of the 62 seabird species from these different sources, including 3 species endemic to southern Africa, i.e. cape cormorant, *Phalacrocorax capensis*, cape gannet, *Morus capensis* and jackass penguin, *Spheniscus demersus*. The database also includes 4 commercially important guano-producing species, i.e. white pelican, *Pelecanus onocrotalus*, white-breasted cormorant, *Phalacrocorax carbo lucidus*, crowned cormorant, *Phalacrocorax coronatus* and bank cormorant, *Phalacrocorax neglectus*. These guano producers are included in the 11 IUCN listed species of seabirds in Namibia. Only 38 of those listed in the database are listed by CITES.

Marine mammals. - Jefferson *et al.* (1993) lists 36 species of cetaceans (8 baleen whales and 28 toothed whales, dolphins and porpoises) and 4 pinnipeds occurring between 20°S and 40°S. Bianchi *et al.* (1999) reported 31 species (8 baleen and 23 toothed whales, dolphins and porpoises) and 3 species of pinnipeds, but provides information on only 12 cetaceans and 2 pinnipeds. The number of cetaceans occurring in Namibian waters represents a considerable 41% of the total number of species of cetaceans worldwide. Of these, only one is endemic, heaviside's dolphin, *Cephalorhynchus heavisidii*, which is a coastal shallow water species found only off southern Africa from about 17°S to the southern tip of Africa (Jefferson et al., 1993). This species is common in Namibian waters and seen mostly about 5 miles from the shore in small groups of 2-7 individuals (Bianchi et al., 1999).

Peddemors (1999), who reported 18 species of delphinids from Africa south of 17°S, concludes that there seems to be little human-induced threat to these species at present and only two inshore species are considered vulnerable, i.e. the Namibian population of heaviside's dolphin, and a localized Namibian population of bottlenose dolphin, *Tursiops truncatus*, which is reported as vulnerable to future coastal development and commercial fishery expansions. The offshore, less common southern right whale dolphin, *Lissodelphis peronii*, also forms a localized population, potentially vulnerable to fishing activities (Peddemors 1999).

The use of stranded marine mammals along the Namibian coast has been recorded since the time when the Khoikhoi, a group of nomadic cattle and sheep herders, occupied the western Namibian coast from 1800 to 1600 years ago, and consumed stranded whale meat and used whale bones to build huts (Boonzaier et al., 1996). Early European visitors recorded an abundance of whales in these waters and 18th century records indicate the dominance of American whalers in the 'Coast of Guinea' in the 1760s and in Angola and Woolwich (Walvis) Bay before 1770 (Best and Ross, 1989). Extensive exploitation of whales in these regions developed rapidly, culminating in the 1930s when whales, notably sperm whales, humpback whales, fin whales and blue whales, were killed and processed in whaling stations along the Namibian coast (Bianchi et al., 1999). Today, though all cetaceans are protected through international and national programmes, most of the once heavily exploited species in this region are still very rare and vulnerable (Bianchi et al., 1999).

Cephalopods. - There are 55 species of cephalopods occurring in Namibia according to Nesis (1991), but Bianchi et al. (1999) include only 19 commercially important or potentially important species in 8 families, though they may be important mainly as bycatch. The database includes 66 species over 26 families and may include species found in South Africa as reported in Branch et al. (1994) and 13 species caught with bottom trawls or as bycatch of bottom trawls. None of the cephalopod species are listed in the IUCN or in the CITES Appendices.

Lobsters. - There are 11 species of lobsters occurring in Namibia (Bianchi et al., 1999) but we found information for only 10 (see database for more details) from 5 families. Four of these have commercial value or of potential commercial interest, i.e., Cape rock lobster, *Jasus lalandii*, royal spiny lobster, *Panulirus regius*, red slipper lobster, *Scyllarides herklotsii*, and scarlet lobsterette, *Nephropsis atlantica*. The most important of these is the Namibian Cape rock lobster stock, which was first exploited in the 1920s, with catches reaching up to 9,000 t in the mid 1960s, decreasing to 1000-3000 t in 1968-1969 with the abolition of the minimum legal size limit, and further decreasing in the 1980s perhaps owing to warm water intrusions from the south adding to the effect of overfishing. The biomass decreased further in 1992 when the TAC was set at 100 t (Grobler and Noli-Peard, 1997). This fishery is now protected by closed seasons (November-May) and size restriction (minimum of 65 mm carapace length; Bianchi et al., 1999).

Crabs. - Bianchi et al. (1999) reported 38 species of crabs occurring in Namibia, of which 9 are either caught by targeted fisheries or frequently occurring in bycatch. Branch et al. (1994) reported 300 species occurring in South Africa and reported a few more species which are known to occur in the southern border of Namibia with South Africa. These are integrated in the online database which lists 43 species, 58% of which are found at depths 0-1000 m.

Shrimps and prawns. - The lists of shrimps and prawns available indicate 54 (Macpherson, 1991) to 56 (Bianchi et al., 1999) species occurring in Namibia. We were able to reconstruct a list of 48 species, 40% of which are swimming and 60% benthic shrimps, and of which 50% are found at 200-1000 m depths, 15% at 0-200 m and about 1% are deepwater species occurring at depths >1000 m. Only a third of these species are of potential commercial interest.

Fishes. - Bianchi et al. (1999) listed a total of 492 fish species from 163 families, consisting of 2 species of jawless fishes from 1 family, 46 sharks from 15 families, 28 species belonging to 7 families of batoid fishes, skates and rays, 6 chimaerids from 3 families and 410 bony fishes from 137 families. Only about 40% of these are discussed in detail by Bianchi et al. (1999) because of their commercial or potential value as exploitable resources. However, the checklist obtained from FishBase (see *www.fishbase.org*) accounts for a total of 515 native and 1 endemic species. Almost 50% of these 516 species are bottom dwellers (20% demersals and 18 benthopelagic; Figure 2) while 36% are deepwater species (18% each bathypelagic and bathydemersals) and only 13% are truly pelagic species. This is not surprising given that the Na-

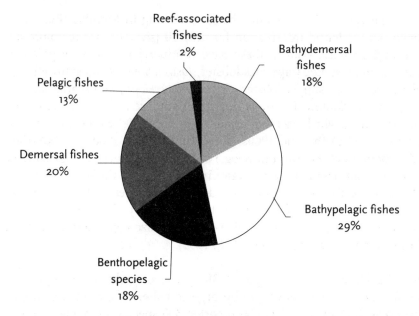

Figure 2. Contribution of marine fish species by habitat in the Namibian marine ecosystem. Data from FishBase list of marine fishes of Namibia (Southeastern Atlantic, FAO Area 47; see www.fishbase.org).

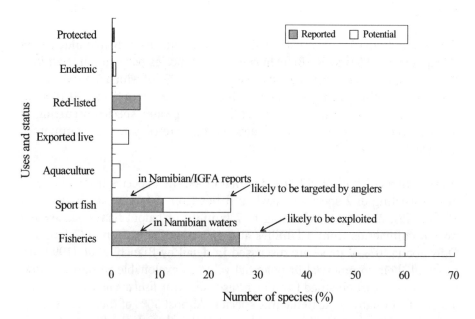

Figure 3. Number of fish species (n=**407**) by commercial use, status of threat and protection in the list of marine fishes of Namibia (Southeastern Atlantic, FAO Area **47**) extracted from FishBase (see www.fishbase.org).

mibian EEZ comprises 64% of waters deeper than 1000 m, 24% between 200 and 1000 m, and 12% between 0 and 200 m (see the *Sea Around Us* Project database at *www.seaaroundus.org*).

The bulk of these fishes, about 65% of the species, are omnivores with trophic levels 3-4, and only about 33% can be considered true carnivores, with trophic levels 4-5. Only about 3% are herbivorous with trophic levels 2-3 (see Willemse and Pauly (this volume) for a trophic level definition and Bianchi *et al.* (this volume) for an analysis of Namibian fish communities). Further analyses of the FishBase data indicate about 34% have commercial use in Namibia and another 42% are reported to have potential commercial value. More than 60% are exploited by commercial and about 25% by the sport fisheries (Figure 3). Three species are currently under the IUCN list of threatened species, i.e. broadnose sevengill shark, *Notorynchus cepedianus* (Data Deficient), yellowspotted catshark, *Scyliorhinus capensis* (low risk/near threatened) and sharptooth houndshark, *Triakis megalopterus* (low risk/near threatened), while two appear in the CITES list of species, i.e., great white shark, *Carcharodon carcharias* (CITES Appendix III for Australian populations) and whaleshark, *Rhincodon typus* (CITES Appendix II).

We do not include in this analysis a discussion of the status of fish stocks and fisheries, as these will be found in other chapters of this book, notably Willemse and Pauly (this volume).

Benthic invertebrates. - Sakko (1998) reports about 200 benthic invertebrate species occurring in Namibia (see Table 1). Our database documents about 70% of these, including 1 Namibian endemic species, i.e. disc lamp shell, *Discinisca tenuis* and 1 endemic to the Benguela: Cape mantis shrimp, *Pterygosquilla armata capensis*. The disc lamp shell is a benthic filter feeder abundant at depths to 25 m off the southern coast of Namibia (Sakko 1998, p. 196). These can attach themselves on top of another to form rafts of shells that are frequently washed up along the driftline (Branch *et al.*, 1994). The Cape mantis shrimp, a stomatopod burrowing in soft terrigenous sediments at depths <300 m, and which occurs between St. Helena Bay and the southern border of Namibia, appears to be endemic to that region (Griffiths and Blaine, 1988). It congregates near the surface in large swarms which are preyed upon by seals, hake and other fish (Branch *et al.*, 1994).

The bulk of these benthic invertebrates occur in depths at 0-200 m; only the sea spider, *Pallenopsis bulbifera*, a species newly described from the Namibian coast by Munilla and Stock (1984) was taken during the cruise Benguela V of the Fisheries Research Institute at a depth of 260-269 m.

Jellyfishes. - The database includes 7 species of jellyfishes. The dominant jellyfish of the Benguela Current system, *Aequorea aequorea*, is a large medusa found usually offshore and to the north of Namibia between 0 and 200 m (Sparks et al., 2001). Another species which was abundant in phytoplankton samples obtained by staff of the University of Namibia from the coastal area off Swakopmund is the moon jelly, *Aurelia aurita* (Senn, 2001). Multi-frequency acoustic data collected by Brierley et al. (2001) indicate that the high densities of gelatinous macrozooplankton, predominantly the scyphozoan, *Chrysaora hysoscella* found in shallow inshore waters (3 individuals per 100 m^3) and *Aequorea aequorea* found in deeper offshore waters (168 individuals per 100 m^3) of the Benguela system off Namibia in 1999 have become a potentially important physical obstruction to pelagic fishing. Moreover, they may bias acoustic estimates of fish abundance.

Macrozooplankton. - Different krill species dominate the different regions of the Benguela current. The shelf region of the southern Benguela current is dominated by *Euphausia lucens* while that the outer shelf is dominated by *Euphausia hanseni*. The neritic region of the northern Benguela is dominated by *Nyctiphanes capensis* (Pillar et al., 1991). These species have high turnover rates, as they reproduce throughout the year with multiple recruitment pulses and large numbers of eggs per brood. Northern Benguela euphausiids have twice as much biomass as those of the southwest Benguela. Adults are opportunistic omnivores, while juveniles are herbivorous. Pillar et al. (1991) report that euphausiids also have a minor impact on the phytoplankton biomass, but have a high predatory impact on other mesozooplankton species. Thus, euphausiids compete directly with various species of fish, which are also the major consumer of euphausiids in the Benguela ecosystem.

Ichthyoplankton. - The eggs and larvae of many fish species are distributed throughout the entire water column, from the surface to depths of 200 m when the upwelling is intense (Olivar et al., 1991). However, ichthyoplankton distribution is clearly segregated between coastal and slope/oceanic areas in intense upwelling seasons in northern Benguela, i.e. the lanternfishes, *Lampanyctodes hectoris* and *Maurolicus muelleri* dominate oceanic areas while *Sufflogobius bibarbatus* dominates the coastal areas. Olivar and Barange (1989) reported that in the 1980s, the most abundant ichthyoplankton species on the Namibian continental shelf and slope were larvae of ringneck blenny, *Trachurus pilicornis*; lanternfishes, *Symbolophorus* sp., *Hygophum macrochir*; and pelagic goby, *Nematogobius bibarbatus* (synonym of *Sufflogobius bibarbatus*; see www.fishbase.org). Larval and egg abundance were different between the southern and northern coasts, i.e. the northern coast apparently acted as a nursery area for the majority of the species. Similarly, vertical distribution showed stratification with highest concentrations above the thermocline and upwelling affected the diurnal distribution of anchovy and sardine eggs and larvae as well as those of king gar, *Scomberesox saurus scombroides* larvae which were most abundant at night. Based on another survey of the area 20-100 miles of the coast between 17°S and 25°S, Belyanina and Stejker (1988) reported that during the non-upwelling period of April-May 1985 and January 1986, ichthyoplankton was scarce with larval oceanic mesopelagic fishes, mainly myctophids, dominating the samples and that at depths shallower than 500 m, the average ichthyoplankton abundance increased with high concentrations of sardine, *Sardinops ocellatus* (synonym of *Sardinops sagax*, see www.fishbase.org) and horse mackerel, *Trachurus* spp.

General results
Reviewing the marine biodiversity of a country proved to be much more than assembling a checklist of species occurring in that country. We were able to assemble, from various sources and websites, a list consisting of 1,053 marine species, of which 47% are non-fish species from 212 families and 53% fish species from 163 families. This list includes roughly 70% of the species reported in Table 1. The list assembled in the online database is in no way complete, but all clearly add to existing databases on Namibian marine biodiversity.

Assuming that the data we used here are representative of the species occurring in the Namibian EEZ, one of the results probably reflecting the particular topography of the Namibian EEZ, i.e. narrow shallow shelf area and large oceanic zone with an upwelling area (Sakko 1998), is the predominance of deep-water fish groups, e.g. bathypelagic fishes (15%) closely followed by demersals (11%) and benthopelagic fishes (10%). The next most important groups are high trophic level groups, which include cephalopods, seabirds and marine mammals, together comprising 15%, while benthic invertebrates together cover 13% of the species composition. Figure 4 gives a bird's eye view of the contribution of each functional group, and implies a larger number of high trophic level species. This is biased however, as large, high trophic level and commercially important species are better documented than the small inconspicuous species that are not commercially exploited.

Table 2 summarizes species counts of endemic, commercial, threatened and protected species, and provides counts of the number of species by depth range and habitat. The number of endemic species accounts for 1% of the total number of species listed in the online database, confirming the relatively low level of endemicity reported in Sakko (1998). Commercially important species account for 14%, with fish groups making up 66% of those that are commercially important.

To compare our results with those that would be obtained by an unwary Internet user, we performed a search for 'Namibia' and 'Southeast Atlantic' area through the IUCN (www.iucn.org) species search. This resulted in 13

Functional group	Number	Endemic	Commer-cial	IUCN	CITES	0 - 200 m	200 - 1000 m	1000 - 4000 m	Pelagic	Demersal	Benthic
Seabirds	58	3	4	11	38	0	0	0	0	0	0
Marine mammals	35	1	4	13	35	16	30	23	0	0	0
Turtles	5	0	1	1	5	5	0	0	5	5	0
Bathydemersal fishes	97	0	17								
Bathypelagic fishes	160	0	4								
Benthopelagic fishes	102	1	24								
Demersal fishes	111	0	27								
Pelagic fishes	70	0	25								
Reef-associated fishes	12	0	3								
Cephalopods	66	0	8	0	0	27	28	5	30	9	5
Lobsters	10	0	4	0	1	3	3	0	0	0	5
Crabs	43	0	9	0	0	11	14	1	0	15	8
Shrimps and prawns	48	0	14	0	0	7	23	3	10	7	11
Invertebrates											
Ascidians	2	0	0	0	0	2	0	0	0	0	2
Brachiopods	1	1	0	0	0	1	0	0	0	0	1
Crustaceans, Cirripeds	3	0	0	0	0	3	0	0	0	0	3
Crustaceans, Amphipods	14	0	0	0	0	14	0	0	1	0	14
Crustaceans, Harpacticoids	2	0	0	0	0	2	0	0	0	0	2
Crustaceans, Isopods	20	0	0	0	0	20	0	0	0	0	19
Crustaceans, Mysids	1	0	0	0	0	1	0	0	1	0	1

Functional group	Number	Endemic	Commer-cial	IUCN	CITES	0 - 200 m	200 - 1000 m	1000 - 4000 m	Pelagic	Demersal	Benthic
Invertebrates											
Crustaceans, Stomatopods	1	1	0	0	0	1	0	0	0	0	1
Echinoderms	5	0	0	0	0	5	0	0	0	0	5
Hexacorals	1	0	0	0	0	1	0	0	0	0	1
Hydroids	7	0	0	0	0	7	0	0	0	0	7
Mollusks, Bivalves	11	0	0	0	0	11	0	0	0	0	11
Mollusks, Chitons	4	0	3	0	0	4	0	0	0	0	4
Mollusks, Gastropods	37	0	1	0	0	37	0	0	0	0	37
Polychaete worms	31	0	1	0	0	31	0	0	0	0	31
Pycnogonid	1	0	0	0	0	0	1	0	0	1	0
Zooplankton											
Crustaceans	10	0	0	0	0	3	3	0	10	0	0
Jellyfishes	7	0	0	0	0	7	2	1	7	0	0
Primary producers											
Macroalgae	41	2	3	0	1	10	0	0	0	0	41
Phytoplankton, Diatoms	26	0	0	0	0	0	0	0	26	0	0
Phytoplankton: Dinoflagellates	7	0	0	0	0	0	0	0	7	0	0
Plankton: Protozoa	3	0	0	0	0	0	0	0	3	0	0
Plankton: Bacteria	1	0	0	0	0	1	0	0	0	0	1
Sub-total: Non-fish groups	501	8	52	25	80	230	104	33	100	37	210

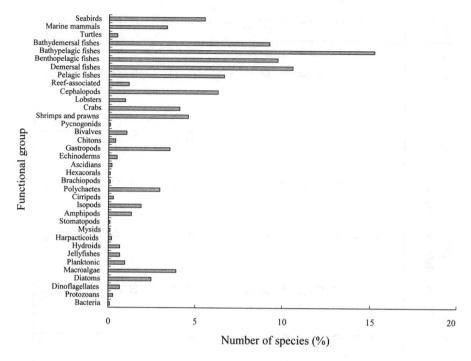

Figure 4. Number of species (n=1053) by functional groups resulting from analysis of the data assembled in the online database for the Namibian marine ecosystem (Southeastern Atlantic, FAO Area 47; see www.seaaroundus.org).

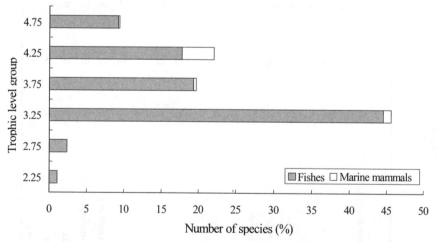

Figure 5. Number of species (n=586) by trophic level groups of fishes (n=552) and mammals (n=34) in the Namibian marine ecosystem. Fish data from FishBase list of marine fishes from Namibia (Southeastern Atlantic, FAO Area 47) (see www.fishbase.org) and the online database of Namibian marine biodiversity at www.seaaroundus.org.

marine species, 9 of which are marine mammals, 3 sharks and 1 turtle. The search did not find seabirds, which should actually have been marked for the Southeast Atlantic as well. Our list contains 28 marine species, i.e. 13 marine mammals, 11 seabirds, 3 sharks and 1 turtle.

A similar search for species listed in the UNEP-WCMC database for Namibia yielded 31 amphibians, 627 birds, 106 fishes, 232 invertebrates, 208 mammals, 216 reptiles, 7 orchids and 459 others. Here, again, habitats were not provided so we examined the list for distinctions by habitat. This yielded 76 species listed in the CITES Appendices I-III, ratified in February 13, 2003 and 6 more which are protected locally by the Namibian government. Of these, 38 are seabirds, 35 marine mammals, 5 turtles, 2 sharks, 1 lobster and 1 macroalga.

In Figure 5, we illustrate the trophic level structure of the Namibian EEZ ecosystem by plotting the number of species (% of total per group) by trophic level group of marine mammals and fishes, considered the top predators of the system. Most marine mammals have trophic levels above 4.0 while fishes tend to have lower trophic levels (see also www.fishbase.org).

Namibian legislature protects all marine mammals, marine turtles and seabirds. This contrasts to the level of protection (and research) focused on invertebrates, of which only few are protected, though many are subject to the direct or indirect effects of fishing. More than 33% of benthic invertebrate species in our checklist (see online database, www.seaaround.org) are potentially affected by commercial fisheries. However, none of them figure in the IUCN nor CITES lists nor appear to receive any specific protection in Namibia. This result, it must be stressed, is similar to the situation in other countries, which tend to protect vertebrate megafauna, but not 'ugly' invertebrates.

As mentioned earlier, a list such as presented here can never be complete and we will endeavor to update and keep this list available at the website of the *Sea Around Us* Project (see above).

- The first lesson was that online resources are still not sufficient to create acceptable marine biodiversity lists for developing countries such as Namibia. Thus, those attempting to create biodiversity lists to meet the requirement of the membership in the convention on biological diversity will generally have to rely on published sources. In this case, the key references were the early work of Bianchi et al. (1999), and gray literature with limited distribution.

- Second, confirming an experience already observed in the process of creating FishBase, we noted that it would be more straightforward for countries interested in creating marine biodiversity lists to team up regionally in the creation of regional or global lists by taxonomic groups, as these can be more efficiently produced, and their species subsequently assigned to countries, than by working country by country.

- Third, the major deficiency and the major need, in this context, is for the creation of global marine invertebrate databases, of which only one so far, 'CephBase', has a global coverage.

References

Belyanina, T.N. and T.N. Stejker (1988): Contribution to the studies of ichthyoplankton from the Benguela upwelling. [Kizucheniyu ikhtioplanktona Bengel'skogo apvellinga]. *Okeanologiya/Oceanology* (MOSC.) 28: 663-666.

Best, P.B. and G.J.B. Ross (1989): Whales and whaling, p. 315-338. In: *Oceans of life off southern Africa* (A.I.L. Payne and R.J.M. Crawford, eds.),. Vlaeberg Publishers, Cape Town.

Bianchi, G., K.E. Carpenter, J.-P. Roux, F.J. Molloy, D. Boyer and H.J. Boyer (1999): Field guide to the living marine resources of Namibia. FAO species identification guide for fishery purposes. FAO, Rome. 265 p.

Birdlife International (2001): Birdlife's online world bird database: the site for bird conservation. Version 1.0 Cambridge, UK: Birdlife International. Available: http://www.birdlife.net (accessed February 10, 2003).

Boonzaier, E., C. Berens, C. Malherbe and A. Smith (1996): *The cape herders: a history of the Khoikhoi of Southern Africa*. D. Philip Publishers (Pty) Ltd. Ohio University Press, Athens, OH, 147 p.

Branch, G.M., C.L. Griffiths, M.L. Branch, and L.E. Beckley (1994): *Two oceans. A guide to the marine life of Southern Africa*. David Philip Publishers (Pty) Ltd., South Africa. 360 pp.

Brierly, A.S., B.E. Axelsen, E. Buecher, C.A.J. Sparks, H. Boyer and M.J. Gibbons (2001): Acoustic observations of jelly fish in the Namibian Benguela. *Marine Ecology Progress Series*. 210: 55-66.

Bustamante, R.H., G.M. Branch, C.R. Velasquez and M. Branch (1993): Intertidal survey of the rocky shores at the Elizabeth Bay area (Sperrgebiet, Namibia). Report to Consolidated Diamond Mines (CDM, Namdeb Diamond Corporation Ltd), 37 pp.

Carola, M. (1994): Checklist of the marine planktonic copepoda of southern Africa and their worldwide geographic distribution. *South African Journal of Marine Science* 14: 225-253.

Devine, D.J. (1993): Marine law developments in Namibia. *International Journal of Marine and Coastal Law*, 8(4):471-495.

FAO (2000): Status and important recent events concerning international trade in fishery products. Minutes of the meeting of the Committee on Fisheries, Subcommittee on Fish Trade, seventh session, Bremen, Germany, 22-25 March 2000. (*www.fao.org/docrep/meeting/x4367e.htm*)

Griffiths, C.L. and M.J. Blaine (1988): Distribution, population structure and biology of stomatopod Crustacea off the west coast of South Africa. *South African Journal of Marine Science* 7:45-50.

Grobler, C.A.F. and K.R. Noli-Peard (1997): *Jasus lalandii* fishery in post-independence Namibia: monitoring population trends and stock recovery in relation to a variable environment. *Marine and Freshwater Research* 48(8): 1015-1022.

Heymans, J.J. and D. Baird (2000): A carbon flow model and network analysis of the northern Benguela upwelling system, Namibia. *Ecological Modelling* 126: 9-32.

IUCN (1994): IUCN Red List Categories. Prepared by the IUCN Species Survival Commission. IUCN, Gland, Switzerland.

Jefferson, T.A., S. Leatherwood and M.A. Webber (1993): FAO Species Identification Guide. Marine Mammals of the World. FAO. Rome. 320 pp.

Karaseva, E.M. and T.A. Shiganov (1993): Species composition and seasonal abundance dynamics of ichthyoplankton over the shelf of Namibia in 1988-1989 [Vidovoj sostav i sezonnaya dinamika chislennosti ikhtioplanktona na shel'fe Namibii v 1988-1989]. *Okeanologiya/Oceanology* (MOSC.) 33(2): 242-247.

Kruger, I. (1980): A checklist of southwest African marine phytoplankton, with some phytogeographical relations. Fisheries Bulletin of South Africa 13:31-53.

Lawson, G.W., R.H. Simons and W.E. Isaac (1990): The marine algal flora of Namibia: its distribution and affinities. *Bulletin of the British Museum (Natural History) Botany Series* 20: 153-168.

Macpherson, E. (1991): Biogeography and community structure of the decapod crustacean fauna off Namibia (Southeast Atlantic). *Journal of Crustacean Biology* 11(3): 401-415.

Macpherson, E and A. Gordoa (1996): Biomass spectra in benthic fish assemblages in the Benguela System. *Marine Ecology Progress Series* 138 (1-3): 27-32.

McGann, J., P. Barnard and S. Shikongo, (2002): National report to the Conference of the Parties on the implementation of the Convention on Biological Diversity in Namibia. Report presented to the Executive Secretary. Secretariat of the Convention on Biological Diversity. April 2002. Ministry of Environment and Tourism, Republic of Namibia. 94 pp.

Munilla, T. and J.H. Stock (1984): A new pycnogonida of the genus *Pallenopsis* off the Namibian coast (SE Atlantic). [Nuevo picnogonido del genero *Pallenopsis* de las costas de Namibia (Atlantico sudoriental)]. *Resultatos Expediciones Cientificas* 12:31-37.

Nesis, K.N. (1991): Cephalopods of the Benguela upwelling off Namibia. *Bulletin of Marine Science* 49(1-2):199-215.

Nielsen, J.G. (1990): Ophidiidae, p. 564-573. In: J.C. Quero, J.C. Hureau, C. Karrer, A. Post and L. Saldanha (eds.): Checklist of the fishes of the eastern tropical Atlantic (CLOFETA). JNICT, Lisbon; SEI, Paris; and UNESCO, Paris. Vol. 2.

Olivar, M.P. and , M. Barange (1989): Vertical distribution of fish eggs and larvae in the northern Benguela region. *Rapports et Procès-verbaux des Réunions du Conseil International pour l'Exploration de la Mer* 191: 454.

Olivar, M.P., P. Rubies and J. Salat (1991): Horizontal and vertical distribution patterns of ichthyoplankton under intense upwelling regimes off Namibia. *South African Journal of Marine Science* 12:71-82.

Peddemors, V.M. (1999): Delphinids of southern Africa: a review of their distribution, status and life history. *Journal of Cetacean Research and Management* 1(2):157-165.

Pillar, S.C., V. Stuart, M. Barange and M.J. Gibbons (1991): Community structure and trophic ecology of euphausids in the Benguela ecosystem. *South African Journal of Marine Science* 12:393-409.

Sakko, A. (1998): Biodiversity of marine habitats, p. 189-226. In: *Biological diversity in Namibia: a country study* (P. Barnard ed.):, Namibian National Biodiversity Task Force, Directorate of Environmental Affairs. Windhoek, Namibia.

Senn, D.G. (2001): A systematic survey on plankton along the coast off Swakopmund (Namibia). Henties Bay Coastal Resource Research Centre *(http://celi.unam.na/research/henties/senn.html)*.

Shannon, L.V. (1989): The physical environment, p. 12-27. In: A.I.L. Payne and R.J.M. Crawford (Eds.): *Oceans of life off southern Africa*. Vlaeberg Publishers, Cape Town.

Shannon, L.V. and S.C. Pillar (1986): The Benguela ecosystem. Part III. Plankton. *Oceanography and Marine Biology Annual Review* 24:65-170.

Sparks, C.A.J, E. Buecher, A.S. Brierley, B.E. Axelsen, H. Boyer and M.J. Gibbons (2001): Observations on the distribution, and relative-abundance of *Chrysaora hysoscella* (Cnidaria, Scyphozoa) and *Aequorea aequorea* (Cnidaria, Hydrozoa) in the northern Benguela ecosystem. *Hydrobiologia* 451:275-286.

4 ON NAMIBIA'S MARINE FISH DIVERSITY

Gabriella Bianchi, Elizabeth Lundsør and Hashali Hamukuaya

Abstract

Conservation of marine diversity is a major challenge worldwide. Some of the difficulties reside in the paucity of baseline information on species diversity at the local level, and the lack of operational indices and reference points that may be useful for management purposes. In this study, various aspects of the diversity of Namibian demersal fish assemblages are explored, based on comprehensive datasets of fish occurrence and abundance obtained from bottom trawl surveys with the Norwegian research vessel Dr. Fridtjof Nansen. The study is meant to provide baseline information on Namibian demersal fish diversity that will hopefully represent a useful reference for future conservation efforts.

INTRODUCTION

Conservation of biological diversity has been high on the international political agenda since the 1992 United Nations Conference on Environment and Development (UNCED). The conference generated the Convention on Biological Diversity (CBD) that entered into force in 1993. Article 7 of the CBD deals with identification and protection of biodiversity, Article 10 deals with its sustainable use while Article 12 encourages the establishment of programmes for scientific research and education '...in measures for the identification, conservation and sustainable use of biological diversity... and provide support for such training and education for specific needs in developing countries'.

Implementing the CBD is a global challenge. Effective biodiversity management requires detailed information on biogeographic distribution of habitats and species and clearly defined objectives and performance indicators. Ten years since the CBD entered into force, this type of information is yet not available for most ecosystems. The challenge is greatest for marine habitats, given the difficulty in studying aquatic systems and the high costs of research on these systems. Although it may be argued that marine species are probably less vulnerable to local extinctions than terrestrial species, given that dispersal and new recruitment are easier than in terrestrial habitats, conservation of marine diversity should still be regarded as very important for the maintenance of ecosystem functions (Warwick and Clarke, 2001).

The only information available on fish diversity off Namibia is the type found in the FAO publications such as the Species Identification Sheets for Fishery Purposes (Fischer et al., 1981), national field guides (Bianchi et al., 1999) and species catalogues. These publications summarise published records of species occurrences and generalised charts are drawn based on these records. Table 1 presents an overview of the taxa included in the FAO national field guide for Namibia (Bianchi et al., 1999), that includes the most comprehensive list of species recorded in this area. Similar information can also be found in FishBase, a global information system on fishes (Froese and Pauly, 2003). The information found in these works, however, does not have the level of detail needed for management purposes given that species distributions are indicated following broad biogeographic patterns and are not based on local habitat distribution. Also, monitoring diversity in a given area should take into account aspects of local abundance and dominance that can only be evaluated by sampling and comparing trends over time.

This study is meant to contribute to the description of biogeographic patterns of the diversity of demersal fish communities off Namibia. Furthermore, conventional and new indices of diversity will be applied to explore different aspects of fish diversity on the continental shelf and upper slope of Namibia.

Main Biogeographic Patterns

Namibia's demersal assemblages have been described in Bianchi et al. (1999) and Hamukuaya et al. (2000). Major changes in the composition of the Namibian demersal fauna take place along the depth gradient and with latitude. A major faunal boundary is found at the shelf edge, between about 300 and 350 m depth, separating shelf from slope communities. Main patterns found

Table 1. Overview of recorded fish taxa, based on Bianchi et al. (1999).

Taxa	Agnatha	Elasmobranchs	Bony fishes
Orders	1	9	24
Families	1	25	129
Genera	2	47	296
Species	2	83	424

on the shelf are associated with latitude and a clear boundary is found at around 27°S, i.e. around the Lüderitz upwelling cell, and another at around 21°S. Main shelf assemblages are therefore three, each characterised by a distinctive species composition and associated with different environmental conditions. The northern shelf assemblage (from the Angolan border to about 21°S) is characterised by a number of tropical/subtropical species that reach here their southernmost distribution (e.g. *Dentex macrophthalmus*, *Synagrops microlepis* and *Pterothryssus belloci*). The central shelf area, between the latitudes of about 21 and 27 °S, is dominated by *Merluccius capensis* (Cape hake) and *Sufflogobius bibarbatus* (pelagic goby) and characterised by being the main oxygen-deficient area in the Benguela system. Near-bottom oxygen levels found here are very low, seldom > 0.5 ml l^{-1}. Diversity here should be expected to be lower and the species found here have special adaptations to withstand these extreme environmental conditions. South of Lüderitz and to the border with South Africa, the fish community is characterised by species that reach their northernmost limit, having their main distribution off South Africa. Near-bottom oxygen levels on this part of the shelf are usually above 2.5 ml l^{-1}. Dominating species are the hakes *Merluccius paradoxus* and *M. capensis*, *Lepidopus caudatus*, *Emmelichthys nitidus*, *Zeus capensis*, *Chelidonichthys capensis*, *Genypterus capensis*, *Thyrsites atun* and *Callorhinchus capensis*.

The upper slope assemblage (between 300 and 600 m depth) includes a few species also found on the shelf but also a number of deep-water benthic fishes. The deep-water hake *Merluccius paradoxus* is the dominating species, accompanied by several species of Macrouridae, *Helicolenus dactylopterus* and several species of deep-water sharks. Figure 1 shows the Namibian coast with the distribution of the main communities.

Demersal fish communities off Namibia have been intensively exploited since the early 1960s and, to the authors' knowledge, no comprehensive diversity studies were carried out at the early stages of exploitation. Bianchi *et al.* (2001) looked at trends in community diversity (richness, Shannon-Wiener and Simpson) and abundance/biomass comparisons (ABC plots) for the decade after Independence. The main feature that was revealed was an increase in diversity due to increased evenness during the study period in the upper slope area. An increase in richness was also observed but was attributed to improved survey protocols. Another important feature was the clear difference in richness and dominance (evenness) patterns between the shelf and the upper slope assemblages, with species richness and evenness increasing with depth.

These conventional measures, however, have some drawbacks (Clarke and Warwick, 2001a) related to their dependence on sample size/effort. Furthermore, they lack a statistical framework to test whether and to what extent a sample departs from the expectation. Finally, they do not reflect phylogenetic diversity and therefore do not take into account to what extent species are taxonomically related. For these reasons Warwick and Clarke (1995) developed a set of indices that also take into account the taxonomic relatedness of the species present in a community. These indices have only been applied to fish communities in two works. Hall and Greenstreet (1998) compared the performance of these indices with conventional diversity indices based on species richness and evenness for a demersal fish commu-

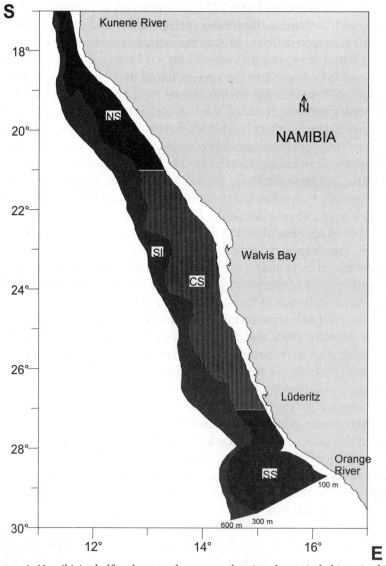

Figure 1. Namibia's shelf and upper slope area, showing the main habitats in this study. SL, Slope; NS, Northern Shelf; CS, Central Shelf; SS, Southern Shelf.

nity of the North Sea. They concluded that there was no difference in the observed trends between the conventional and the new 'taxonomic' indices but they argued that this was due to the fact that major changes had taken place in the community structure and that these were therefore well reflected in both types of indices.

Rogers et al. (1999) applied the new taxonomic indices on bottom-dwelling fish off NW Europe and found that greater resolution was achieved by using these indices as compared with conventional ones. They also noted that the taxonomic range, including both species redundancy and range of taxa with species fulfilling key functional roles, may be important for maintaining the ecosystem stability.

MATERIALS AND METHODS

The data used in the current analysis were collected through bottom trawl surveys by the Norwegian research vessel Dr Fridtjof Nansen off Namibia, from 1990 to 2000 (Table 2). The stations were sampled randomly along transects perpendicular to the coast separated from each other by about 20 nautical miles and data collected from a total of 2309 stations were used for this analysis. The sampling gear was a high-opening shrimp and bottom fish trawl, with a headline of 31 m, footrope of 47 m, roller disks of 12 cm in diameter. The estimated headline height during towing was 5 m. The codend was lined with fine meshes of 20 mm. Towing lasted about 30 minutes at a speed of 3 knots.

Table 2. Overview of the surveys and number of trawl hauls used in the analysis. Project codes and station numbers, as used in the database from which the data were extracted, are also included.

Year	Start date	End date	Project code	St. Nr.	N Stations
1990	26/01	18/03	NA	2 - 242	241
1991	27/01	20/03	NA	501 - 710	210
1992	24/04	19/05	NA	1061 - 1225	165
1993	21/01	23/02	NA	1566 - 1759	194
1994	21/01	21/02	N1	1 - 152	152
1995	22/04	28/05	N1	920 - 1104	185
1996	14/01	17/02	N1	1185 - 1427	243
1997	12/01	19/02	N1	1857 - 2124	267
1998	13/01	20/02	N1	2229 - 2442	214
1999	13/01	18/02	NC	2600 - 2814	215
2000	17/01	22/02	NC	2815 - 3037	223

The catches were sampled and all species identified, on the basis of existing literature (Fischer et al., 1981; Bianchi et al., 1993; Smith and Heemstra, 1991). Each species was weighed and counted and catches standardised to catch per hour tow. Pelagic species belonging to the families Clupeidae, Engraulididae and Carangidae were not included in the analyses.

Data analysis

The continental shelf and upper slope of Namibia, between 100 and 600 m, were subdivided into four areas (Figure 1 and Table 3) based on Hamukuaya et al. (2000). For each area, analyses were performed both on individual samples and on the average catches over the whole sampling period.

Data were exported from the NANSIS database (Strømme, 1992) and matrices were generated with species abundance by sampling station, for each of the four areas over the whole study period. Another table was generated including pooled samples representing the average species abundance in each of the four areas described above. The former were used for calculating α- diversity while the latter would produce γ- diversity indices. α- diversity is the diversity at the local scale while γ- diversity describes diversity on a larger scale, here the total diversity of a given area (Whittaker, 1960).

All the calculations were done using the Primer-e programme (Clarke and Warwick, 2001b) for Windows.

Diversity indices

Since no single biodiversity index covers all aspects of biodiversity, a range is needed to describe marine communities (Jennings and Reynolds, 2000). In this work we have chosen to include the Shannon~Wiener index, species richness, Simpson's indices of dominance and evenness and newly developed indices for taxonomic diversity and distinctness. The higher taxa used for the analyses are based on the taxonomy presented by Moyle and Cech (1982), for the higher taxa (phylum to suborder) and Bianchi et al. (1999) for lower taxa (family to species). Fishbase (Froese and Pauly, 2003) was used for those species not included in Bianchi et al. (1999).

Table 3. Summary of spatial distribution of assemblages by latitude (Ang, border with Angola; SA, border with South Africa).

Area	Depth (m)	Latitude	N Stations
Upper slope	301-600	Ang-SA	1010
Northern shelf	100-300	Ang- 21°S	315
Central shelf	100-300	21- 27°S	513
Southern shelf	100-300	27°S-SA	180

The Shannon-Weiner index, H', is a commonly used index that expresses both species richness and evenness. It gives the share of N (individuals) per S (species).

$$H' = - \sum p_i \log p_i \quad (1)$$

where p_i is the percentage of species i in relation to the total number (S).

Species richness (S) is often given as total number of species. Richness is dependent on sample size so that it is important to show the relationship between the total number of species versus the number of samples or individuals sampled (rarefaction curves). We have chosen to display 'area plots' to show richness (cumulative number of species sampled) as a function of number of samples collected for each of the assemblages considered. For this purpose all samples taken in a given area have been used.

Simpson's indices, D and E, express dominance and evenness of species:

$$\text{Dominance: } D = \sum p_i^2 \quad (2a)$$

$$\text{Evenness: } E = 1\text{-}D \quad (2b)$$

None of the above indices, however, includes the higher taxonomic relationship between species. They may therefore show the same diversity values for assemblages that comprise species closely related to each other as for assemblages where the species are taxonomically more distinct. Intuitively we would consider the latter case more diverse that the former. Warwick and Clarke (1995) developed biodiversity indices that express taxonomic diversity.

Average taxonomic diversity Δ can be considered as an extension of the Simpson diversity with the inclusion of taxonomic separation. It is described as the mean hierarchical path length between any randomly chosen individuals (x) (Jennings and Reynolds, 2000) while reflecting dominance patterns among species.

$$\Delta = [\sum\sum_{i<j} \omega_{ij}\, x_i\, x_j] / [N(N\text{-}1)/2] \quad (3)$$

where $N = \sum_i X_i$ i.e. all the individuals sampled and ω_{ij} is the taxonomic distances through the classification tree, between every pair of individuals.

Taxonomic distinctness Δ^* is the expected taxonomic distance between two randomly chosen individuals under the condition that they represent two different species (Jennings and Reynolds, 2000). The dominance effect is reduced as compared with Δ. This is considered by far the most sensitive univariate measure of community structure by Warwick and Clarke (1995).

$$\Delta^* = [\sum\sum_{i<j} \omega_{ij}\, x_i\, x_j] / [\sum\sum_{i<j} x_i\, x_j] \quad (4)$$

Average taxonomic breadth Δ^+ is the taxonomic path length for presence/absence data and expresses the average taxonomic distinctness (Warwick and Clarke, 1995)

$$\Delta^+ = [\sum\sum_{i<j} \omega_{ij}] / [S(S-1)/2] \qquad (5)$$

where S is the number of species in the sample.

Clarke and Warwick (1998) have shown that these indices are largely independent of sampling effort.

Weighing of samples

ω is the path length between two species. ω can be equal for each step in the taxonomic tree, (100/number of steps) or it can be based on taxon richness (Clarke and Warwick, 1999). The last approach was used in this study.

The values obtained in taxonomic breadth for samples within each community were tested against the average taxonomic breadth for all groundfish species sampled and identified during the study period, as described in Clarke and Warwick (2001b).

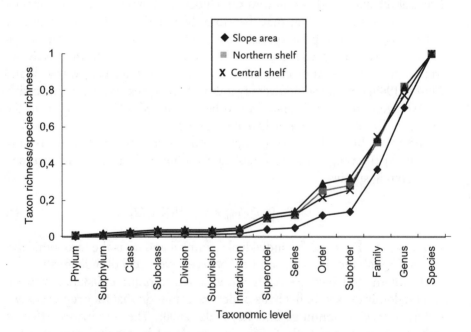

Figure 2. Profiles of ratios between richness at each taxonomic level and number of species in each community. Diamonds: Slope community; crosses: Southern shelf community; squares: Northern shelf community; triangles: Central shelf community.

RESULTS

Appendix 1 shows the list of species and families of bottom-dwelling fishes caught in all trawl stations used in this study. Table 4 shows the number of taxa for each level of classification by main areas. The upper slope holds the highest number of species, genera and families, while at higher taxonomic levels the taxon richness values are comparable between the various communities/areas. In Figure 2 this is shown by comparing the ratio between taxon richness at all taxonomic levels used and the number of species in each area/community. The profiles show lower ratios for the slope community at higher taxonomic levels.

α- diversity

A measure of local species richness is provided by the average number of species caught per trawl haul (Figure 3). There are significant differences, with the upper slope area showing the highest value, followed by the southern shelf, while the central and northern shelves have the lowest values. Figure 3 also shows trends in dominance, with the shelf communities clearly characterised by higher dominance levels (lower evenness) and the upper slope with lower dominance (higher evenness).

Figure 4 shows means and variances for the Shannon~Wiener index (H'), average taxonomic diversity (Δ), distinctness (Δ^*) and average taxonomic breadth (Δ^+) for each of the communities at the α- diversity level. Two of the

Table 4. Number of groups in each taxonomic level, by main area.

	Slope	Northern shelf	Central shelf	Southern shelf
Phylum	1	1	1	1
Subphylum	2	1	2	1
Class	3	2	3	2
Subclass	4	3	4	3
Division	4	3	4	3
Subdivision	4	3	4	3
Infradivision	5	4	5	4
Superorder	11	11	12	10
Series	13	13	14	12
Order	30	27	29	21
Suborder	35	30	32	25
Family	93	55	53	53
Genus	178	87	81	75
Species	252	106	99	97

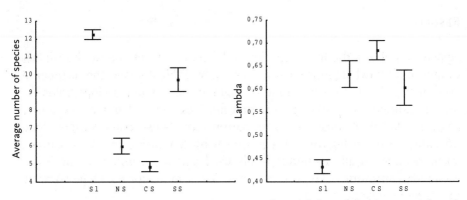

Figure 3. α – Diversity: Average species richness (left) and dominance (right). Sl, Slope; NS, Northern Shelf; CS, Central Shelf; SS, Southern Shelf. Mean ±0.95 Confidence Interval.

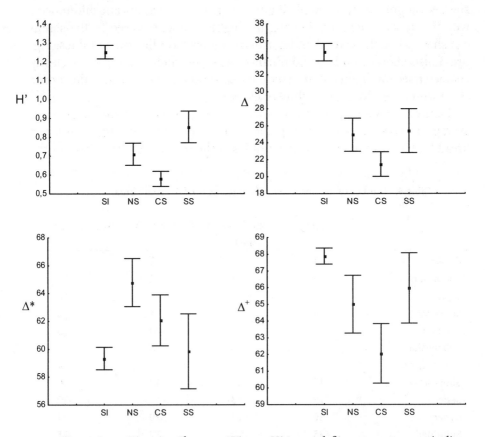

Figure 4. α – Diversity. Shannon–Wiener, H',(upper left); average taxonomic diversity, Δ (upper right); average taxonomic distinctness, Δ* (lower left); and average taxonomic breadth, Δ+ (lower right). Sl, Slope; NS, Northern Shelf; CS, Central Shelf; SS, Southern Shelf. Mean ±0.95 Confidence Interval.

indices (H' and Δ) show a pattern of higher diversity in the upper slope area as compared with the shelf communities, with the central slope having the lowest diversity values. This result is consistent with the patterns in richness and dominance described above. As regards taxonomic indices where the dominance component has been removed (Δ* and Δ+), differences are not as marked and trends different. Remarkably, taxonomic distinctness is significantly higher in the northern and central shelves as compared with the slope community while taxonomic breadth is still highest in the slope area. The step lengths applied for the taxonomic indices are shown in Table 5.

γ- diversity

Figure 5 shows the species-area plots by major assemblage while Figure 6 presents the species-area plot for the three shelf communities combined.

Number of stations varies from area to area but clearly species richness is higher on the upper slope than in any of the other regions. Despite the very high number of samples taken here (> 1000) the cumulative species curve does not seem to reach the asymptote, and therefore the total number of species may be expected to increase as sampling goes on. Comparison between shelf and upper slope (Figures 5, upper left and 6) shows how species richness is higher on the upper slope also when the three shelf communities are combined and number of samples is comparable between the two regions.

Table 5. Levels of classification used (k) and weights (step lengths) applied using the 'standard ' weighting scheme (ω_k) and an alternative scheme (ω_k^0) that uses an accumulation of the proportional decrease in taxon richness values (s_k).

k	Level	s_k	ω_k	ω_k^0
1	Species	308	7.14	6.02
2	Genus	232	14.29	18.88
3	Family	112	21.43	34.87
4	Suborder	40	28.57	38.03
5	Order	35	35.71	52.52
6	Series	15	42.86	56.04
7	Superorder	13	50	70.42
8	Infradivision	6	57.14	75.35
9	Subdivision	5	64.29	75.35
10	Division	5	71.43	75.35
11	Subclass	5	78.57	75.35
12	Class	5	85.71	87.68
13	Subphylum	2	92.86	100
14	Phylum	1	100	100

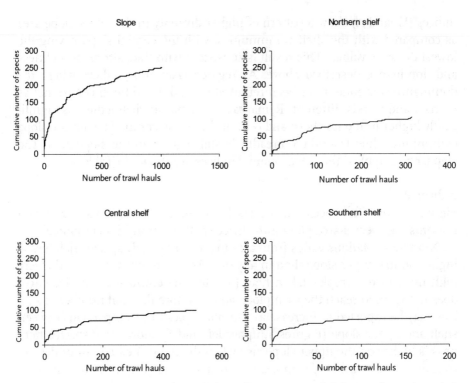

Figure 5. Species-area plot including all stations sampled during the study period, by main habitat type.

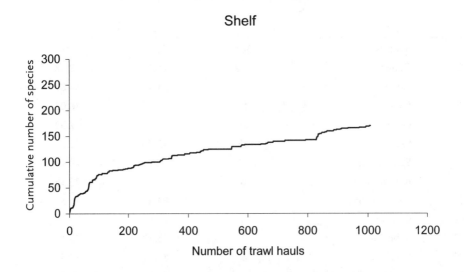

Figure 6. Species-area plot for the three shelf communities combined.

Table 6 shows a summary with the γ- diversity indices calculated for each of the assemblages and including all the stations sampled during the study period. These values provide therefore an average total diversity for the communities over a 10 year period.

γ- diversity calculated by the Shannon-Wiener index gives the highest value for the slope area and the lowest for the central shelf area. This correlates with a high dominance (lambda) on the central shelf and a high evenness on the upper slope and is consistent with the results obtained at the α-diversity level. The same pattern is shown in Δ, but for Δ^* the central shelf has the highest value and the three other areas are not significantly different from each other.

Figure 7 shows the 95% confidence intervals for the four regions of the mean taxonomic breadth. The mean is the theoretical mean from the overall value of Δ^+ for the global species list from the whole region while the variances are obtained by sampling a varying number of species (without replacement) from the total list of species. This produces an overall 'confidence funnel' that allows to check to what extent taxonomic breadth of a given set of samples falls within the expected values. The results show that most of the samples fall within the expected range but that most of them were below the expected mean.

Discussion

Data limitations

Fish diversity studies have important limitations due to the difficulties of sampling a three-dimensional environment. Trawl gear, although among the least selective gears, can only sample a small fraction of the water column at the time, is size selective (but possibly this problem is less important on the RV Dr. Fridtjof Nansen, given that double lining in the codend and very small mesh sizes are used) and is species selective (some species living very close to the bottom might not be sampled properly). It is difficult to evaluate how this affects the various measures of diversity. It may be particularly im-

Table 6. γ-diversity with taxonomic diversity indices weighed for species richness.

Area	S	H'(loge)	Lambda	Δ	Δ^*	Δ^+
Slope	252	3.12	0.07	57.11	61.25	72.77
Northern shelf	106	2.01	0.17	50.91	61.65	71.21
Central shelf	99	1.19	0.40	41.83	69.45	72.34
Southern shelf	97	2.13	0.18	50.36	61.75	72.40

portant for the measures that have an equity component given that the relative abundance of species found in the catches does not necessarily reflect the actual relative abundance. These inaccuracies are further exacerbated by fish behaviour. Although the various species are described in the broad categories "demersal" or "pelagic", most carry out vertical diel migrations, more or less extensive depending on the species and/or the environmental conditions. Most of the species included in the analyses show this behaviour, but species that are mainly pelagic have been excluded from the analyses. For this reason, the study does not provide a comprehensive description of the total fish fauna found in the area.

Patterns of fish species diversity on the Namibian shelf and upper slope areas

The Namibian shelf and upper slope belong to the temperate South African region. Its northern limit coincides with the Angola~Benguela frontal area, situated off southern Angola where the tropical Eastern Atlantic zoogeographic region has its southern limit. This region is the poorest in fish species of all tropical regions, with 434 shore fishes (as compared, for example, with 900 on the other side of the Atlantic). The South African region, which includes both the Indian Ocean and Atlantic sides of southern Africa, has a higher number of species given the presence of warm currents that

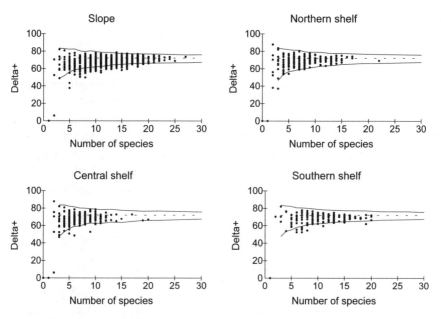

Figure 7. Funnel plots showing the average taxonomic distinctness of samples within each community, as compared with the 95% probability limits for Δ^+ and the mean taxonomic distinctness for the whole area.

extend the tropical Indo-Pacific fauna off the eastern and southern coasts of South Africa. Given its long isolation, this region has a high level of endemism. Namibia is part of the Southwestern Africa province (Briggs, 1974) that extends from southern Angola to the Cape of Good Hope.

Shelf diversity off Namibia is relatively low, with a total of 170 species recorded from 1008 trawl hauls. Only an average of 5 species per trawl haul is caught in the central shelf where near-bottom oxygen levels are lowest. Species richness at α-diversity level on the shelf reflects the poor ecological conditions in the central shelf, while the trends in the northern and southern shelves reflect the diversity of adjacent areas. Species richness measured as average number of species per trawl haul is highest on the slope. Both of the above trends seem to be in contrast with two accepted rules in marine zoogeography: 1) that fish richness increases with decreasing latitude and; 2) that richness decreases with depth. Macpherson and Duarte (1994) have shown for the Eastern Atlantic that species richness declines by about 1% for each degree of latitude. Obviously this seems not to apply locally for the Namibian marine fishes. Richness and dominance are negatively correlated (Figure 3), with high and low richness corresponding to low and high dominance, respectively. High dominance and low diversity are typical of 'stressful' environments and the low oxygen levels found in large parts of the shelf area off Namibia are likely the reason for the observed patterns. Other studies have shown how fish diversity is strongly affected by oxygen levels < 1 ml l^{-1}. (Bianchi 1991 and 1992).

Taxonomic indices showed similar trends to the species indices. In particular, the average taxonomic diversity (Δ) and the average taxonomic breadth (Δ^+) showed similar patterns to the Shannon-Wiener index, thus apparently not adding more information and insight, similarly to what was concluded by Hall and Greenstreet (1998) for patterns of change in the fish communities of the northern North Sea. However, taxonomic indices present a new aspect of diversity that seems worthwhile pursuing. Despite the higher species richness of the slope, its taxonomic range is similar to the shelf regions. Furthermore, diversity of higher taxa (expressed as ratio of number of higher taxa divided by the number of species in the assemblage) is in fact higher on the shelf than in the upper slope area (Figure 2). Two conclusions can be derived from this. The first is that relatively higher taxonomic diversity on the shelf may be related to higher ecological diversity. On the other hand, high taxonomic diversity associated with low species diversity at taxon level, may also result in ecologically unstable conditions. For example, the central shelf area holds relatively few species and has high levels of dominance. This may indicate the presence of (keystone?) fish species particularly well adapted to the given conditions but whose function in the ecosystem may be difficult to replace by other species in the case of population crashes due, for example, to overexploitation. In these cases of low species diversity in stressful environments, resilience of the system may be low and the precautionary approach to fisheries management should be more strongly emphasized.

Of the three taxonomic indices used in this study (Δ, Δ^* and Δ^+), Δ^* has been difficult to interpret. It displays a quite different pattern as compared with all the other indices, but is consistent with the analysis of diversity of higher taxa shown in Figure 2.

Visualization of trends of taxonomic breadth within each community as compared with the theoretical mean is indicative of a lower local taxonomic breadth at community level as compared with the total. This confirms the distinctness of the four regions. It is suggested that diversity studies in a given geographic area should be preceded by analyses that test the homogeneity of the area, by, for example, multivariate analyses. Correlation with environmental variables leads to the definition of habitats, thus allowing separation of environmental effects from anthropogenic effects in understanding patterns of diversity.

Demersal fish diversity and fisheries management - A pragmatic approach toward monitoring fish diversity

Clearly diversity has many facets and can hardly be represented by a single index. The question still remains as to which indices should be used to indicate ecosystem health as related to possible effects of fishing, and what are the reference points for these measures that fisheries management should aim at. Based on the experience made through the work of the SCOR working group on ecosystem effects of fishing (ICES, 2000) and the present study, the following is an outline for a pragmatic approach to monitoring demersal marine fish diversity.

Bottom trawl surveys, usually carried out under controlled conditions and with appropriate sampling schemes, are performed in areas exploited by the industrial fisheries. This type of data could be used to study diversity patterns in space and time, according to the following analytical steps:
- Stratification of a given area into communities by multivariate analyses and habitat definition (by, for example, main environmental variables). This would provide a definition of ecological and management units.
- Analysis of community properties as regards diversity, taking into account species richness, dominance and measures of taxonomic diversity.
- Monitoring trends over time. Development of reference points should be done in an adaptive way, at least to start with.

Methods should be standardized so that experiences gained in a given area could more easily be compared. Standardization of methods is a key issue and should be sought for all the analytical steps. For example, when looking at taxonomic indices, it is essential that the number of taxonomic levels and the type of taxa utilized are the same given that the choice of number and type of taxa affects the taxonomic index. Furthermore, there are other features of diversity that may be important to evaluate (e.g. functional diversity) and studies should be carried out to find which of the indices are most informative in relation to assessing ecosystem health.

It is recommended that international (e.g. FAO) and regional fisheries organizations should take the lead as regards promoting the standardization of methods for describing and monitoring marine fish diversity.

References

Bianchi, G. (1991): Demersal assemblages of the continental shelf and upper slope edge between the Gulf of Tehuantepec (Mexico) and the Gulf of Papagayo (Costa Rica). *Marine Ecology Progress Series* 73:121-140.

Bianchi, G. (1992): Demersal assemblages of the continental shelf and upper slope of Angola. *Marine Ecology Progress Series* 81:101-120.

Bianchi, G., Carpenter, K.E., Roux, J-P., Molloy, F.J., Boyer, D. and Boyer, H. (1993): FAO species identification field guide for fisheries purposes. The living marine resources of Namibia. Rome, FAO. 250 pp., 8 colour plates.

Bianchi, G., Carpenter, K.E., Roux, J-P., Molloy, F.J., Boyer, D. and Boyer, H. (1999): *FAO species identification field guide for fisheries purposes. The living marine resources of Namibia.* 2nd edition. Rome, FAO. 265 pp., 11 colour plates.

Bianchi, G., Hamukuaya, H. and Alvheim, O. (2001): On the dynamics of demersal fish assemblages off Namibia in the 1990s. In: A.I.L. Payne, S.C. Pillar, and R.J.M. Crawford (eds.). *South African Journal of Marine Science* 23: 419-428.

Briggs, J.C. (1974): *Marine Zoogeography.* McGraw-Hill Book Company. 475pp.

Clarke, K.R. and Warwick, R.M. (1998): A taxonomic distinctness index and its statistical properties. *Journal of Applied Ecology* 35:523-531.

Clarke, K.R. and Warwick, R.M. (1999): The taxonomic distinctness measure of biodiversity: weighing of step lengths between hierarchical levels. *Marine Ecology Progress Series*, 184: 21-29.

Clarke, K.R. and Warwick, R.M. (2001a): A further biodiversity index applicable to species lists: variation in taxonomic distinctness. *Marine Ecology Progress Series* 216: 265-278.

Clarke, K.R. and Warwick, R.M. (2001b): *Change in marine communities: an approach to statistical analysis and interpretation,* 2nd edition. PRIMER-E: Plymouth.

Fischer, W.G., Bianchi, G. and Scott, W.B. (eds.) (1981): FAO species identification sheets for fisheries purposes. Eastern Central Atlantic; fishing areas 34, 47 (in part). Rome, Food and Agricultural Organisation of the United Nations, vols 1-7: pag. var.

Froese, R. and Pauly, D. (eds.) (2003): FishBase. World Wide Web electronic publication: *www.fishbase.org*, version 26 February 2003.

Hall, S.J. and Greenstreet, S. P. (1998): Taxonomic distinctness and diversity measures: responses in marine fish communities. Marine Ecology Progress Series 166:227-229.

Hamukuaya, H., Bianchi, G. and Baird, D. (2000): The structure of demersal assemblages off Namibia in relation to abiotic factors. *South African Journal of Marine Science* 23:397-417.

ICES (2000): Ecosystem Effects of Fishing. Proceedings of an ICES/SCOR Symposium held in Montpellier, France 16-19 March 1999. ICES Marine Science Symposia Vol. 57, no. 3.

Jennings, S. and Reynolds, J.D. (2000): Impacts of fishing on diversity: from pattern to process. In: *The Effects of Fishing on Non-target Species and Habitats* (M.J. Kaiser and S.J. de Groot, eds.), pp. 235-250.

Macpherson, E. and Duarte, C.M. (1994): Patterns in species richness, size, and latitudinal range of East Atlantic fishes. *Ecography* 17:242-248.

Moyle, P.B. and Cech, J.J. (1982): *Fishes. An Introduction to Ichthyology*. Prentice-Hall International (UK) Limited, London. 612 pp.

Rogers, S.I., Clarke, K.R. and Reynolds, J.D. (1999): The taxonomic distinctness of coastal bottom-dwelling fish communities of the North-east Atlantic. *Journal of Animal Ecology* 68: 769-782.

Smith, M. M. and Heemstra P.C. (eds.) (1991): *Smiths' Sea Fishes*. Southern Book Publishers, Johannesburg. 1048 pp.

Strømme, T. (1992): NAN-SIS: Software for fishery survey data logging and analysis. User's manual. FAO. Computerised information services (Fisheries) 4. Rome. 103 pp.

Warwick, R.M. and Clarke, K.R. (1995): New 'biodiversity' measures reveal a decrease in taxonomic distinctness with increasing stress. *Marine Ecology Progress Series* 129, 301-305.

Warwick, R.M. and Clarke, K.R. (2001): Practical measures of marine biodiversity based on relatedness of species. *Oceanography and Marine Biology: an Annual Review* 39: 207-231.

Whittaker, R.H. (1960): Vegetation of the Siskiyou Mountains, Oregon and California. *Ecological Monographs* 30: 279-338.

APPENDIX 1. List of fish families and species collected by bottom trawl by the RV DR. F. Nansen (1990-2000).

Family	Species	Family	Species
HAGFISHES			*Scymnodon squamulosus*
Myxinidae			*Isistius brasiliensis*
	Myxine capensis		*Oxynotus centrina*
		Centrophoridae	
CARTILAGINEOUS FISHES			*Centrophorus* sp.
Lamnidae			*Centrophorus uyato*
	Isurus oxyrinchus		*Centrophorus granulosus*
	Apristurus sp.		*Centrophorus squamosus*
	Apristurus saldanha		*Deania profundorum*
	Galeus polli		*Deania calcea*
Scyliorhinidae			*Deania quadrispinosum*
	Holohalaelurus regani	Squalidae	
	Scyliorhinus capensis		*Squalus blainvillei*
	Galeorhinus galeus		*Squalus megalops*
Triakidae			*Squalus acanthias*
	Mustelus palumbes		*Squalus mitsukurii*
	Mustelus mustelus	Rajidae	
Carcharhinidae			*Raja* sp.
	Carcharhinus signatus		*Raja miraletus*
	Prionace glauca		*Raja caudaspinosa*
Chlamydoselachidae			*Raja alba*
	Chlamydoselachus anguineus		*Raja straeleni*
			Raja wallacei
	Heptranchias perlo		*Raja clavata*
Hexanchidae			*Raja stenorhyncus*
	Hexanchus griseus		*Raja leopardus*
Echinorhinidae			*Raja confundens*
	Echinorhinus brucus		*Raja doutrei*
Dalatiidae			*Raja pullopunctata*
	Etmopterus sp.		*Raja springeri*
	Etmopterus pusillus		*Raja spinacidermis*
	Etmopterus lucifer		*Cruriraja parcomaculata*
	Etmopterus spinax		*Bathyraja smithii*
	Etmopterus brachyurus	Torpedinidae	
	Centroscymnus crepidater		*Torpedo nobiliana*
	Centroscymnus coelolepis	Dasyatidae	
	Centroscyllium fabricii		*Dasyatis* sp.

Family	Species	Family	Species
Myliobatidae			Xenodermichthys copei
	Myliobatis aquila	Platytroctidae	
Callorhinchidae			Maulisia microlepis
	Callorhinchus capensis	Gonostomatidae	
Rhinochimaeridae			Diplophos sp.
	Neoharriotta pinnata		Diplophos maderensis
Chimaeridae			Triplophos sp.
	Hydrolagus sp.		Triplophos hemingi
	Hydrolagus mirabilis	Sternoptychidae	
			Argyropelecus aculeatus

BONY FISHES

Family	Species	Family	Species
Albulidae			Argyropelecus affinis
	Pterothrissus belloci		Maurolicus muelleri
Colocongridae		Phosichthyidae	
	Coloconger cadenati		Phosichthys sp.
Muraenidae			Phosichthys argenteus
	Uropterygius wheeleri		Yarrella blackfordi
Ophichthidae		Stomiidae	
	Mystriophis rostellatus		Melanostomias sp.
	Ophisurus serpens		Odontostomias micropogon
Nemichthyidae			Photonectes braueri
	Nemichthys scolopaceus		Stomias boa boa
	Nemichthys curvirostris		Astronesthes sp.
Congridae			Bassanago albescens
	Bathyuroconger vicinus	Ateleopodidae	
Nettastomatidae			Guentherus altivela
	Venefica proboscidea	Chlorophthalmidae	
Synaphobranchidae			Chlorophthalmus atlanticus
	Synaphobranchus kaupii		Chlorophthalmus agassizi
Halosauridae			Chlorophthalmus punctatus
	Halosaurus ovenii	Notosudidae	
Notacanthidae			Scopelosaurus meadi
	Notacanthus sexspinis	Paralepididae	
Microstomatidae			Lestidium atlanticum
	Nansenia sp.		Lestrolepis intermedia
Bathylagidae			Lestidiops sp.
	Bathylagus glacilis		Macroparalepis macrogeneion
Alepocephalidae			
	Alepocephalus sp.		
	Alepocephalus rostratus		Macroparalepis affinis

Family	Species	Family	Species
Neoscopelidae			*Physiculus capensis*
	Neoscopelus macrolepidotus		*Tripterophycis gilchristi*
Myctophidae		Bregmacerotidae	
	Diaphus sp.		*Bregmaceros* sp.
	Diaphus dumerili		*Merluccius* sp.
	Diaphus hudsoni		*Merluccius polli*
	Lampadena sp.		*Merluccius capensis*
	Lampadena pontifex	Merlucciidae	
	Lampanyctodes hectoris		*Merluccius paradoxus*
	Lampanyctus australis		*Lyconus pinnatus*
	Symbolophorus boops	Batrachoididae	
Ophidiidae			*Batrachoides* sp.
	Ophidion sp.		*Chatrabus melanurus*
	Dicrolene intronigra		*Perulibatrachus rossignoli*
	Brotula barbata	Lophiidae	
	Genypterus capensis		*Lophius piscatorius*
	Lamprogrammus exutus		*Lophius vaillanti*
	Selachophidium guentheri		*Lophius vomerinus*
Macrouridae		Chaunacidae	
	Coelorinchus sp.		*Chaunax pictus*
	Coelorinchus fasciatus	Ogcocephalidae	
	Coelorinchus coelorhinc.		*Dibranchus atlanticus*
	poll	Melanocetidae	
	Coelorinchus braueri		*Melanocetus johnsoni*
	Coelorinchus matamua	Diceratiidae	
	Coelorinchus acanthiger		*Phrynichthys wedli*
	Hymenocephalus italicus	Ceratiidae	
	Malacocephalus laevis		*Ceratias holboelli*
	Malacocephalus		*Cryptopsaras couesii*
	occidentalis	Scomberesocidae	
	Nezumia sp.		*Scomberesox saurus*
	Nezumia aequalis	Trachipteridae	
	Nezumia leonis		*Trachipterus trachypterus*
	Nezumia milleri		*Trachipterus jacksonensis*
	Nezumia micronychodon		*Zu elongatus*
	Trachyrincus scabrus	Regalecidae	
	Gadella imberbis		*Regalecus glesne*
	Laemonema laureysi	Trachichthyidae	
Moridae			*Hoplostethus cadenati*
	Lepidion capensis		*Hoplostethus mediterraneus*

Family	Species	Family	Species
	Hoplostethus atlanticus	Acropomatidae	
	Hoplostethus melanopus		Synagrops microlepis
Berycidae		Polyprionidae	
	Beryx splendens		Polyprion americanus
Barbourisiidae		Callanthiidae	
	Barbourisia rufa		Callanthias legras
	Cyttus traversi	Serranidae	
Zeidae			Anthias anthias
	Zeus faber	Epigonidae	
	Zeus capensis		Epigonus sp.
	Zenopsis conchifer		Epigonus telescopus
Oreosomatidae			Epigonus pandionis
	Allocyttus verrucosus		Epigonus denticulatus
	Neocyttus rhomboidalis	Bramidae	
	Oreosoma atlanticum		Brama brama
Grammicolepididae			Taractichthys longipinnis
	Xenolepidichthys dagleishi		Taractes sp.
Centriscidae		Caristiidae	
	Macrorhamphosus scolopax		Caristius groenlandicus
	Notopogon macrosolen	Emmelichthyidae	
Congiopodidae			Emmelichthys nitidus
	Congiopodus spinifer	Sparidae	
	Congiopodus torvus		Dentex macrophthalmus
Liparidae			Spondyliosoma cantharus
	Careproctus griseldea		Argyrosomus hololepidotus
Scorpaenidae		Sciaenidae	
	Scorpaena sp.		Umbrina canariensis
Psychrolutidae			Atractoscion aequidens
	Ebinania costaecanarie	Percophidae	
	Psychrolutes macrocephalus		Bembrops heterurus
Setarchidae		Callionymidae	
	Ectreposebastes imus		Paracallionymus costatus
Sebastidae		Gobiidae	
	Helicolenus dactylopterus		Sufflogobius bibarbatus
	Sebastes capensis	Sphyraenidae	
	Trachyscorpia capensis		Sphyraena guachancho
Triglidae		Gempylidae	
	Chelidonichthys capensis		Paradiplospinus gracilis
	Chelidonichthys queketti		Ruvettus pretiosus
	Trigla lyra		Thyrsites atun

Family	Species	Family	Species
Trichiuridae		Tetragonuridae	
	Aphanopus sp.		*Tetragonurus cuvieri*
	Benthodesmus tenuis		*Tetragonurus atlanticus*
	Lepidopus caudatus	Bothidae	
	Trichiurus sp.		*Arnoglossus imperialis*
	Trichiurus lepturus		*Arnoglossus capensis*
	Scomber japonicus		*Monolene microstoma*
Scombridae		Soleidae	
	Allothunnus fallai		*Austroglossus microlepis*
Centrolophidae			*Austroglossus pectoralis*
	Centrolophus niger		*Dicologoglossa cuneata*
	Hyperoglyphe moselii		*Synaptura kleini*
	Schedophilus pemarco	Cynoglossidae	
	Schedophilus huttoni		*Cynoglossus capensis*
	Schedophilus ovalis		*Cynoglossus zanzibarensis*
Nomeidae		Diodontidae	
	Cubiceps caeruleus		*Chilomycterus reticulatus*

5 RECONSTRUCTION AND INTERPRETATION OF MARINE FISHERIES CATCHES FROM NAMIBIAN WATERS, 1950 TO 2000

*Nico E. Willemse and Daniel Pauly**

Abstract
Time series of catches taken by local and foreign fleets from Namibian waters were reconstructed for the period from 1950 to 2000 from a variety of published and unpublished sources. Examination of these time series demonstrated a gradual shift in catches from larger, long-lived piscivorous fish species to smaller, short-lived planktivorous fishes and invertebrates, which, as would be expected, reflects similar changes in the relative abundance of these species in the Namibian marine ecosystem. Similar results were obtained from the examination of trends in trophic-based indicators, which confirmed that the fishing down marine food web phenomenon occurs in Namibian waters. More precisely, three periods marking different trends and developmental stages of the fishery were identified, viz, an 'undeveloped' stage (1950-1964), marked by the dominance of southern African sardine, a 'developing/mature' stage, characterised by increased fishing effort, landings and diversification of target species (1965-1969), and a 'senescent' stage with declining ratios of piscivorous to planktivorous fishes and, more ominously, declining total landings (1970-2000). Some of the ecological ramifications of these findings are discussed, along with the need for ecosystem-based management of Namibia's marine fisheries.

INTRODUCTION

The concept of 'ecosystem-based fisheries management' is about two decades old, and widely discussed in the scientific literature, but it is still not

---

* We would like to thank the National Marine Information and Research Centre (NatMIRC), Ministry of Fisheries and Marine Resources (MFMR), Swakopmund for data and invaluable assistance. We would also like to thank M.L. Deng Palomares, D.C. Boyer, P. Cury and G. Bianchi for useful comments on the draft version. Nico E. Willemse thanks Jorge Santos, Norwegian College of Fishery Science, Tromsø, Norway, for his contribution toward the M.Sc. thesis upon which this chapter is based. Daniel Pauly thanks the Pew Charitable Trust for their support of the Sea Around Us Project.

part of the practice of fisheries management. One reason, clearly, is that it seems to imply a vast knowledge about the ecosystems in which the fisheries are embedded, and an excessive emphasis on conservation, to the detriment of a still perceived need for fisheries 'development', not only in mature fisheries such as Namibia, but particularly in developing countries.

Here we show that, on the contrary, inferences on the state of an exploited marine ecosystem can be derived from data that are not only straightforward to obtain, but are in fact required for the more traditional single-species approaches. Moreover, we suggest that by not considering ecosystem effects, fisheries managers increase the risk of deleterious changes in the ecosystem within which fisheries are embedded, all the way to their collapse (Pauly, 1998; Pauly et al., 1998a).

The key data for both single-species and ecosystem-based approaches are long time series of catch data. These time series must refer to catches (i.e., the biomass of all organisms that are killed by fishing operations), rather than only landings (i.e., the part of the catch that is brought ashore), because it is through their catches that fisheries impact ecosystems. Further, the time series must be long, in order to incorporate contrast, and thus help identify the cause for change (Hilborn and Walters, 1992).

Namibia's history, in the last 50 years, was not conducive to the accumulation of such long time series of fisheries catch data. Hence, this contribution starts with a brief account of how such time series were assembled, both to support our studies, and to support other analyses of Namibian fisheries. The biology, and especially the trophic role, of major fish species in the Namibian marine ecosystem are then summarized. We conclude with a preliminary analysis of the resulting time series, i.e., we test for the occurrence in Namibian waters of the 'fishing down marine food webs' phenomenon originally described by Pauly et al. (1998a), through the examination of times series of mean trophic levels in the catches, as corroborated by an analysis of the ratio of piscivorous to planktivorous fishes (Caddy and Garibaldi, 2000), and of the FiB (fishing-in-balance) index of Pauly et al. (2000).

RECONSTRUCTION OF CATCHES FOR NAMIBIAN MARINE WATERS, 1950-2000

Data on landings (in metric tons, or 'tonnes') from Namibian waters were extracted primarily from Statistical Bulletins and other documents of the Madrid-based International Commission for the South Eastern Atlantic Fisheries (ICSEAF), which reported from 1971 on the activities of distant water fleets (DWF) operating off Namibia. Other data, notably early catch figures from 'South West Africa,' then under the administration of South Africa, were obtained from the Food and Agriculture Organisation (FAO) and South African fisheries statistics. Also, reports of the National Marine Information and Research Centre (NatMIRC; Ministry of Fisheries and Marine Resources, Namibia) and articles in scientific journals were searched for information. For example, Crawford et al. (1987) published landings data extracted from ICSEAF Statistical Bulletins, which also formed the basis of many subsequent studies (including by NatMIRC staff).

The fish species whose catch is documented here largely overlap with the species listed in statistical reports of the Ministry of Fisheries and Marine Resources (MFMR), Namibia, and include all groups that can be expected to impact, either as prey or as predator, on the structure and functioning of the Namibian marine ecosystem (see also Palomares and Pauly, this volume). Only those species whose landings are negligible (e.g., orange roughy, alfonsino, jacopever) were not explicitly considered here, though they are considered implicitly in the analysis of catch trends, through their inclusion in the 'mixed species' category.

Landings for the two *Merluccius* species caught off Namibia, *Merluccius capensis* and *M. paradoxus*, were combined, and treated as a single entity ('Cape hakes'), because the two species are not distinguished in catch records. The various species of tuna caught in Namibian waters were also aggregated (see below).

As the catch data assembled here largely resembled those published by Boyer and Hampton (2001), we abstain from presenting them here. They may be found, however, under 'Namibia' in the global catch database of the *Sea Around Us* Project (see *www.seaaroundus.org*).

BIOLOGY AND TROPHIC LEVELS OF MAJOR FISH SPECIES IN NAMIBIAN MARINE WATERS

We present here key features of major commercial and other species in Namibian waters, each of the brief accounts ending with the trophic level estimate (TL; see below for definition) we used to compute time series of mean TL, and with a one letter assignment (for fishes) to either piscivory (P) or planktivory (Z, see below). The TL values used here are updated, based on FishBase (Froese and Pauly, 2000; version of February 2004; see *www.fishbase.org*), and other sources (notably Heymans and Baird, 2000; Jarre-Teichmann et al., 1998; and contributions in Payne et al., 2001), and the preliminary estimates used in Willemse (2002) and Willemse and Pauly (2004).

The trophic levels of lobsters and crabs are based on estimates for related species in Eastern Canadian waters (Pauly et al., 2001).

Tunas. - Seven species of tuna occur in Namibian waters, but only two are identified as important to the fisheries: albacore *(Thunnus alalunga)* dominate the pole fishery, while bigeye *(Thunnus obesus)* dominate the long-line fishery. The high market prices of tuna make up for their relatively low catches off Namibia (Manning, 1998).

Albacore, which can attain a fork length of 130 cm, are found between 10°S and 40°S in the South Atlantic, and spawn off Brazil just south of the equator and in the Central Atlantic, where the surface temperature of the water exceeds 24°C (Manning, 1998). Bigeye attain a fork length of 200 cm, and range across the Atlantic between 45°S and 45°N. Spawning occurs in the east central Atlantic, north of 5°N in the warmest season when the season surface temperature is above 24°C, and in the Gulf of Guinea (Manning, 1998).

Yellowfin *(Thunnus albacares)* can reach a fork length of over 200 cm and are preyed upon by toothed whales and billfishes (Bianchi *et al.*, 1993). This pelagic oceanic species is caught mainly by longlines at depth ranging from 200 to 300 m. Skipjack *(Katsuwonus pelamis)* attain a fork length of 108 cm and are distributed widely in all oceans. They are taken with purse seines, pole-and-line and longlines.

The food of these four tuna species is rather similar, and consists of a wide range or fish, notably sardine, anchovy, lanternfishes, cephalopods and crustaceans; cannibalism is also common (Bianchi *et al.*, 1993); TL = 4.4, P.

Southern African sardine. - The southern African sardine *(Sardinops sagax)* is a relatively short-lived species with a high reproductive capacity. Thus, like other small pelagics, this sardine fluctuates strongly in response to environmental changes, which impact on the early life history (Boyer, 1994). Spawning occurs north or south of the Lüderitz upwelling cell, as the cold, highly turbulent, weakly stratified water of this area is not conducive for larval development (Crawford *et al.*, 1987).

Younger sardine spawn in the warmer waters of the north, while the older ones spawn in the vicinity of Walvis Bay (Boyer and Hampton, 2001). Tagging studies demonstrated no movement of sardine from the Western Cape to Namibia and minimal movement of sardine to the Western Cape coast (Boyer *et al.*, 2001).

Juvenile sardine feed on zooplankton while the adults feed on phytoplankton and zooplankton (Boyer and Hampton, 2001; Bianchi *et al.*, 1993; van der Lingen, 1998); TL = 2.5, Z.

Cape anchovy. - The distribution and movement of Cape anchovy *(Engraulis capensis)* off Namibia used to be similar to that of sardine, but spawning was only significant north of Walvis Bay, with dense concentrations of larvae occurring beyond 100 km from the coast (Boyer and Hampton, 2001). The northern limit of anchovy is usually the southern edge of the Angola-Benguela front. Cape anchovy are caught in the purse-seine fishery (Manning, 1998).

Anchovy feed primarily on zooplankton, both as juveniles and adults (Boyer and Hampton, 2001); TL = 3.0, Z.

Cape horse mackerel ('Maasbanker'). - Juvenile horse mackerel (*Trachurus capensis*) are pelagic and live in waters <200 m deep to the age of two years (<20 cm) before they are recruited to the adult, semi-pelagic stock (Klingelhoeffer, 1994; Manning, 1998). They then move to deeper water, mainly north of Walvis Bay with dense concentrations between Cape Cross and the Kunene River. Adult horse mackerel, which may reach a length of up to 60 cm, spawn during both summer and winter, with peak activity between January and April (Klingelhoeffer, 1994).

While a small portion of juvenile horse mackerel are used for human consumption, most are reduced to fish meal (Manning, 1998). On the other hand, 60% of the catch of adults is frozen whole, while the remainder is processed into fish meal or dried and salted.

Juvenile horse mackerel feed on zooplankton, while the adults feed mainly on euphausiids shrimp (krill), with pelagic goby, lantern fish and juvenile horse mackerel making up the remainder of the diet (Klingelhoeffer, 1994). Juvenile TL = 3.1, Z; adults: TL = 3.5, P.

Pelagic goby. - The 3 to 6 cm juveniles of the Pelagic goby (*Sufflogobius bibarbatus*) live close to the sea surface, while the adults, ranging from 7 to 15 cm, live in deeper water. According to Crawford *et al.* (1987), pelagic gobies are consumed by most piscivorous fish species, and by seabirds and marine mammals, suggesting that pelagic goby play a major role as prey species in Namibian waters. This species is occasionally targeted by purse-seines, and is sometimes caught in bottom trawls (Bianchi *et al.*, 1993), though not in quantities reflecting their abundance in the ecosystem, a theme to which we return further below.

Pelagic goby are phytoplankton feeders; TL = 2.2, Z.

Round herring. - Round herring (*Etrumeus whiteheadi*) also known as redeye, reach about 20 cm, and are caught in purse seines and trawls at depths which may range from 10 to 200 m, but usually less than 100 m.

Round herring feed mainly on large zooplankton; TL = 3.4, Z.

Angelfish. - The Angelfish (*Brama brama*) occur from the sea surface to a depth of 1000 m, where it is preyed upon by billfish and tuna, and commonly caught by trawling (Bianchi et al., 1993).

Angelfish feed on myctophids and other small fishes, euphausiids and cephalopods; TL = 4.1, P.

Cape hakes. - Two species of hake are caught in Namibian waters: *M. capensis*, or Cape hake (also known as shallow water hake), occurring mainly between the 200 and 350 m isobaths and constituting the bulk of the hake catch; and *M. paradoxus* or deepwater hake, occurring typically deeper, around the 350 m isobath in the south, and making up an average of thirty per cent (30%) of the catch between 1990 and 1996. A third species, of no commercial importance, *M. polli* or Benguela hake, is found in northern Namibian waters (Manning, 1998).

Young hake occur in the inshore waters at depths between 25 and 100 m for close to one year before migrating to deeper waters (Hamukuaya, 1994). Hake are targeted by the bottom trawl fishery, Namibia's most important fishery in terms of value since the mid-1990s.

Hake are predominantly piscivorous, and feed on both demersal and semi-pelagic fishes, which they catch staying close to the bottom during the day, and in mid-water at night; TL = 4.75, P.

John Dory. - John Dory (*Zeus faber*) is a demersal species ranging from coastal water to depths of 400 m; they are caught mainly by bottom trawls.

The food of John Dory consists mainly of schooling bony fishes, with occasional crustaceans and cephalopods (Bianchi et al., 1993); TL = 4.5, P.

Snoek. - Snoek (*Thyrsites atun*) are pelagic in coastal waters, at temperatures between 13°C and 18°C. They are one of the dominant commercial fish species in Namibia and are fished with line gear (Bianchi et al., 1993).

Snoek feed mainly on anchovy and sardine, but also on a variety of other organisms when these two species are scarce (Bianchi et al., 1993); TL = 4.0, P.

Silver kob. - Silver kob (*Argyrosomus inodorus*) also known as kabeljou, can attain a length of 130 cm. They are primarily a coastal species, ranging from the surface to 150 m, but also occur at 400 m. Kob feed at night and/or in turbid waters, on a variety of crustacean species (Bianchi et al., 1993).

Here, the method described in Froese and Pauly (2000, Box 26) was applied to obtain an estimate of TL (via a size adjustment) based on that of a close relative, *Argyrosomus hololepidotus*, for which it has often been mistaken. As well, the food items for *A. inodorus* and *A. hololepidotus* in FishBase and in Bianchi et al. (1993) resembled each other sufficiently for the trophic level estimate obtained for the latter to be used for the former species. Thus, TL = 4.3, P.

West Coast steenbras. - The West Coast steenbras (*Lithognathus aureti*) can reach up to 100 cm, and live in schools near the coast usually over sandy sea bottoms.

Steenbras feed on sand mussels, crabs and worms (Bianchi *et al.*, 1993); TL = 3.4, Z.

Cape monk. - Monk (*Lophius vomerinus*) reach maturity at 4 years at about 40 cm, but can reach a maximum length of 100 cm (Bianchi *et al.*, 1993). They occur on the deeper continental shelf and upper slope at depths between 200 m and 400 m, and are caught by bottom trawls and longlines. Monk are, with the tuna, one of the most valuable fish species of Namibia in terms of price per unit weight.

Monk ambush any small and large prey, up to their own size, but hake are their main food items (Boyer and Hampton, 2001); TL = 4.8, P.

Cape gurnard. - Cape gurnard (*Chelidonichthys capensis*) occur on the continental shelf between depths of 10 and 390 m, and are caught mainly in bottom trawls.

Cape gurnard feed mainly on small fishes (Bianchi *et al.*, 1993); TL = 4.3, P.

Other species. - We used the following settings for the other species in the catch database: chub mackerel: TL = 4.3, P; large eye dentex: TL = 3.7, Z; kinglip: TL = 4.4, P; Westcoast sole: TL = 3.5, Z (here treating small zoobenthos as if it were zooplankton); silver kob: TL = 4.3, P; panga and reds:TL = 3.9, Z; rock lobster: TL = 2.6 (not considered for P/Z computation); crabs: TL = 2.3 (not considered for P/Z computation); and 'Mixed species': not used in computation of mean TL.

DERIVATION OF TIME SERIES OF INDICES OF ECOSYSTEM STATES

Using catch data as a proxy to describe the underlying structure and dynamics of marine systems is not perfect. However, in the absence of more direct data, catch data are believed to broadly reflect the ecosystem from which the catches were made. Such data are frequently used for single-species stock assessment purposes (e.g. catch per unit of effort data) and have previously been used in other ecological studies (see Pauly, 1998). Therefore, while not perfect, we believe that these data are informative for our purposes and as such represent a cost-effective information for use in descriptions of the ecosystem status of the Namibian marine system.

Trophic levels (TL) express the number of steps a consumer organism is removed from the primary producers at the base of a food web, and can be defined by the equation:

$$TL_j = 1 + \sum_{j=1}^{n} DC_{ij} \cdot TL_j \quad (1)$$

where i is the predator, j the nth prey, and DC_{ij} is the fraction of j in the diet of i. TL assignment starts with detritus and plants, both with definitional TL value of 1.

The mean trophic level of the catch from Namibian waters was computed for each year from 1950 to 2000 landings, using

$$\overline{TL_k} = \cdot \sum_{i=1}^{m} Y_i k \cdot TL_i \bigg/ \sum_{i=1}^{m} Y_i k \quad (2)$$

where Y_{ik} is the landings of species i in year k and TL_i is its trophic level, implying that the mean trophic level thus obtained is weighted by the catch.

The FiB index (Pauly et al., 2000) was computed from:

$$FiB = log\left[\left(\sum_i Y_{ik} \cdot 10^{TL_i}\right) \bigg/ \left(\sum_i Y_{i0} \cdot 10^{TL_i}\right)\right] \quad (3)$$

where i and k are defined above, and where the subscript '0' refers to the year at the start of a series, which serves as anchor. The FiB index changes its value only when a decrease in TL is not matched by a corresponding increase in catch, and conversely for increasing TL. Here, 'corresponding' is defined as a 10-fold increase for a decline of one trophic level, as implied by a 10% transfer rate between trophic levels (Pauly and Christensen, 1995).

Finally, a piscivory index, the ratio of the sum of catches of piscivorous fishes (P) to the sum of catches of piscivorous (P) plus planktivorous fish, was computed. Here, the planktivores are identified by the letter Z (for zooplankton, although phytoplankton is also consumed, notably by the southern African sardine), to match the nomenclature of Caddy and GaribaldI (2000), who proposed this index. The text above gives our P, Z assignments, which are based on the diet information data also used to estimate TL values. The species consuming finfish or pelagic cephalopods were regarded as 'piscivorous,' while the rest were defined as '(zoo-) planktivorous' (which excluded detritivores and benthic-feeding fish that eat invertebrates off the sea bottom, and all invertebrates from this analysis). More precisely, the index in question was computed for each year from 1950 to 2000, using

$$\textit{Piscivory index} = \sum_{j=1}^{n} P_k \bigg/ \left(\sum_{j=1}^{n} P_k + \sum_{j=1}^{n} Z_k \right) \tag{4}$$

where P_k represents the reported catch of piscivorous finfish in year k and correspondingly for Z_k.

RESULTS AND DISCUSSION

Based on the analyses of total landings from 1950 to 2000 (Figure 1) and of the time series of indices of ecosystem status (Figures 2-5), three periods with distinct trends can be readily identified:
- the 'undeveloped' stage of the fishery (1950-64), when sardine dominated the catches in landed weight, and during which large fisheries, particularly for horse mackerel, Cape hake and chub mackerel developed;
- the 'developing/mature' stages of the fishery (1965-69), when landings and mean trophic level increased rapidly, followed by the collapse of the sardine stock; and
- the 'senescent' stage (1970-2000), during which the mean TL of catches stagnated (Figure 2), while the Piscivory index (Figure 3) and the FiB index (Figure 5) declined.

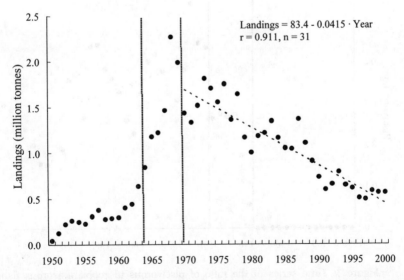

Figure 1. Analysis of catch time series from Namibian waters, 1950 to 2000, by major species groups. Note occurrence of three distinct periods, discussed in the text.

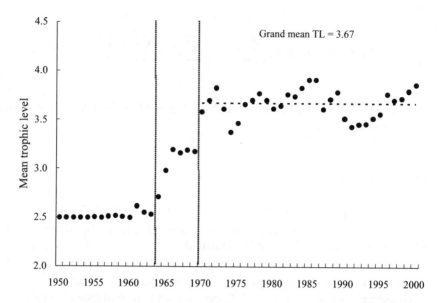

Figure 2. Time series of mean trophic level of catches from Namibian waters, 1950 to 2000. Note occurrence of three distinct periods, during the last of which stagnating mean trophic levels mask ecosystem changes detected by two other ecosystem status indices (see text).

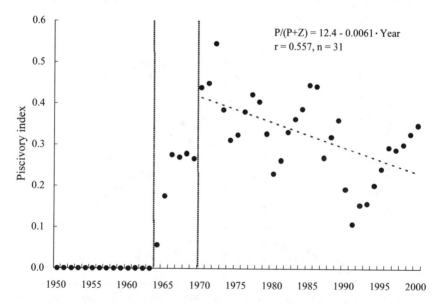

Figure 3: Time series of the ratio of piscivorous to zooplanktivorous fishes in catches from Namibian waters, 1950 to 2000. Note occurrence of three distinct periods, as in Figure 1.

In 1950-64, southern African sardine dominated the fishery with peak catches of 1.4 million tonnes in 1968, followed by a collapse in the early 1970s. During this period, the landings showed an increasing trend (Figure 1), while the mean TL (Figure 2) and the piscivory index showed no distinct trends (Figure 3), and the FiB index showed an increasing trend (Figure 5). Fisheries for Cape hake, horse mackerel and chub mackerel commenced in 1964, 1955 and 1971 respectively. Cape hake and horse mackerel landings peaked during the early to late 1970s. The second period, 1965-69, saw an increase in the impact of distant water fleets (DWF), which began operating off Namibia in the early 1960s (Bonfil, 1998). During this 5-year period, the rapid increases in landings (Figure 1), notably on shallow- and deep-water Cape hake, Cape anchovy, horse mackerel and rock lobster, were reflected as increases of all three ecosystem status indices.

These trends turned around during the 1970-2000 period, where declining trends are observed for total landings (Figure 1), piscivory index (Figure 3), and FiB index. Of these only the piscivory index indicates that fishing has impacted the ecosystem structure (the declining trend in the FiB index is trivial, as it reflects only declining catches). Here, contrary to our earlier analyses, which were based on preliminary estimates of trophic levels (Willemse, 2002; Willemse and Pauly, 2004), the impact of fisheries on the ecosystem is not reflected in the mean TL of catches, which stagnates during the period in question. We attribute this to the fact that off Namibia there is no

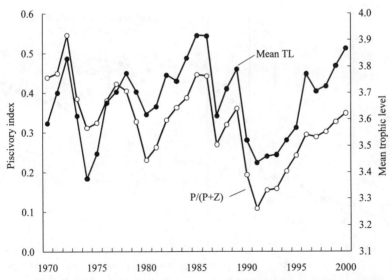

Figure 4: Mean trophic levels and piscivory index of fisheries catches from Namibian waters, 1970 to 2000. Note that the fluctuations of these two indices of ecosystem states in response to long-term environmental events are largely in phase, suggesting that they both reflect the same underlying phenomena (see text).

directed fishery for pelagic goby and other low level organisms (incl. jellyfish), which are now extremely abundant in the ecosystem. We are here reminded that for the 'fishing down marine food web' phenomenon to be detected, the assumption must be met that the relations between species in the fisheries catches used for the analysis broadly reflect their relative abundances in the ecosystem (Pauly et al., 1998b).

The choice of trophic levels is also important: the use of different values in Willemse (2002) and Willemse and Pauly (2004), notably of a high value for southern African sardine (3.4 instead of the 2.5 value used here) generated for the 1970-2000 period what may have been a spurious downward trend of mean TL, not recorded here (see Figure 2).

Environmental fluctuations do matter in an ecosystem such as that occurring in Namibian waters, but they cannot be evoked to explain the 30 year trend in Figure 1, 2, and 5. Indeed, as might be seen from Figure 4, these fluctuations, while inducing the two ecosystem state indices in Figure 4 to vary in unison, do not prevent a long-term trend from appearing that can be readily explained by the gradual removal of large, piscivorous, high-TL species with benthic affinities, and their replacement, in the ecosystems and the catches, by predominantly planktivorous, small, low-TL species.

Thus, we can conclude that the fisheries off Namibia have induced changes of ecosystem structure mainly as a result of pre-independence overfishing. This implies increased management effort by the national govern-

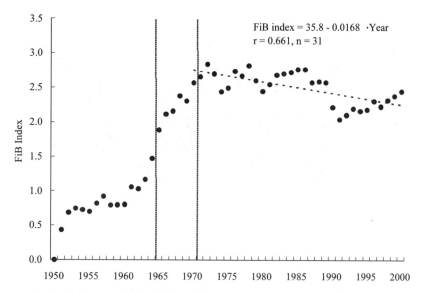

Figure 5. Time series of the FiB (fishing-in-balance) index for Namibian waters, 1950-2000. This series fails here to indicate anything beyond what a perusal of the catch trends (Figure 1) would suggest, owing to the catches, in this case, not reflecting changes in underlying ecosystem structure (see text).

ment. Also the research needs to be continued which has so far contributed to a much increased understanding of the Benguela ecosystem (Iyambo, 2001). However, this will require more emphasis on ecosystem-based indicators such as illustrated here.

REFERENCES

Bianchi, G., Carpenter, K.E., Roux, J.P., Molloy, F.J., Boyer, D. and Boyer, H.J. (1993): The living marine resources of Namibia. FAO species identification field guide for fishery purposes. Food and Agricultural Organisation, Rome. 265 pp., 11 colour plates. (This guide was revised in 1999.)

Bonfil, R. (1998): Case study: distant water fleets off Namibia. In: *Distant water fleets: an ecological, economical and social assessment* (R. Bonfil, G. Munro, U.S. Sumaila, H. Valtysson, M. Wright, T. Pitcher, D. Preikshot, N. Haggan and D. Pauly. Eds.), Fisheries Centre Research Report 6(6). University of British Columbia. Vancouver. 77 pp.

Boyer, D. (1994): Research provides information for management. *Namibian Brief: Focus on Fisheries and Research* 18: 75-78.

Boyer, D.C. and Hampton, I. (2001): An overview of the living marine resources of Namibia. In: *A decade of Namibian Fisheries Science. South African Journal of Marine Science* 23: 5-35.

Boyer, D.C., Boyer, H.J., Fossen, I. and Kreiner, A. (2001): Changes in abundance of the northern Benguela sardine stock during the decade 1990-2000, with comment on the relative importance of fishing and the environment. In: *A decade of Namibian Fisheries Science. South African Journal of Marine Science* 23: 67-84.

Caddy, J.F. and Garibaldi, L. (2000): Apparent changes in the trophic composition of world marine harvests: perspective from the FAO capture database. *Ocean and Coastal Management* 43: 615-655.

Crawford, R.J.M., Shannon, L.V. and Polloch, D.E. (1987): The Benguela Ecosystem VI: the major fish and invertebrate resources. In: *Oceanography and Marine Biology: An Annual Review* (M. Barnes, ed.), Vol. 25, pp. 353-505. Aberdeen, University Press.

Froese, R. and Pauly, D. (eds.) (2000): FishBase 2000: concepts, design and data sources. ICLARM, Los Baños, Philippines. 344 pp. [Distributed with four CD-ROM; updates at *www.fishbase.org*].

Hamukuaya, H. (1994): Research to determine the biomass of hake. *Namibian Brief: Focus on Fisheries and Research* 18: 73-74.

Heymans, J.J. and Baird, D. (2000): Network analysis of the northern Benguela ecosystem by means of NETWRK and ECOPATH. *Ecological Modelling* 131: 97-119.

Hilborn, R. and Walters, C.J. (1992): *Quantitative Fisheries Stock Assessments: Choice, Dynamics and Uncertainty*. Chapman and Hall, New York and London.

Iyambo, A. (2001): A decade of Namibian fisheries science: an introduction. *South African Journal of Marine Science* 23: 1-4.

Jarre-Teichmann, A., Shannon, L.J.,

Moloney, C.L. and Wickens, P.A. (1998): Comparing trophic flows in the southern Benguela to those in other upwelling ecosystems. In: S.C. Pillar, C.L. Moloney, A.I.L. Payne and F.A. Shillington (eds.): *South African Journal of Marine Science* 19: 391-414.

Klingelhoeffer, E. (1994): Emphasis on distribution and abundance. *Namibian Brief: Focus on Fisheries and Research* 18: 79-81.

Manning, P.R. (1998): Managing Namibia's marine fisheries: optimal resource use and national development objectives. Doctoral thesis. London School of Economics and Political Science, London. 313 pp.

Palomares, M.L.D. and Pauly, D. (2004): Biodiversity of the Namibian exclusive zone: a brief review with emphasis on online databases. In: *Namibian Fisheries: Ecological, Economic and Social Aspects.* (U.R. Sumaila, D. Boyer, M. Skogen and S.I. Steinshamn, eds.), pp. 53-74. Eburon, Delft.

Pauly, D. (1998): Large marine ecosystems: analysis and management. In: *Benguela dynamics: impacts of variability on shelf-sea environments and their living resources* (S.C. Pillar, A.I.L. Payne and F.A. Shillington, eds.), *South African Journal of Marine Science* 19: 487-499.

Pauly, D. and Christensen, V. (1995): Primary production required to sustain global fisheries. *Nature* 374: 255-257. [Erratum in Nature (376): 279.]

Pauly, D., Christensen, V., Dalsgaard, J., Froese, R. and Torres, F.C. Jr. (1998a): Fishing down marine food webs. *Science* 279: 860-863.

Pauly, D., Froese, R. and Christensen, V. (1998b): How pervasive is "Fishing down marine food webs": response to Caddy *et al. Science* 282: 183 [full text (p. '1383a') on *www.sciencemag.org/cgi/content/full/2 82/5393/1383*].

Pauly, D., Christensen, V. and Walters, C.J. (2000): Ecopath, Ecosim and Ecospace as tools for evaluating ecosystem impact of fisheries. *ICES Journal of Marine Science* 57: 697-706.

Pauly, D., Palomares, M.L., Froese, R., Sa-a, P., Vakily, M., Preikshot, D. and Wallace, S. (2001): Fishing down Canadian aquatic food webs. *Canadian Journal of Fisheries and Aquatic Science* 58: 51-62.

Payne, A.I.L., Pillar, S.C. and Crawford, R.J.M. (eds.) (2001): *A Decade of Namibian fisheries science. South African Journal of Marine Science* 23. 466 pp.

van der Lingen, C.D. (1998): Gastric evacuation, feeding periodicity and daily ration of sardine *Sardinops sagax* in the southern Benguela upwelling ecosystem. In: *Benguela dynamics: impacts of variability on shelf-sea environments and their living resources* (S.C. Pillar, A.I.L Payne and F.A. Shillington, eds.), *South African Journal of Marine Science* 19: 305-316.

Willemse, N. (2002): Major trends in the marine fisheries catches of Namibia, 1950-2000. MS Thesis, Department of Biology, Norwegian College of Fishery Science, University of Tromsø.

Willemse, N. and Pauly, D. (2004): Ecosystem overfishing: a Namibian case study. In: P. Chavance *et al.* Eds.): Proceedings of the international Symposium on "Marine fisheries, Ecosystems, and Societies in West Africa: Half a Century of Change" held in Dakar, Senegal on 24-28 June 2002. [In press]

6   MANAGEMENT REGULATIONS OF NAMIBIAN ANGLING FISH SPECIES

*Johannes Andries Holtzhausen and Carola Heidrun Kirchner**

Abstract

An economic survey in 1997 showed that recreational angling is the most important sector of the Namibian inshore linefish fishery. Prior to the year 1995, fisheries regulations for Namibian angling fish species were based on regulations adapted from South African legislation enforced prior to Namibia's Independence in 1990, rather than the results of formal stock assessments, because some biological parameters and the total annual catches of shore anglers for the two most important species, silver kob and West Coast steenbras, were not available. Two separate studies were launched to obtain the necessary biological parameters as input for age-based assessment models. To obtain the total annual catch of recreational anglers, a roving-roving creel survey design was developed whereby anglers were interviewed while they were fishing. Their catches were also measured during the interview. Based on some of the results of the assessments, previous management regulations were revised and/or new regulations introduced in December 2001. These include a reduced daily bag per angler, the re-introduction of minimum size limits, and the introduction of a daily limit per species that may be retained over a certain maximum size limit. For transportation purposes, an angler may now accumulate three days' catch, but a limit on the number per species that may be transported, was also introduced. The chapter gives an overview of why and how these new regulations were formulated.

INTRODUCTION

The Republic of Namibia is situated in south-western Africa and is famous for its excellent shore angling opportunities. The Namibian fishing industry,

---

[*] The Director of the Ministry of Fisheries and Marine Resources is thanked for permission to publish the data. We also thank all private anglers for their efforts in the tagging project. The technical assistance of our colleagues Messrs Stefanus Voges, Shaun Wells and Brian Louw is greatly appreciated.

well known for its rich stocks of demersal and small pelagic species, is based on the Benguela Current, one of the four major eastern boundary upwelling systems of the world. What these systems lack in terms of species diversity is more than compensated for in abundance. Consequently, for angling enthusiasts, the Namibian coast has for decades been synonymous with large catches of linefish species such as kob (*Argyrosomus* spp.), West Coast steenbras (*Lithognathus aureti*), galjoen (*Dichistius capensis*), blacktail (*Diplodus sargus*) and various shark species. However, recreational anglers share the resource with a commercial sector, which makes management complex.

Namibia's inshore linefish fishery can best be described as the fishery that uses rods-and-reels or handlines with baited hooks to catch various angling species either for recreation from the shore or from skiboats, or commercially from skiboats and lineboats (Table l). The linefish fishery therefore supports three different sectors, namely a) the recreational rock-and-surf sector, b) the recreational skiboat fishery and c) the commercial skiboat and lineboat sector. Lineboats are approximately 20 m in length and carry up to 16 fishermen who each operates a handline with two hooks. In order to land fresh fish on ice, these boats are able to put to sea for a maximum of one week and each boat has a carrying capacity of 12 tonnes of fish. Currently, two freezer lineboats are operating in Namibian waters; this type of boat can spend longer times at sea and the fish are also processed and packed onboard. A skiboat is approximately 5-6 m long and can carry between four and six fishermen. Fishing is done with boat-rods and reels and normally only one rod per fisherman is used. Skiboats usually fish just behind the breaker-zone and most skiboat fishing is concentrated in the vicinity of Swakopmund.

Silver kob and West Coast steenbras were identified as the two most important species because these are the two most popular recreational angling

Table 1: Fish species caught by the multi-sector Namibian linefish fishery.

Species	Shore-angling Sector	Lineboat sector	Skiboat sector
Silver kob	x	x	x
West Coast steenbras	x	x	
Snoek		x	x
Galjoen	x		x
Blacktail	x		x
Barbel	x		x
Spotted gullyshark	x		x
Coppershark	x		x
Cowshark	x		x
Smooth houndshark	x		x

species and both are also commercially exploited by other sectors of the linefish fishery (Holtzhausen et al., 2001a). However, these species were never assessed owing to lack of information on the necessary biological parameters and the total annual catch of anglers, used as input for fishery assessment models. Two separate studies were thus launched in 1995 to obtain the necessary biological data for each species. To obtain the total annual catch of recreational anglers, a roving-roving creel survey design was developed (described by Kirchner and Beyer, 1999) whereby anglers were interviewed while they were fishing. Their catches were also measured during the interview.

For historical reasons South African angling regulations were enforced here before Namibia's Independence in 1990. These management regulations, such as minimum size limits and daily bag limits, were not based on scientific investigations on local species, but were formulated for South African species. After Independence, Namibia amended or abolished some of these regulations (Ministry of Fisheries and Marine Resources (MFMR), 1993); for instance an arbitrarily chosen daily bag limit of 30 fish per angler came into effect while the current roving-roving creel surveys show that an angler catches on average no more than 2.6 fish per day (Kirchner et al., 2000). Thus these bag limits had no effect as a management tool. Minimum size limits were also abolished for all angling species (Table 2). Thus, there

Table 2: Some of the old angling regulations enforced from 1993 and new angling regulations enforced after December 2000.

Old regulations	New regulations
No size limits	Size limits on angling species
	40 cm TL kob and West Coast steenbras
	30 cm galjoen and 25 cm blacktail
	Only 2 kob ≥ 70 cm TL and
	2 West Coast steenbras ≥ 65 cm TL
	per angler per day.
Daily bag per angler	Daily bag per angler of
30 fish or 30 kg fillets per angler	10 fish.
For transportation	For transportation
A maximum of 60 fish or 60 kg fillets per vehicle, minimum of two anglers in vehicle.	An angler may accumulate his catch of 3 days of which only 10 may be of any one species – no restriction on anglers per vehicle.
Fish may be filleted for transportation.	Fish may only be transported in a whole state.
	This is to be able to enforce the size limits.

was a need to start managing the resource basing management decisions on scientific evidence.

The annual contribution of the linefish fishery to Namibia's Gross Domestic Product (GDP) has not been estimated prior to 1997. Kirchner *et al.* (2000) conducted an economic survey in 1996/97, which showed that recreational angling earned the country approximately N$30 million compared with only N$5 million for the commercial sector. A similar study conducted in 1998 by Zeybrandt and Barnes (2001) confirmed the figure for the recreational shore fishery. It is therefore clear that the recreational sector is the most important of the three sectors.

The movements of a fish throughout its range and in the various stages of its life may have implications for its management as a population or stock. Migratory patterns of adult fish, especially of the spawner stock, and the dispersal of individuals into areas adjacent to where they have spent their early life, must be known if a stock is to be assessed or managed as a unit. The identification of discrete stocks is also basic to the conservation and rational exploitation of all fisheries resources (Holtzhausen *et al.*, 2001a).

Throughout the world, the method of tagging and recapturing specific fish species is used to study their migrations. The main aim of the Namibian tagging programme, launched in the late 1980s, was to investigate the possibility that linefish move out of the Namib Naukluft Park (NNP) and Skeleton Coast Park (SCP) to the West Coast Recreational Area (WCRA; see Figure 1). The NNP is closed to all shore-angling and the SCP is partly closed to shore-angling. The hypothesis was that such areas serve as reserves seeding other parts of the coast, including the angler-accessible WCRA. The possibility also existed that there might be one or more populations of each species occurring in Namibian waters.

Prior to 1995, it was thought that kob (*Argyrosomus hololepidotus*) ranged from northern Natal, on the east coast of South Africa, to around Congo, on the west coast of Africa. As such, all kob in Namibian waters were classified as the species *A. hololepidotus*. However, in the early 1990s, Griffiths and Hecht (1993) suggested that there may in fact be two species of kob off southern Africa, referring to them as kob A and kob B. Specimens of these fish were obtained during March 1995 along the Namibian coastline during a linefish tag-and-release excursion. Taxonomic investigations indicated that the Namibian kob A was a different species from the South African kob A, *Argyrosomus japonicus*. The Namibian kob A has subsequently been described as a new species and named the West Coast dusky kob, *A. coronus*, whereas kob B was described as the same in both South African and Namibian waters, reclassified as silver kob *A. inodorus* (Griffiths and Heemstra, 1995).

Tag-recapture results proved the existence of a separate, closed population of West Coast steenbras in the vicinity of Meob Bay, and a northern population off central and northern Namibia (WCRA and SCP). Also, distinct differences in growth rates, otolith morphology, size at maturity, sex ratios and length-at-age were found between the Meob Bay and the more northern population. Electrophoretic analysis on samples from the two populations showed significant genotypic differentiation at two loci, indicating that effective barriers exist to isolate them (Van der Bank and Holtzhausen, 1998/99). Thus, it was shown that two distinct populations of West Coast steenbras occur in Namibian waters, and they might require different management measures as one population is only accessible to lineboats while the other population is accessible to lineboats and recreational anglers.

The two separate studies on silver kob (Kirchner, 1998) and West Coast steenbras (Holtzhausen, 1999) that were launched in 1995 were aimed at

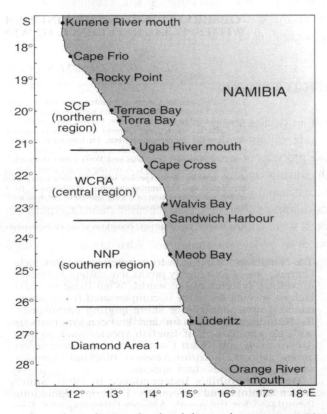

Figure 1: Map of Namibia indicating the closed diamond-mining area (Diamond Area 1) and conservation areas, the partially restricted Skeleton Coast Park (SCP), the closed Namib Naukluft Park (NNP), and the West Coast Recreational Area (WCRA) that is open to angling. It should be noted that lineboats may operate along the entire coastline.

investigating their life histories, obtaining the necessary biological parameters and the total annual catches of recreational anglers for assessing both stocks, and finally at recommending management options to harvest the resources on a sustainable basis. Furthermore, characteristics of sparids that may have serious implications for their management are hermaphroditism, longevity, slow growth and high rates of natural mortality. West Coast steenbras have all these characteristics and it is therefore crucial that all these issues be considered when developing an effective management strategy for the species. This chapter describes the methods that were used to obtain the necessary data and how management options were formulated ~ mostly from the results, but sometimes contrary to the results due to e.g. socio-economic implications. Most of the information in this chapter has been published before and should thus be seen as a review of these. The three most important publications to be consulted for more information on migration, biology, stocks assessments and economics are 1) by Holtzhausen (1999) on the West Coast steenbras, 2) by Kirchner (1998) on silver kob, and 3) by Kirchner *et al.* (2000) on the economics of the linefish fishery.

MATERIALS & METHODS

Migration studies and genetics

Fish were caught with rod and reel from the shore, and standard tag-and-release procedures were followed (Botes, 1994; Kirchner, 1998; Holtzhausen, 1999). During tagging excursions fish were normally tagged over a period of four days, usually making use of the same anglers to keep the effort constant for calculating CPUE at various fishing stations. Approximately 50 such excursions were conducted, mostly in the closed southern and northern areas (Table 3). A yellow monofilament T-bar anchor tag with an alpha-numeric code and "Fisheries, Namibia" imprinted on it, was inserted with a Banox applicator into the muscle posterior and on the top-left side of the dorsal fin.

Table 3: Numbers of silver kob and West Coast steenbras tagged and released and the numbers recaptured in each region along the Namibian coastline.

Area	Silver kob				West Coast steenbras			
	Number tagged	Number recaptured			Number tagged	Number recaptured		
		South	Central	North		South	Central	North
South	7 888	34	44	0	23 834	329	29	0
Central	2 795	0	236	3	8 555	2	260	4
North	9 337	2	31	87	3 189	0	67	77

Biochemical genetic studies (starch-gel electrophoresis) were performed on both species a) to determine if the silver kob and dusky kob were of the same gene pool, and b) to test the hypothesis that the southern West Coast steenbras population at Meob Bay was a separate and closed population to the stock that occurs in the central and northern shore areas.

Biological data

Biological samples were also collected during these tagging excursions. For example, otoliths were extracted for determination of age and growth rate and to formulate age-length keys, and gonads were weighed and classified to develop gonadosomatic indices (GSI) to determine breeding seasons, and macroscopically staged (in the case of West Coast steenbras) to confirm that the species is a protandrous hermaphrodite. Growth parameters were derived for both species from otolith readings and the results were verified with mark-recapture data (Kirchner and Voges, 1999, Holtzhausen and Kirchner 2001b). Size-specific natural mortality was determined for West Coast steenbras by developing a length-based catch curve, and this method was also adapted for silver kob (Beyer et al., 1999). Age-length keys were constructed for each species and used to transform length- to age-frequency distributions. Table 4 depicts the biological parameters used for modeling.

Table 4: Parameter estimates and the range/best estimates of values used in the models for assessment of the northern West Coast steenbras stock (Holtzhausen and Kirchner, 2001b) and the silver kob stock (Kirchner, 2001) off Namibia.

Parameter	Range / best estimate	
	West Coast steenbras	Silver kob
K	0.088 year^{-1}	0.136 year^{-1}
L_∞	84.6 cm	103 cm
T_0	-2.756	-1.58
W_∞	14.19 kg	11.38 kg
M	0.23 year^{-1}	0.15 year^{-1}
F	0.11 year^{-1}	0.22 year^{-1}
Z	0.35 year^{-1}	0.365 year^{-1}
M_∞	0.21	0.19
F_{term}	0.11	
T_r	1 year	0.75 years
T_c	2 years	1 years
A	0.00003	4.8×10^{-5}
B	2.9444	2.71
T_m	5 years	5 years
T_f	10 years	10 years

Table 5: Estimated values for depletion of the stock, long-term biomass, MSY and expected catches of Namibian silver kob in 1997/98 determined using the Thompson and Bell model, with catches per age-class = N[mean, SD], M year^{-1} = U[0.15–0.25] and F_{term} year^{-1} = U[0.17–0.27] for scenarios 1–4. Percentiles (95%) are given with the average value in parentheses (after Kirchner, 2001).

Scenario	Depletion (%)	Biomass (1000tons)	MSY (1000tons)	Catches (1000tons)
1 Current	30–47 (39)	7.8–10.6 (9.1)	1.14–1.36 (1.24)	1.00–1.17 (1.13)
2 Minimum size of 40 cm	36–54 (46)	9.7–12.3 (10.9)	1.05–1.15 (1.10)	0.95–1.05 (1.00)
3 Cut F of linefish boats by 25%	34–52 (43)	8.7–11.6 (10.0)	1.10–1.34 (1.20)	1.00–1.10 (1.05)
4 40 cm size limit + cut F of linefish boats by 25%	41–58 (50)	10.0–13.2 (11.9)	1.00–1.12 (1.07)	0.88–1.02 (0.95)

Table 6: Estimated values for depletion, biomass (tons), maximum sustainable yield and catches in tons of the silver kob stock for the years 2000/01. Percentiles (95%) are given in parentheses except for yield.

Parameter	Best estimate	Percentiles
Depletion (%)	40	(29-51)
Biomass (tons)	7 175	(5 734–9 125)
MSY (tons)	955	(855–1 114)
Catches (tons)	856	(817–896)

Table 7: Estimated values for depletion, long-term spawning biomass (tons), proportion of females in the spawner stock biomass (SSB) and the total biomass, maximum sustainable yield and catches in tons of the northern stock of West Coast steenbras for the years 2000/01. Percentiles (95%) are given in parentheses except for yield (after Holtzhausen, 1999).

Parameter	Best estimate	Percentiles
Depletion (%)	53	(37-67)
SSB (tons)	2006	(1 294–2 942)
% Females in SSB	44	(38–50)
% Females in biomass	22	(19–26)
MSY (tons)	278	(202–384)
Catches (tons)	191	

Models and assessments

In Namibia, silver kob and West Coast steenbras were first assessed in 1998 (Kirchner, 1998) and 1999 (Holtzhausen, 1999) respectively. The results of these assessments were based on only three years' data and therefore the results are not shown in this chapter but merely mentioned where appropriate.

Two yield-per-recruit approaches were used for silver kob and West Coast steenbras (Tables 5, 6 and 7).

In 1997/98, a Beverton and Holt (1957) yield-per-recruit model was used to investigate the potential effect of different fishing mortalities, natural mortality and age-at-first-capture schedules on silver kob and West Coast steenbras. In a Thompson and Bell (1934) yield-per-recruit model, fishing mortality arrays and recruitment, estimated by cohort analysis and size-specific natural mortality, were used to study different scenarios that could be implemented as management measures for the sustainable exploitation of the two species (Kirchner *et al.*, 2001; Holtzhausen and Kirchner, 2001c).

To estimate the total annual catch and effort of shore-anglers, a roving-roving creel beach-survey design, through which data were collected by intercepting and interviewing anglers while they were fishing, was used (Kirchner and Beyer, 1999). Skiboat and lineboat catches were sampled routinely at offloading sites and the data raised to represent the total annual catch (Kirchner, 1998). The combined catches of the three fishery sectors for six consecutive fishing seasons are illustrated in Figures 2 and 3 for silver kob, and that of the recreational anglers in Figure 4 for West Coast steenbras.

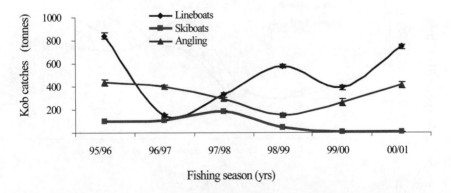

Figure 2: Catches in tonnes of the three forms of kob fisheries for six consecutive fishing seasons off Namibia.

Economics

The economic value of the Namibian recreational rock-and-surf fishery was determined by surveying a stratified sample of 240 anglers over a period of one year (Kirchner *et al.*, 2000). Skiboat owners and lineboat skippers were also surveyed to estimate their annual contributions to Namibia's Gross Domestic Product (GDP) (Kirchner *et al.*, 2000). Zeybrandt and Barnes (2001) conducted a similar survey in 1998 of 626 coastal recreational anglers aimed at measuring further economical characteristics of demand in the fishery. In particular, consumer surpluses and value added for the different market segments, as well as elasticities of demand, were investigated.

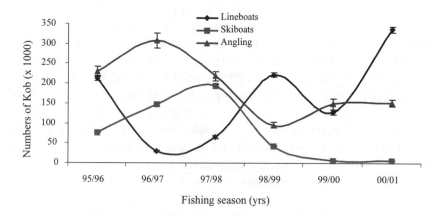

Figure 3: Catches in numbers of the three forms of kob fisheries for six consecutive fishing seasons off Namibia.

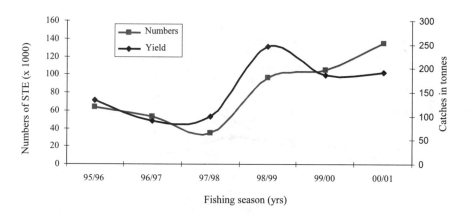

Figure 4: Numbers and catches (in tonnes) of West Coast steenbras of recreational shore anglers for six consecutive fishing seasons off Namibia.

RESULTS

Migration and genetics
Silver kob. - In all, 20 020 kob were tagged and released, 437 (2.18%) subsequently being recaptured (Table 3). The recapture results demonstrate that there is only one stock of silver kob, ranging from Cape Frio in the north to Meob Bay in the south (Kirchner and Holtzhausen, 2001; see Figure 1). Biochemical genetic studies (starch-gel electrophoreses) on specimens from two Namibian populations of *A. inodorus* and one population of *A. coronus* further confirmed that the two species were not from the same gene pool (Van der Bank and Kirchner, 1997). The southern distribution of the latter species ends about 55 km north of Cape Frio (Van der Bank and Kirchner, 1997; Figure 1). As <10% of the annual kob catch (by number) of recreational shore-anglers in Namibia is of this species (Kirchner, 1998), research effort was directed towards silver kob.

West Coast steenbras. - Of the 35 578 West Coast steenbras tagged and released, 768 (2.15%) were recaptured (Table 3). The results strongly indicate the existence of a separate, closed population of West Coast steenbras in the vicinity of Meob Bay, and a northern population off central and northern Namibia (WCRA and SCP). Also, distinct differences in growth rates, otolith morphology, size at maturity, sex ratios and length-at-age were found between the Meob Bay and the more northern population. Electrophoretic analysis on samples from the two populations showed significant genotypic differentiation at two loci, indicating that effective barriers exist to isolate them (Van der Bank and Holtzhausen, 1998/99).

Models and assessments
Silver kob. - Kirchner (2001) calculated that by introducing a minimum size limit of 40 cm TL, the long-term biomass could increase by approximately 10%. By further reducing the catches of lineboats by 25% (i.e. implement a total allowable catch of 350 tonnes gutted and headed weight) it was calculated that the long-term biomass could increase to 11 900 (10 700, 13 200) tonnes. For ages 1-5 only, c. 10% of silver kob mortalities are attributable to fishing (Figure 5). In contrast, fishing mortalities are closer to 50% for ages 6, 10, 11 and 13. Lineboats fish heavily on age-classes 6-7 and, when a Thompson and Bell analysis was run, lowering the fishing mortality for lineboats by c. 25%, the total biomass would reach 50% of its pristine level under steady state conditions. The target reference level of depletion for the silver kob stock is set at 50%. In 2000/01, the stock was assessed with the maximum sustainable yield (MSY) for silver kob estimated at 955 tonnes with 95 percentiles of (855, 1 114) tonnes at a biomass of 7 175 (5 734, 9 125) tonnes

and a level of depletion of 40% (29, 51; Table 6). The high fishing mortalities of ages 10, 11 and 13 are the result of shore-anglers and skiboats targeting the large spawners when they aggregate in shallow water during the breeding season (Figure 5). It is clear that some sort of protection should be offered to these spawners to ensure sustainable recruitment.

Kirchner *et al.* (2001), using two different empirical equations, calculated for silver kob the length and age (L_{opt} and t_{opt}) at which the biomass should be at its highest level. These produced similar results, where L_{opt} and t_{opt} were estimated as 68 cm (7.8 yr) and 67 cm (7.5 yr) respectively. These are the lengths (TL) at which silver kob first start to take part in a spawning migration. The highest spawning potential is usually reached at the length of L_{opt}. Therefore the daily bag limit was lowered to 10 fish in total, of which only two may be above 70 cm TL.

Table 8: Catches and yields of the three forms of kob fishery (1 October 1995 to 30 September 1996.

Fisheries	Numbers ±SE (× 1 000)	Mass ±SE (tonnes)
Anglers	230 ± 13	361 ± 22
Skiboats	75 ± 4	97 ± 4
Lineboats	219 ± 6	728 ± 22
Total	524 ± 15	1 187 ± 32

Estimation of catch and effort (Table 8) showed that, in the 1996/97 season, the three kob fisheries combined harvested approximately 1 187 (±32) tonnes of silver kob; most was taken by recreational shore-anglers. Compared to the

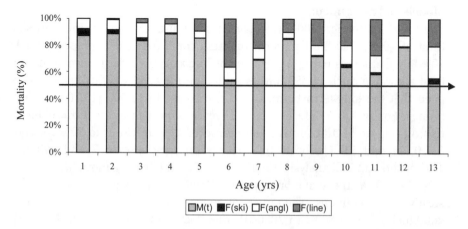

Figure 5: Mortality of silver kob of different ages attributable to natural causes $M(t)$ and fishing mortality by the skiboat fishery $F(ski)$, shore-anglers $F(angl)$ and the lineboat fishery $F(line)$. The arrow indicates 50% mortality.

current MSY of 955 tonnes (Table 6), the stock is evidently optimally utilised. The Thompson and Bell yield-per-recruit model indicated that, if a minimum size limit of 40 cm TL were to be introduced, the long-term biomass could increase by approximately 10% (Table 5). However, Kirchner *et al.* (2001) argued that implementing a minimum size limit for the silver kob fishery would have serious economic implications for coastal communities. Nevertheless, something needs to be done, and therefore a total allowable catch (TAC) of 350 tonnes (headed and gutted) has been proposed for the lineboat fishery. By selectively targeting bigger silver kob with big hooks, this fishery has a large impact on spawner biomass, so effort has to be reduced.

Figure 2 shows that the kob catches of recreational shore-anglers doubled in the last three fishing seasons. However, catches in numbers have remained constant over the last two seasons (Figure 3). This might be as a result of the skiboat sector, which has collapsed during the same period. On average, skiboats have caught smaller kob of about 35 cm TL. Recreational anglers catch larger kob on average, which is reflected in a higher tonnage being caught but in numbers that remain relatively constant.

West Coast steenbras. - In 2000/01, the northern West Coast steenbras population was assessed and the MSY was estimated at 278 (202, 384) tonnes with a spawning stock biomass level of 2006 (1 294, 2 942) tonnes and a level of depletion of 53% (37, 67; Table 7). The target reference level of depletion for the silver kob stock is set at 50%. The proportion of females in the spawning biomass was estimated at 38–50%, with the best estimate at 44% (Thompson

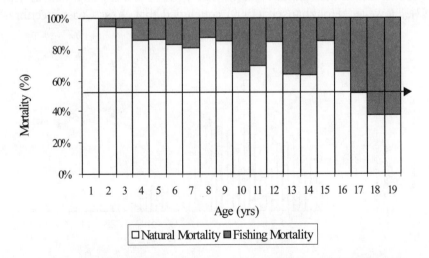

Figure 6: Mortality of northern West Coast steenbras of different ages attributable to natural causes and fishing mortality by recreational rock-and-surf anglers off the Namibian coast. The arrow indicates 50% mortality.

and Bell model). The MSY has been denoted as a limit reference point, in other words catches should not be allowed to exceed this value. In the 1998/99 fishing season, the recreational catch was 245 tonnes, only 33 tonnes lower than the MSY.

Fishing and natural mortality per age group are shown in Figure 6, the latter obtained from values of age-specific natural mortality. Occasionally, for poorly understood fish stocks, a management procedure keeping F ≈ M can be advocated (Gulland, 1970). This is possible because it is an adequate approximation of the optimal $F_{0.1}$ criteria in cases when $1 < M/K < 4$ (Deriso, 1987; M/K for northern West Coast steenbras is approximately 2.6). Mortality attributable to fishing for ages 1–8 does not exceed 20% for each of the age-classes. Between ages 10 and 16, mostly females, the mortality attributable to fishing increases to 40%. The high fishing mortality of West Coast steenbras older than 16 years is the result of increased availability of large fish in Terrace Bay, where they aggregate in shallow waters, hence presenting an easy target for shore-anglers. Similarly to silver kob, only two West Coast steenbras larger than 65 cm FL (16 yr) may be retained per angler per day. This management regulation will offer some protection to these large spawners which contribute most to reproduction. The Thompson and Bell yield-per-recruit model indicated that, if a minimum size limit of 40 cm TL were to be introduced, the long-term biomass could increase by approximately 8%.

The southern West Coast steenbras stock could not be assessed with conventional fishery models, because recreational rock-and-surf anglers are not allowed to fish in the Namib Naukluft Park, which is a closed area (MFMR, 1993). However, Holtzhausen (1999) concluded that West Coast steenbras

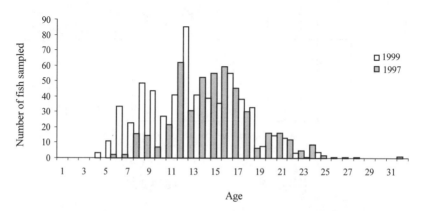

Figure 7: Age composition of West Coast steenbras sampled from commercial catches landed by lineboats at the Walvis Bay harbour, Namibia (n_{1997} = 466, n_{1999} = 600; after Holtzhausen, 1999).

are protandrous hermaphrodites, in other words that fish first function as males but then change sex to become females, at a FL of approximately 40 cm in the southern population. Female West Coast steenbras from the southern population reach 50% maturity at about 43.2 cm FL, an age of about 10 years. Commercial lineboat catches of West Coast steenbras from the southern population only resumed in 1997 as no West Coast steenbras were harvested in this area from 1985 to 1996. Lineboat catches during 1997 and 1999 are presented in Figure 7 (Holtzhausen and Kirchner, 2001c); 82% of the catch was of fish older than 10 years (93% of the 1997 catch and 73% of the 1999 catch). This means that lineboats target the female component of the population, ultimately leading to a change in the sex ratio of the population that could impair its reproductive potential.

Shore-angling catches of West Coast steenbras, estimated from the roving-roving creel survey data, declined over two years but increased significantly during the 1998/99 season. This could have been due to increased catches in the Torra/Terrace Bay area (Figure 1). Usually, catches from that area contribute up to 20% of the total catches of West Coast steenbras for the northern region, but in 1998/99 the proportion of the catch caught there increased to 50%. These were mostly large fish which must have moved into the area from the northern Skeleton Coast Park area.

It should be noted that there is a high degree of uncertainty in these assessments, which should be expected when dealing with a paucity of data. This was only the second time that these stocks were assessed. This is captured in the formal statistical uncertainty estimates, where the 95% percentiles are fairly wide = e.g. the depletion for silver kob in the 2000/01 assessment ranges from 29% to 51% (Table 6), and that for West Coast steenbras ranges from 31% to 67% (Table 7). In addition, given the uncertainties in these assessments, the changes predicted in biomass, yield per recruit and depletion rates are fairly small and might not be detectable over a period of time. However, the topic of a precautionary approach in fisheries research, management and technology has become an important issue globally (Gabriel and Mace, 1999). Therefore, although it seems that both stocks are in good condition, one has to be careful of implementing radical management measures when there is some uncertainty in the results.

Economics

Results from the economic survey indicated that, between October 1996 and September 1997, some 8 800 recreational shore-anglers went on 173 000 angling outings along the Namibian coast and had direct expenditure amounting to N$29.7 million (US$3.7 million in mid 2001). Similarly, Zeybrandt and Barnes (2001) found that the aggregate angler expenditures per annum amount to about N$30 million. Foreign visitors, mostly South Africans, contributed 55% of the expenditure (Kirchner *et al.*, 2000). The skiboat fishery contributes ±N$2 million annually to Namibia's GDP and the lineboat fishery another ±N$3.4 million (Kirchner, 1998). In total, the linefishery contributes approximately N$35 million annually to Namibia's GDP. The estimated expenditure for each fish (all species) caught by a shore-angler, a lineboat, and a skiboat is given in Table 9.

Table 9: Estimated cost of each fish (all species) caught by population group, area, shore-anglers, lineboats and skiboats off the coast of Namibia during the 1996/97 fishing season (after Kirchner, 1998).

Source	N$/fish
Coastal Namibian	34.00
Inland Namibian	94.10
Foreigner	96.40
Terrace Bay	45.80
Torra Bay	44.20
Recreational anglers (average)	64.30
Skiboats	17.70
Lineboats	24.30

Discussion

Daily bag limit

Bag limits are the one management measure mostly used to regulate a recreational fishery. Results from the roving-roving creel survey (Kirchner and Beyer, 1999) indicated that in Namibia, recreational anglers land on average 2.6 fish per day. Thus, as a management measure, the previous daily bag of 30 fish per angler had no effect as a management measure on Namibian anglers. For a bag limit to have an effect in increasing the biomass, it should thus be set at between two and five fish, but this is unrealistic and would destroy the recreational fishery (Kirchner, 2001). For this reason a daily bag limit of 10 fish per angler was introduced for the four important linefish species that include silver kob and West Coast steenbras.

Silver kob

Tag-recapture results support the assumption that only one stock of silver kob is found between Cape Frio and Meob Bay. As only two silver kob tagged in Namibia were recaptured in South African waters and one off Oranjemund (the border between the two countries, but Namibian territory), it is considered that emigration of silver kob to South Africa is minimal. This theory is further supported by the fact that none of the silver kob tagged in South African waters, 6 904 up to 30 April 2000 (Tagging News **13**, July 2000: p. 5), has been recaptured in Namibian waters to date. Therefore, the Namibian silver kob stock is currently assessed using cohort analysis under the assumption that it is a discrete stock targeted in three different areas by the three Namibian line-fisheries.

Tag-recapture results of silver kob were further used to demonstrate the migratory cycle of adult fish (Kirchner and Holtzhausen, 2001). Spawning adults start migrating southwards against the north-westerly surface currents at the beginning of the austral summer, from the northern end of their distributional range to their spawning grounds, Sandwich Harbour and Meob Bay (NNP), at the southern end of their distributional range. After spawning, larvae probably drift north with the current to the nursery area in the WCRA. When juveniles reach the age of approximately 2 years, they gradually move north towards SCP waters. At the end of the spawning season when the surf-zone water temperature decreases to about 15°C, adult silver kob complete their spawning cycle by returning to the same SCP waters, probably moving slightly offshore and with the current.

The daily bag limit of 30 fish per angler was too high to be effective, and has been reduced to 10. Of these 10, only two (2) should be silver kob >70 cm TL (Kirchner *et al.*, 2001). This is an adaptive management approach (Walters and Hilborn, 1978), whereby a deliberate attempt is made to learn about the effectiveness of a management measure and to improve the knowledge about the system. Introducing the minimum size limit of 40 cm TL could increase the long-term biomass by only 10%, but it might enhance survival of recruits into the age classes that contribute to the spawning potential of the stock (Kirchner *et al.*, 2001). However, after a few years the stock will be re-assessed and the size limits and daily bag limits adjusted accordingly.

Lineboat catches of kob (tonnes) have also increased sharply from 1999/00 to 2000/01 while the numbers that were caught have more than doubled (Figures. 2 and 3). This could be explained by the fact that four freezer lineboats are now employed. These boats can stay at sea for long periods as they do not rely on ice to keep the headed and gutted catch fresh, but instead all fish are being packed and frozen at sea. Therefore, they target all sizes of kob hence the dramatic increase in numbers caught. It was also recommended to reduce the catches of lineboats by 25% (i.e. implement a total allowable catch of 350 tonnes gutted and headed weight). It was calculated that the long-term biomass could increase to 11 900 (10 700, 13 200) tonnes. Due to the negative impact that such a management measure might have upon the socio-economic well-being of some coastal communities, it has not yet been implemented.

West Coast steenbras

Recapture results for West Coast steenbras showed that those from the northern region (SCP) (northern population) move over large distances to the WCRA, whereas the southern population (Meob Bay area) constitutes a closed and separate population. Evidence to support the hypothesis of an isolated population is provided by Agenbag and Shannon (1988), who suggested that the combined effect of changes of circulation and turbulence/stratification causes a biological discontinuity in the vicinity of Meob Bay, so providing a barrier to interchange of biota. The biochemical genetic study by Van der Bank and Holtzhausen (1998/99) confirmed that the southern population is a closed population, separate from the northern population. Also, different life histories and tag-recapture data show that the populations are isolated, and the negative effects of local overfishing will not be cancelled out by immigration from less exploited areas. Therefore, it is proposed that the two populations be managed separately, while taking cognizance of the geographical structure of the resource. West Coast steenbras from the northern population that move from the northern region to the WCRA are mostly males in a reproductive stage. The hypothesis is therefore that these males disperse southwards to find gravid females with which to mate. However, no spawning migrations of large West Coast steenbras, as suggested by anecdotal evidence, were found during the study period (Holtzhausen, 1999).

A daily bag limit of only two (2) large West Coast steenbras >65 cm FL per angler has been implemented in order to protect the female component of the population. At such a size they are 16+ years old, and fishing mortality appears to be too high for this age group (Figure 6). By introducing a minimum size limit of 40 cm FL the spawning stock biomass would increase by approximately 8%, so improving the spawning potential of the population in the long term, e.g. more males would survive to become females. The level of depletion of the stock would also decrease to c. 56%, with 95% percentiles of 43–68%. For the southern West Coast steenbras population, it is clear that the older fish need protection to provide a stable stock structure for optimal recruitment. Although the southern West Coast steenbras population is currently not exploited by shore-anglers and only intermittently by lineboats, as a unique population it needs to be protected.

The yield in tonnes for West Coast steenbras harvested has decreased from 245 tonnes in 1998/99 to 191 tonnes in 2000/01 while the numbers harvested have increased significantly (Figure 4). In 1998/99, more than 50% of the harvest consisted of large spawners caught in the Terrace Bay area. Since then, fewer of these large fish were caught and the average size West Coast steenbras caught is around 45 cm FL.

Economics

The average expenditure per fish caught is highest for visiting foreign anglers (mostly South Africans), approximately N$96 (Table 9). Inland Namibian anglers almost matched that figure, at an average of N$94 per fish. Coastal resident Namibian anglers spent far less, just N$32 on average, to catch each fish. Visitors to Terrace Bay and Torra Bay had the highest expenditures of any angling group but, because their catches there were much bigger, the cost per fish caught was on the order of N$45 (Kirchner, 1998). Kirchner *et al.* (2000) argued that these values could be sustainable if policies to reduce fish mortality without affecting angler numbers were implemented. Therefore, smaller but realistic bag limits need to be set. At N$17.70 per fish landed, the catches of the skiboat anglers are the cheapest of the various fisheries. Approximately N$24.28 is the cost of each fish (excluding snoek) caught by a lineboat, and a fish caught by a shore-angler (all angler categories and localities combined) costs approximately N$64.30. These results indicate that the recreational shore-angler is by far the most valuable (to the Namibian economy) user of the linefish resource. The commercial linefish fishery is worth only a fraction (1/7) of the value of the recreational fishery. Therefore, the ongoing conflict between the different user groups where one group accuses the other for dwindling catches in this multi-user fishery needs resolution with these facts in mind.

Before this economic survey was conducted, the linefishery was not considered an important component of the Namibian fishery as a whole. The results of the survey indicated that the annual revenue derived from utilising the linefish resource is approximately N$35 million. Foreign (mainly South African) recreational anglers buy food, fuel, bait and refreshments from various Namibian suppliers during their holidays in the country. As most of these visitors travel by vehicle and, because Namibian towns are spaced at great distances apart, they stop overnight on their way to and from the coast at various guest farms, game lodges or accommodation establishments in towns. Also, some of the visitors make use of the camping facilities available at various campsites along the coast, which are administered by the Ministry of Environment and Tourism (MET), which collects daily camping and entrance fees. Some visitors hire private houses at the coast for the duration of their stay, some make use of municipal bungalows or caravan parks, and others stay in hotels, at bed-and-breakfast establishments, or with friends and relatives. Therefore, the revenue indirectly derived through the linefish resource is reflected in contributions by other Ministries to Namibia's GDP.

Adding to this, the sociological and psychological benefits to the recreational angler and his family cannot be measured in terms of hard cash, a fact that eludes most economists. However, it is the duty of the Government of any country to guard its natural resources, the ownership of which belongs to its inhabitants. This fact becomes even more relevant if the resource contributes to the economy, small as it may be.

Based on the stock assessment results of the two important linefish species, minimum size limits for these two species were introduced (40 cm TL), a restriction was placed on the daily allowable take of specimens over a certain size (70 cm TL for silver kob and 65 cm TL for West Coast steenbras), and a drastically reduced daily bag for recreational anglers was implemented (30 to 10 fish). It was further recommended that a TAC for the lineboat sector be introduced, but since this would have far reaching socio-economic implications for some coastal communities, this has not been implemented. Table 2 shows some of the new regulations for recreational shore-anglers that have been accepted and implemented in December 2000.

REFERENCES

Agenbag, J.J. and Shannon, L.V. (1988): A suggested physical explanation for the existence of a biological boundary at 24°30'S in the Benguela system. *South African Journal of Marine Science* 6: 119-132.

Beverton, R.J.H. and Holt, S.J. (1957): *On the Dynamics of Exploited Fish Populations*. Fishery Investigations, Series 2, 19. London; Her Majesty's Stationery Office. 533 pp.

Beyer, J.E., Kirchner, C.H. and Holtzhausen, J.A. (1999): A method to determine size-specific natural mortality applied to westcoast steenbras *Lithognathus aureti* in Namibia. *Fisheries Research* 41(2): 133-153.

Botes, F. (1994): Extensive research on angling fish resource. *Namibia Brief* 18: 82–83.

Deriso, R.B. (1987): Optimal $F_{0.1}$ criteria and their relationship to maximum sustainable yield. In: Proceedings of an International Symposium on Stocks Assessment and Yield Prediction (W.J. Christie and G.R. Spangler, eds.), *Canadian Journal of Fisheries and Aquatic Sciences* 44(Suppl. 2): 339–348.

Gabriel, W.L. and Mace, P.M. (1999): A review of biological reference points in the context of the precautionary approach. In: *Proceedings of the Fifth National NMFS Stock Assessment Workshop*: Providing Scientific Advice to Implement the Precautionary Approach Under the Magnuson-Stevens Fishery Conservation and Management Act (V.R. Restrepo, ed.), pp. 34-45. U.S. Dep. Commer., NOAA Tech. Memo. NMFS-F/SPO-40.

Griffiths, M.H. and Hecht, T. (1993): Two South African *Argyrosomus hololepidotus* species: implications for management. In: *Fish, Fishers and Fisheries*. Proceedings of the Second South African Marine Linefish Symposium, Durban, Ocober 1982 (L.E. Beckley and R.P. van der Elst, eds.), Special Publications of the Oceanographic Research Institute South Africa 2: 19–22.

Griffiths, M.H. and Heemstra, P.C. (1995): A contribution to the taxonomy of the marine fish genus *Argyrosomus* (Perciformes: Sciaenidae), with descriptions of two new species from southern Africa. Ichthyology Bulletin J.L.B. Smith Institute of Ichthyology 65: 40 pp.

Gulland, J.A. (1970): The fish resources of the ocean. F.A.O. Fisheries Technical Paper 97: 425 pp.

Holtzhausen, J.A. (1999): Population dynamics and life history of westcoast steenbras (*Lithognathus aureti* (Sparidae)), and management options for the sustainable exploitation of the steenbras resource in Namibian waters. PhD thesis, University of Port Elizabeth: 213 pp.

Holtzhausen, J.A, Kirchner, C.H. and Voges, S.F. (2001a): Observations on the linefish resources of Namibia, 1990–2000, with special reference to West Coast steenbras and silver kob. In: *A Decade of Namibian Fisheries Science* (A.I.L. Payne, S.C. Pillar and R.J.M. Crawford, eds.), *South African Journal of Marine Science* 23: 124-135.

Holtzhausen, J.A. and Kirchner, C.H. (2001b): Age and growth of West Coast steenbras *Lithognathus aureti* in Namibian waters, based on otolith readings and mark-recapture data. In: *A Decade of Namibian Fisheries Science* (A.I.L. Payne, S.C. Pillar and R.J.M. Crawford, eds.), *South African Journal of Marine Science* 23: 169-179.

Holtzhausen J.A and Kirchner, C.H. (2001c): An assessment of the current status and potential yield of Namibia's northern West Coast steenbras *Lithognathus aureti* population. In: *A Decade of Namibian Fisheries Science* (A.I.L. Payne, S.C. Pillar and R.J.M. Crawford, eds.), *South African Journal of Marine Science* 23: 158-169.

Kirchner, C.H. (1998): Population dynamics and stock assessment of the exploited silver kob (*Argyrosomus inodorus*) stock in Namibian waters. PhD thesis, University of Port Elizabeth. 276 pp.

Kirchner C.H. (2001): Fisheries regulations based on yield-per-recruit analysis for the linefish silver kob *Argyrosomus inodorus* in Namibian waters. *Fisheries Research* 12(3): 155-167.

Kirchner, C.H. and Beyer, J.E. (1999): Estimation of total catch of silver kob *Argyrosomus inodorus* by recreational shore-anglers in Namibia using a roving-roving creel survey. *South African Journal of Marine Science* 21: 191-199.

Kirchner C.H. and Holtzhausen, J.A. (2001): Seasonal movements of silver kob *Argyrosomus inodorus* in Namibian waters. *Fisheries Management and Ecology* 8(3): 239-251.

Kirchner, C.H. and Voges, S.F. (1999): Growth of Namibian silver kob *Argyrosomus inodorus* based on otoliths and mark-recapture data. *South African Journal of Marine Science* 21: 201-209.

Kirchner, C.H., Holtzhausen, J.A. and Voges, S.F. (2001): Introducing size limits as a management tool for the recreational line fishery of silver kob, *Argyrosomus inodorus*, in Namibian waters. *Fisheries Management and Ecology* 8: 227-237.

Kirchner, C.H., Sakko, A.L. and Barnes, J.I. (2000): An economic valuation of the Namibian recreational shore-angling fishery. *South African Journal of Marine Science* 22: 17–25.

MFMR (1993): Sea Fisheries Regulations. Government Gazette of the Republic of Namibia. Act 29 of 1992 (Section 32). 55 pp.

Thompson, W.F. and Bell, F.H. (1934): Biological statistics of the Pacific halibut fishery. 2. Effect of changes in intensity upon total yield and yield per unit of gear. Rep. int. Fish. (Pacific Halibut) Commn 8: 49 pp.

Van der Bank, F.H. and Holtzhausen, J.A. (1998/1999): A preliminary biochemical genetic study of two populations of Lithognathus *aureti* (Perciformes: Sparidae). *Southern African Journal of Aquatic Science* 24(1/2): 47–56.

Van der Bank, F.H. and Kirchner, C.H. (1997): Biochemical genetic markers to distinguish two sympatric and morphologically similar Namibian marine fish species, *Argyrosomus coronus* and *Argyrosomus inodorus* (Perciformes: Sciaenidae). *Journal of African Zoology* 111(6): 441–448.

Walters, C.J. and Hilborn, R. (1978): Ecological optimization and adaptive management. *Annual Review of Ecological Systems* 9: 157–188.

Zeybrandt F. and Barnes, J.I. (2001): Economic characteristics of demand in Namibia's marine recreational shore fishery. In: *A Decade of Namibian Fisheries Science* (A.I.L. Payne, S.C. Pillar and R.J.M. Crawford, eds.), *South African Journal of Marine Science* 23: 145-156.

7 AGGREGATION DYNAMICS AND BEHAVIOUR OF THE CAPE HORSE MACKEREL (*Trachurus trachurus capensis*) IN THE NORTHERN BENGUELA – IMPLICATIONS FOR ACOUSTIC ABUNDANCE ESTIMATION

Bjørn Erik Axelsen, Jens-Otto Krakstad and Graça Bauleth-D'Almeida[*]

Abstract
The horizontal and vertical distribution patterns in pelagic fish vary greatly between species and ecosystems. Such variability can be seen in both the spatial (distribution area, densities, patchiness, clustering) and temporal (diurnal, seasonal, annual, inter-annual) dimensions. Knowledge of the aggregation dynamics of a species therefore greatly influences our ability to understand the systems and to estimate abundance by means of hydro-acoustic surveys. This chapter reviews the behaviour, aggregation dynamics and migration patterns of Cape horse mackerel (*Trachurus trachurus capensis*) in the northern Benguela, with special emphasis on their impacts on acoustic estimates of abundance.

INTRODUCTION

The Benguela is one of the world's four major western boundary upwelling regions, and one of the most productive marine ecosystems in the world

---

[*] The research reported here originates to large extents from the Benguela Environment Fisheries Interaction and Training Programme (BENEFIT) and the Nansen Programme, funded by the Norwegian Agency for Development Cooperation (NORAD). Fellow scientists, crew, technicians and technical assistants from the National Marine Information and Research Centre (NatMIRC) in Namibia, Instituto de Investigação Marinha (IIM) in Angola, Marine and Coastal Management (MCM), and the Institute of Marine Research (IMR) in Norway, are all thanked for excellent cooperation and support. A special thanks goes to Mr. Dave Boyer at Fisheries Resource Surveys, Cape Town (Formerly, Chief Scientist at NatMIRC) for his keen interest and for his valuable and constructive comments to several earlier versions of this chapter.

(Shannon, 1985). In the south it borders the Agulas bank region in South Africa at about 35° S, while the northern border of the system is the Angola – Benguela frontal zone between 14° S and 17° S (Shannon, 1985). The Benguela is split into two ecologically separate sub systems, the northern and southern Benguela, by an upwelling cell off Lüderitz (26° S – 27.5° S) that brings cold nutrient-rich water up along the coast and acts as a semi-permanent barrier for pelagic fish (O'Toole, 1977; Agenbag, 1980; Boyd and Cruickshank, 1983; Agenbag and Shannon, 1988).

The Cape horse mackerel, *Trachurus trachurus capensis* (Carangidae), is currently the most abundant of the commercially important fish species in the northern Benguela, with annual catches exceeding 300 000 tonnes since the mid-1970s (Figure 1), and plays a central role in the ecosystem. The species has a continuous distribution from Port Alfred on the south-eastern coast of South Africa (Kerstan and Leslie, 1994) to the northern border of the Benguela, Tombwa in southern Angola (16°00'S). Historically, three stocks have been identified (Hecht, 1990), but it has been suggested that the populations on the east coast and south-west coast of South Africa belong to the same stock (Hecht, 1990; Naish *et al*,. 1991; Barange *et al.*, 1998), with a distribution consisting mostly of adults on the south and south-west coast, and younger cohorts (<10 cm) dominating on the north-west coast (Barange *et al.*, 1998). The Namibian and South African Cape horse mackerel populations are managed as separate stocks (Zenkin and Komarov, 1981; Hecht,

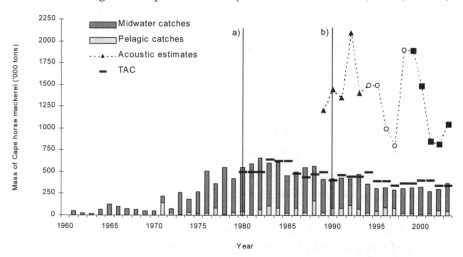

Figure 1. Commercial catches, Total Allowable Catch (TAC) and acoustic abundance estimates (since 1990) of Cape horse mackerel in Namibia in the period 1961-2002 (▲: Old "Dr F. Nansen"; O: New "Dr F. Nansen"; ■: "Welwitchia"). a) Introduction of TAC, protection zone (12 nmi from shore) and mesh size regulations (60 mm) (ICSEAF). b) Expansion of protection zone for the mid-water fleet (200 m bottom depth) (Krakstad and Kanandjembo, 2001). Note that the landings of the midwater fleet are estimated for 2003.

1990) due to biological differences and the presence of the Lüderitz upwelling cell (O'Toole, 1977; Agenbag, 1980; Boyd and Cruickshank, 1983; Agenbag and Shannon, 1988). Recent genetic studies involving mitochondrial DNA have, however, found no genetic indication that the Namibian and South African Cape horse mackerel stocks are biologically separate (Lovell, 2000).

The Namibian Cape horse mackerel is mainly pelagic, but adult fish become increasingly demersal with age. Adults generally form monospecific schools, but juveniles can be found in mixed schools with clupeoids like sardine (*Sardinops sagax*), anchovy (*Engraulis capensis*) and round herring (*Etrumeus whiteheadii*) (Cruickshank, 1983). Both juvenile and adult Cape horse mackerel in Namibia feed mainly in the pelagic zone, but while the juvenile (<20 cm) diet consists of small zooplankton such as copepods (*Calanus* spp) (Venter, 1976; Konchina, 1986), there is a gradual transition towards larger prey in bigger fish (Andronov, 1983). Adults mainly prey on euphausiids, copepods and, to lesser extents, on lantern fish (Myctophiidae) and gobies (Gobidae) (Andronov, 1985). Benthic food items such as polychaetes have also been recorded (J.-P. Roux, *pers. comm.*). Most authors identify the daytime and, in particular, dusk, as the most active feeding periods (i.e. Andronov, 1985). Important predators of Cape horse mackerel in the northern Benguela upwelling region include hake (*Merluccius capensis* and *Merluccius paradoxus*) (Andronov, 1983; Konchina, 1986) and other predatory fish, marine mammals like the Cape fur seal (*Arctocephalus pussilus pussilus*) (David, 1987, 1989) and dolphins (Sekigushi *et al.*, 1992), and a number of sea bird species (Crawford, 1999).

Several authors have suggested that there is a negative relationship between the abundance levels of the sardine and Cape horse mackerel stocks. The hypothesis sprang from the large increase in the horse mackerel population during the mid 1970s, coinciding with the crash in sardine stocks (Boyer *et al.*, 2001; Fossen *et al.*, 2001). Although similar shifts between pelagic species have been shown in other upwelling regions (Jacobson and MacCall, 1995; Wada and Jacobson, 1998), this idea is not generally accepted. The two species occupy ecological niches with limited overlap, and another explanation for the changes in stock abundance could therefore be environmental changes like increase in temperatures, comprising a regime-shift (Boyer *et al.*, 2001).

Acoustic biomass surveys of the Cape horse mackerel in the northern Benguela have been conducted since the 1980s, annually since 1995 (Bauleth-D'Almeida, 2001). The primary objective of the surveys has been to estimate an annual biomass index of the Cape horse mackerel stock in the northern Benguela. Prior to 1995, acoustic surveys where either multispecies surveys or targeted on clupeids, and consequently did not cover the

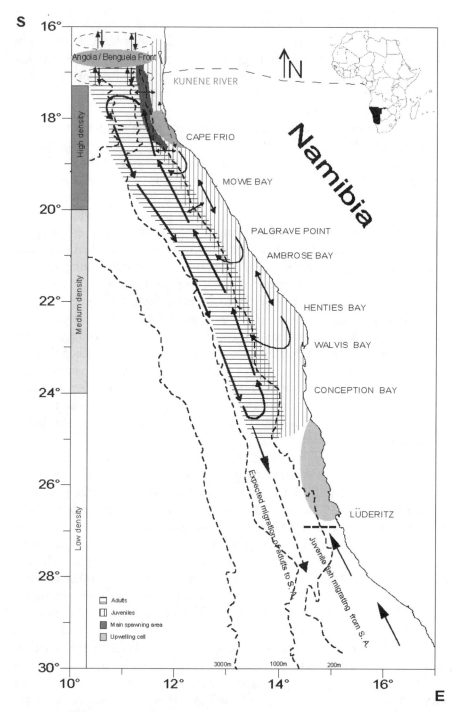

Figure 2. Generalized distribution (shaded areas) and migration (arrows) of Cape horse mackerel in Namibia.

entire distribution area of the Cape horse mackerel. From 1995 to 1998, Cape horse mackerel surveys where conducted using the Norwegian R/V Dr. Fridtjof Nansen during the winter season. The Namibian R/V Welwitschia took over in 1999, and has since conducted annual summer surveys. Disregarding the change of seasons, the survey strategies have been similar during the time series, covering the northern Benguela from 25° or 26° S in the south and northwards to the Namibian-Angolan border at 17°15 'S, and in some years extending northwards to Tombwa in southern Angola (16° S). The transects have systematically been perpendicular to the coastline, with the distance between transect lines adjusted to the density of the Cape horse mackerel in different parts of the survey area. The biomass index has, alone or in combination with virtual population analysis (VPA), been used to set an annual total allowable catch (TAC) for Cape horse mackerel in Namibia (Butterworth et al., 1990; Boyer and Hampton, 2001). The acoustic estimates represent the only fisheries-independent cue of the stock level, and therefore represent vital inputs to the assessment of the Cape horse mackerel.

Acoustic surveys are conducted in order to provide independent abundance indices of most commercially important pelagic fish stocks in the northern Benguela, including Cape horse mackerel (Boyer and Hampton, 2001), and are advantageous in that large sampling volumes are sampled with relatively low effort, and that high sample resolutions can be obtained in both the horizontal and vertical planes (e.g. MacLennan and Simmons, 1992). However, the aggregation dynamics and behaviour of the fish may significantly influence biomass estimates, and it is therefore vitally important to understand these processes and their implications on acoustic estimates of abundance in the next two sections we address the aggregation and migration dynamics in the horizontal and vertical planes, outlining implications for acoustic abundance estimation. Then we summarize the sources of survey errors, including a quantitative synopsis of error sources in acoustic surveys of Cape horse mackerel in the northern Benguela.

HORIZONTAL DYNAMICS

Distribution area

A basic understanding of the distribution and migrations of the surveyed stock is essential to ensure adequate coverage and appropriate survey design, which are required for obtaining reliable abundance estimates. The main distribution and migration patterns of the Cape horse mackerel in Namibia are illustrated Figure 2. The Angola–Benguela front marks the northern boundary of the Benguela current, which shifts seasonally between 14° S in the winter and 17° S in the summer (Shannon, 1985), and the northern boundary of the Cape horse mackerel stock largely follows that of the Benguela current. The northern part of the Namibian continental shelf (17°15' S - 20° S) has the highest aggregation densities and can be considered the core area of the Cape horse mackerel distribution in Namibia (Assorov *et al.*, 1988; Anon., 1998; 2003a). Smaller, more scattered concentrations found further south, between 19° S and 23° S (Anon., 1998, 2003a), and south of 23° S only small quantities of older, more demersal fish have been found. This general distribution pattern does, however, change seasonally, with patchier and denser aggregations during the summer months (December - February) than in winter (June-August) (Krakstad and Kanandjembo, 2001).

Migrations

There is a size separation of horse mackerel across the northern Namibian continental shelf, with adults primarily distributed over the outer shelf and shelf break (250-300 m depth), especially in the north, and juvenile fish predominantly occurring inshore (Figure 2). Spawning is mostly confined to the

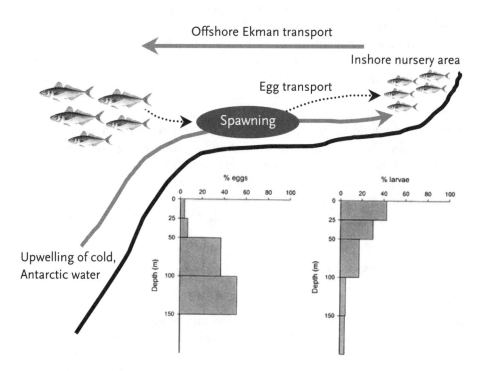

Figure 3: Cross-shelf distribution of horse mackerel during summer. Adult horse mackerel spawn on the shelf. The eggs, being slightly buoyant and drifting upwards, follow the undercurrent in a slightly southwards direction until reaching the inshore nursery areas as larvae. Ekman transport in the upper 25 m transports the warm surface water off the shelf. Data from Anon. (2001a).

continental shelf during summer, with the main spawning grounds in the north. Smaller fish are believed to start spawning first. The eggs are slightly buoyant and drift upwards, following the undercurrent in a slightly southwards direction until reaching the inshore nursery areas as larvae (Figure 3) (Anon., 2001a). During winter, the juveniles (>14 cm, 1.5 year old) leave the coastal areas to join the adult stock (Assorov et al., 1988). The average size of adults increase southwards along the shelf, with the largest Cape horse mackerel usually found south of the Lüderitz upwelling cell. The distribution is very dynamic, however, with daily alongshore and cross-shelf migrations that could be related to feeding.

Relatively large concentrations of juvenile Cape horse mackerel are found in the southern Benguela, off the west coast of South-Africa, and some of these fish are though to undertake migrations northwards past the Orange River and into Namibian waters (Anon., 1993). This is consistent with size compositions in catches from this area (Namwandi, 2002). Northward migration is restricted by the Lüderitz upwelling cell, which acts as a geographical barrier for juvenile pelagic fish (O'Toole, 1977), and these juveniles do not mix with the Namibian Cape horse mackerel stock. It has been suggested, however, that adults from the northern Benguela migrate to deep, offshore waters, and take advantage of the southward-flowing undercurrent to migrate past the upwelling cell and into South-African waters (Anon., 1998). The hypothesis is based on observations from the Namibian bottom trawl surveys where adult Cape horse mackerel often has been observed throughout the Lüderitz upwelling cell area, although in small quantities (Namwandi, 2002).

Survey design
The decision to carry out the surveys for Cape horse mackerel in the northern Benguela during the summer months (Anon., 1999) was based on knowledge of the aggregation patterns and the position of the Benguela front, with associated denser aggregations and more southern distribution during summer, respectively. Furthermore, the commercial fisheries have experienced higher catchability rates (CPUE) during the summer season. The current survey design takes the separation of the two stock components during summer (Anon., 2003a) into account by covering the juvenile (inshore) and adult (offshore) stock components separately. Parts of the nursery areas along the coast are inaccessible as they are too shallow to be surveyed (<15 m depth), and this may represent a small negative bias in the abundance estimates (Anon., 2003a). The southern limit of the surveys was set north of the Lüderitz upwelling cell (25° S), because the concentration of Cape horse mackerel south of this has been considered insignificant. With this coverage, it is believed that the majority of the stock is being covered,

and that any negative bias due to incomplete coverage of the stock is fairly small (Anon., 2003b).

VERTICAL DYNAMICS

Vertical movements of fish during acoustic surveys have important implications for the biomass estimates. First of all, survey practitioners should have an idea about the depth and aggregation pattern of the Cape horse mackerel in order to correctly identify the acoustic traces. Besides, acoustic transducers on research vessels typically operate at distance from the surface of about 5 m, and therefore cannot detect anything above this depth. Additionally, interference from the transducer face in the acoustic "near field" (Medwin and Clay, 1998) makes target detection dubious at short transducer ranges, and integration can therefore only be carried out meaningfully at a certain range. The surface blind zone is typically about 10-12 m from the surface. Similarly, fish cannot be detected near the bottom, in the acoustic "dead zone", due to interference from bottom echoes (Ona and Mitson, 1996). The bottom dead zone would usually correspond to less than 1 m during acoustic surveys of Cape horse mackerel. Acoustic blind zones represent a potential source of availability errors, and it is essential that survey practitioners know to what extent the Cape horse mackerel occupy these zones during the surveys. Compression of the closed, gas-filled swimbladder during descents reduces the buoyancy of the Cape horse mackerel, creating a need for compensation that increases with depth. Compensation may occur either through production of gas in the *rete mirabilae* in the swimbladder, or through increased swimming activity, both of which involve greater metabolic costs (Blaxter and Tytler, 1978). Changes in the acoustic scattering properties following uncompensated vertical movements (Harden Jones and Scholes, 1985) and changes in angular orientations during vertical excursions (Foote, 1978) both affect the acoustic scattering properties of the fish, ultimately affecting the abundance estimates.

Diel vertical migration
Diel vertical migration (DVM) is a common feature in many pelagic fish (e.g. Pearre, 2003). In the southern Benguela, Cape horse mackerel generally form dense shoals in mid-water or close to the bottom during the day, migrate to midwater at night and disperse, before aggregating and descending again at dawn (Pillar and Barange, 1998). The same general pattern is seen in the northern Benguela, although the horse mackerel there often remain in mid-water during the day, ascending to near the surface at night.

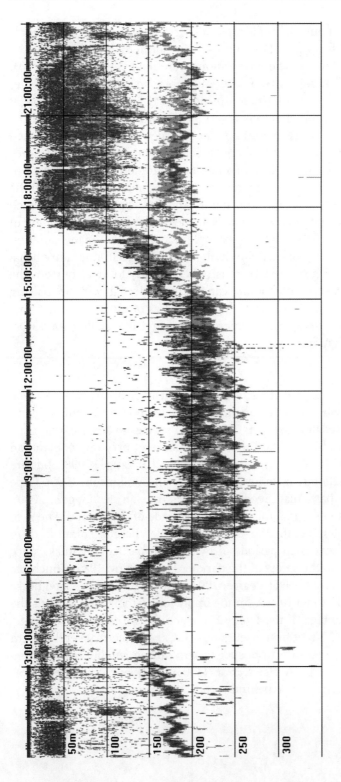

Figure 4. Composite echogram (38 kHz) showing a 24-hour cycle of acoustic recordings of Cape horse mackerel (17 cm mean length) offshore (1000 m bottom depth) near the Namibian-Angolan border. The euphausiid layer (blue/grey) and horse mackerel layer (yellow/green) stayed in mid-water (150-200 m) at daytime and migrated to near-surface depths (0-50 m) at night. The grey layer that did not lift at night consisted mainly of red shrimp (*Aristeus varidens*). The observations were made in August 2001, using the R/V Dr. Fridtjof Nansen.

Figure 4 shows a 24-hour cycle of acoustic observations of pelagic Cape horse mackerel located offshore at about 17°20' S in northern Namibia, near the Namibian-Angolan border. The observations were made in September 2001, using the R/V Dr. Fridtjof Nansen. The Cape horse mackerel can be seen in mid-water at a depth of about 200 m during the day and near the surface at night. The vertical movements took about 30 min. for both upward and the downward migrations, corresponding to an average vertical migration speed of about 6.7 m min^{-1}. Layers of krill and copepods, which are important prey for the Cape horse mackerel, followed similar migration patterns to that of the Cape horse mackerel, although generally staying slightly deeper than the horse mackerel during both day and night. Another scattering layer, which consisted of red shrimp (*Aristeus varidens*), stayed in mid-water throughout the diel cycle (Anon., 2004). In this example, identification of the different groups was fairly straightforward due to their differences in migratory behaviour. Although Cape horse mackerel is generally easy to identify from the acoustic traces, target identification can be complicated with increasing numbers of co-occurring species and higher degrees of mixing, and it is therefore essential that survey practitioners understand the driving forces behind the vertical migrations.

Light and feeding

The main driving force in diel vertical migration in fish can be said to be light, as it makes fish and other organisms visible and hence vulnerable to visually mediated predators (Pearre, 2003). Most species in any ecosystem act as both predators and prey. The Cape horse mackerel prey on zooplankton and krill (Venter, 1976; Konchina, 1986; Pillar and Barange, 1998), but are also subject to predation (Andronov, 1983; Pillar and Wilkinson, 1995). Cape horse mackerel are particulate feeders (Twatwa and van der Lingen, 2000), and light is therefore essential for feeding. The trade-off between the need to eat and not to be eaten, i.e. the life-dinner principle, therefore often leads to peaks in feeding activities in periods of twilight, in particularly dusk, since this transition leads to the safety of the darkness (Andronov, 1985; Konchina, 1986; Pillar and Barange, 1998; Pearre, 2003; Anon., 2004). The Cape horse mackerel illustrated in Figure 4 seem to simply follow their prey towards the surface, taking advantage of the twilight to catch them (Pillar and Barange, 1998; Anon., 2001b). Night-time excursions of Cape horse mackerel (15 cm total length) near the surface to deeper waters during conditions of strong moonlight were noted by Axelsen *et al.* (2003), demonstrating the importance of light as a driving force behind vertical migrations.

Boundaries and preferences

The surface and bottom are obvious obstacles to vertical movements in fish. Cape horse mackerel have been found to remain at similar depths in shallow waters (50 m) while in adjacent, deeper waters, taking on extensive vertical migrations (Anon., 2003b). Figure 5 shows a scattering layer of Cape horse mackerel aggregated at about 200-250 m near the bottom along the continental slope, while fish further offshore were found at similar depths (250-300 m). While the surface and the bottom always represent absolute barriers, hydrographical and environmental conditions sometimes constitute boundaries in similar ways. This is illustrated in Figure 6, where a steep oxycline is shown to prevent Cape horse mackerel from diving into waters inhabited by krill, which is one of their main prey items. In this situation, the hypoxic waters (dissolved $O_2 < 0.7$ ml l^{-1}) formed a spatial refuge for the prey, effectively acting as a physiological barrier for the horse mackerel. The Namibian shelf is known for its prevalence of hypoxic bottom waters (Chapman and Shannon, 1985; Dingle and Nelson, 1993), and this could be one of the reasons why Cape horse mackerel are less commonly found close to the bottom in Namibia than in South Africa (Pillar and Barange, 1998), since Cape horse mackerel cannot sustain long periods in hypoxic water (Anon., 2001b). Temperature and salinity variations play important roles in terms of limiting the distribution, both horizontally and vertically. The temperature determines the metabolic levels in fish, and fish therefore often have temperature preferences (e.g. Boyd and Cruickshank, 1983). Pillar and Barange (1998) noted that Cape horse mackerel that ascended to warmer waters enjoyed higher metabolic rates, i.e. maximizing growth rates by means of vertical migration. Similar relations have not been established for salinity levels.

Predator-prey interactions

Fish exhibit strong avoidance reactions when attacked by predators, and such predator responses may significantly affect the shape, size and density of fish schools (e.g. Pitcher and Wyche, 1983; Axelsen *et al.*, 2001). Changes in aggregation patterns following predator avoidance may have a significant bearing on recorded acoustic densities, potentially affecting acoustic estimates. Figure 7 shows acoustic recordings of 14-20 cm Cape horse mackerel under attack by dusky dolphins (*Lagenorhynchus obscurus*) in coastal waters in the border area between Namibia and Angola in September 2002. Observations were recorded both night and day, as the vessel was drifting across aggregations of Cape horse mackerel. Prior to the attacks shown in Figure 7, the horse mackerel were confined to a homogeneous, pelagic layer at about 120 m as the dolphins attacked the layer from underneath. The Cape horse mackerel reacted by diving 60 m in about 2 min., i.e. at about 0.5 m s^{-1} vertical diving speed, or 4.5 times faster than during the natural vertical migra-

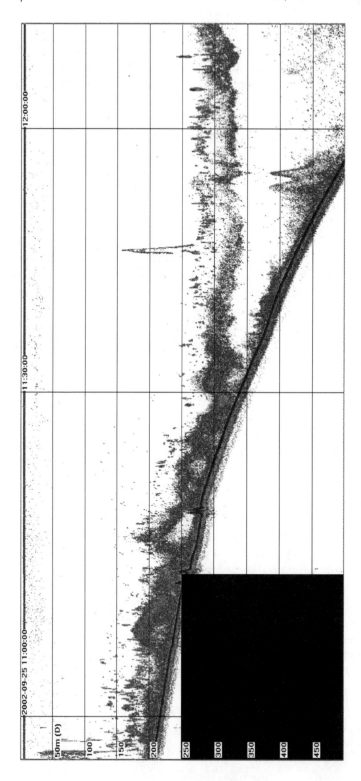

Figure 5. Echogram (38 kHz) showing adult Cape horse mackerel (red/green schools) and mesopelagic fish (lanternfish, blue layer) distributed over the shelf-break of northern Namibia, near the Namibian-Angolan border. The observations were made in September 2002, using the R/V Dr. Fridtjof Nansen.

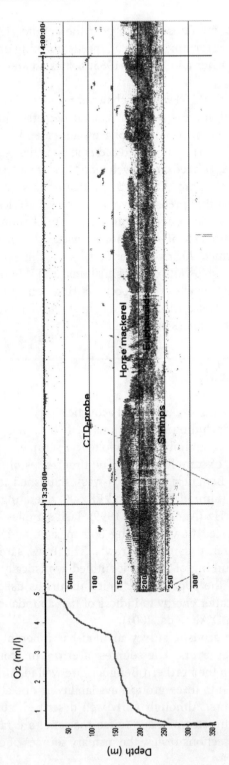

Figure 6. Echogram (38 kHz) showing a daytime aggregation of Cape horse mackerel (17 cm total length) at about 170-190 m depth, located immediately above a layer of krill (200-250 m). The oxygen profile shows a steep oxycline from 150 m to 190 m, at which depth the dissolved oxygen (DO) corresponded to the tolerance limit of the Cape horse mackerel (DO= 0.7 ml l^{-1}), as determined *in situ* during the experiments. The bottom depth in the area was about 1000 m.

tion. The horse mackerel stayed in a dense school near the bottom (170 m depth) for another 2 min., before returning at similar speeds, forming dense, smaller schools, or tight balls (Pitcher and Wyche, 1983; Nøttestad and Axelsen, 1999), at 120-140 m depth.

The recorded acoustic densities (38 kHz) also changed considerably during the 10 min. attack period, with increased mean acoustic densities during the descent (Sv= -48.2 dB) and bottom (-45.3 dB) phases compared with the initial undisturbed phase (-53.6 dB). The integrated acoustic densities (s_A, m^2 nmi^{-2}), following the same trend, peaked at a level more than eight times higher while the fish were at the bottom than in the undisturbed situation. The example reported here shows that predator avoidance may significantly affect acoustic measurements of density, representing a potential source of error to acoustic abundance estimates of Cape horse mackerel. Predator avoidance events are rarely identified during conventional surveying, simply because survey data only provide snapshots of the situations, and it is therefore difficult to evaluate how often such events occur during conventional acoustic surveys. Consequently, it is difficult to evaluate to which extents they represent a source of error to biomass estimates.

ABUNDANCE AND ERROR ESTIMATION

Acoustic surveying involves recording backscattered acoustic signals from all scattering objects in the water, including aggregations of fish targets, and zooplankton and other non-targets, and obtaining biological samples of the acoustic targets using some type of sampling gear. The time series of Cape horse mackerel in the northern Benguela (Figure 1) was established using vertically oriented echosounders (Simrad EK 400 and EK 500) operating at a discrete carrier frequency of 38 kHz (Anon., 1998, 1999). Fish samples have been collected by means of targeted trawling using both pelagic and, particularly in shallow waters, demersal sampling trawls. The trawl samples serve the functions of ground-truthing of targets identified acoustically and estimating the size structures of the populations, which in turn is used for conversion of the acoustic backscatter energy to indices of fish abundance in the sampled size groups (Bauleth-D'Almeida, 2001).

As for any other method, the acoustic survey method has inherent systematic (bias) and variable (noise) errors. The sources of errors in acoustic surveys of Cape horse mackerel in the northern Benguela are well identified, and can conveniently be divided into three groups: availability errors, acquisition errors and conversion errors. Although fairly well described, survey errors in acoustic abundance estimates of Cape horse mackerel are rather poorly quantified. The following sections review the primary sources of error in acoustic abundance estimates of Cape horse mackerel, including a summary of the current knowledge of error types and ranges.

Availability errors
Incomplete coverage. - A certain proportion of the Cape horse mackerel stock may be distributed in Angolan waters, also during summer. Systematically extending the surveys into southern Angola would have reduced such errors, which may vary between years due to changing position of the front. Errors related to incomplete survey coverage as a whole, i.e. loss of fish on the Angolan side of the border in the north, inshore fish in shallow waters, and fish distributed south of 25° S, are, however, considered comparably low. A negative bias of about 10% has been suggested as a result of incomplete coverage (Anon., 2003b).

Blind zones. - Although the Cape horse mackerel in the northern Benguela are known to sometimes enter into the acoustic blind zones, it is not believed that this contributes any major errors to the estimates. Blind zone errors do, however, affect age groups differently. Juvenile Cape horse mackerel (<10 cm) can sometimes be seen schooling immediately below the surface at night and can in such cases be caught with a surface trawl, although unavailable to the echosounder beam. Older size groups, on the other hand, stay deeper in the water column, where they are acoustically detectable. At a certain age, however, the Cape horse mackerel in the northern Benguela tend to adopt a more demersal lifestyle, thus entering into the bottom dead zone. Evidently, a negative bias due to blind zone errors should be expected, but the bias is believed to be relatively small as the Namibian Cape horse mackerel surveys only target fish >10 cm and the proportion of large fish (fish >30 cm) is relatively small, contributing accordingly little to the total estimates (Bauleth-D'Almeida, 2001; Anon., 2003a).

Acquisition errors
Weather. - The weather effect is a consequence of signal loss caused by air bubbles near the surface during bad weather. The problem has been estimated to be about 5% in the acoustic surveys of Cape horse mackerel in Namibia, and is primarily due to conditions of strong wind in southern Namibia. The problem may in principle be avoided simply by interrupting the survey whenever weather conditions are unfavourable for acoustic surveying, but this is often not practically feasible, and extensive delays may represent even greater errors. Mounting the acoustic transducers on a protruding keel that can be submerged below the air bubbles significantly reduces weather errors, and this has been done on the R/V Dr. Fridtjof Nansen. Installing protruding keels on ships that do not have this feature originally does, how-

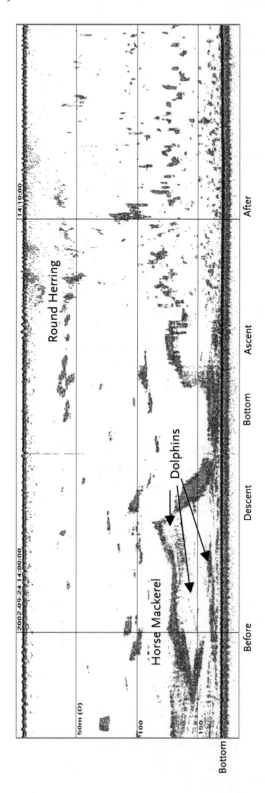

Figure 7. Echogram (38 kHz) showing Cape horse mackerel under attack by dusky dolphins (*Lagenorhynchus obscurus*). The horse mackerel (red/green/blue schools) are diving more then 60 m in about 2 min when attacked by the dolphins (blue, almost straight lines), before returning to mid-water and forming small, dense schools. The observations were made in September 2002, using the R/V Dr. Fridtjof Nansen.

ever, require considerable technical modifications to vessels. Although significantly reducing weather errors, operating with a protruded instrument keel increases the upper blind zone of the vessel, which in principle may lead to availability errors.

Calibration and equivalent beam factor. - Acoustic calibration errors follow the accuracy of the reference sphere calibration technique (Foote, 1987), and are normally low. Typically, the accuracies of the calibrations represent errors smaller than ± 0.1 dB if conditions are favourable, but can be as high as ± 0.3 dB in cases of less favourable conditions, corresponding to a random error of maximally ± 5%. Nevertheless, it is very important to conduct acoustic calibrations when conditions are good, i.e. when the weather is calm and there are few scattering objects in the water near the reference sphere.

Embedded noise in the acoustic signal converted to abundance estimates may constitute another instrument error that is referred to as equivalent beam factor error. The effect is proportional to carrier frequency, and constitutes a relatively small problem at 38 kHz with regard to abundance estimation of Cape horse mackerel. The BENEFIT Survey Error workshop in Cape

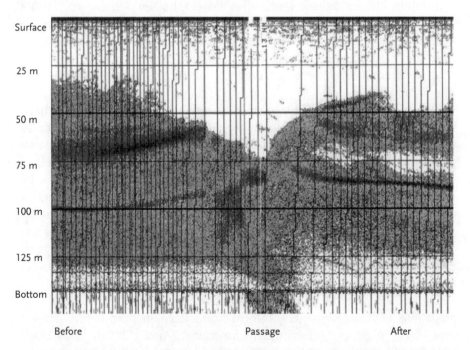

Figure 8. Echogram showing vessel avoidance in dense aggregations of juvenile Cape horse mackerel (17-21 cm total length) recorded using a self-contained acoustic buoy (38 kHz) during passage of the R/V Welwitschia. The vessel passed at standard survey speed (about 10 knots) at a distance of about 10 m from the buoy.

Town 2000 noted, however, that the error could be as large as 1 db, consequently contributing up to 10% to the error of the overall abundance estimates (Simmonds et al., 1992), depending on a number of factors such as bottom depth, weather conditions, transducer arrangement and vessel noise. Beam factor errors may be quantified and subsequently corrected for (Korneliussen, 2000).

Target identification. - Target identification involves allocation of acoustic backscatter to target groups, i.e. separating acoustic traces originating from Cape horse mackerel from those originating from non-target groups. Target identification requires knowledge about the acoustic characteristics of the fish aggregations and of the location of the fish in the water column. Separation of echoes from targets and non-targets becomes increasingly difficult when the targets and other scatterers are mixed. Horse mackerel are generally fairly easy to identify on the acoustic properties alone. Juvenile Cape horse mackerel along the coast of the northern Benguela are, however, often found in mixed aggregations with anchovy, round herring and sardine (Cruickshank, 1983), which complicates identification. Adults tend to mix with layers of mesopelagic lanternfish (Myctophiidae). On the continental shelf in the border area between Namibia and Angola, Cape horse mackerel of all size groups also mix with Cunene horse mackerel (*Trachurus trecae*) from the Angola current system. Plankton layers, in particular at the surface, tend to sometimes obscure the presence of horse mackerel, especially in mid-water at night.

Although targeted trawling helps to identify the relative species composition of recorded shoals, the catches are not always representative of the actual species composition in the surveyed area. Target identification errors are generally random, and for Cape horse mackerel in the northern Benguela an error range of about ± 10% has been suggested. There is inevitably some subjectivity in the scrutinizing process, as different survey practitioners examining the same acoustic data may allocate different proportions to the target groups. There are therefore allocated considerable regional efforts into developing automated tools for species identification. Two main developments are taking place, one exploring differences in target strength between frequencies (multi-frequency target identification) (Korneliussen and Ona, 2002), and the other technique combining acoustic characteristics of the aggregations with supplementary information such as bottom depth, latitude, time of day/night etc. (discriminant functions) (Anon., 2003b, 2004). Both techniques have the overall goal of producing automated, objective predictions of likely identities of encountered targets in order to assist survey practitioners in the scrutinizing process.

Sv-threshold. - Plankton and other weak scatterers often co-occur with fish targets, contributing to the recorded target densities. It is therefore customary to apply a minimum threshold in order to reduce erroneous contribution from plankton and other non-targets (S_V threshold errors) (MacLennan and Simmonds, 1992; Anon., 2003b). Two principally different thresholding practices have been followed during acoustic surveys of Cape horse mackerel in Namibia: a fixed minimum S_V threshold of -65 dB (R/V Welwithchia) and a variable threshold (R/V Dr. Fridtjof Nansen).

The problem is that there does not exist a cut-off level that removes all non-target contributions while keeping all target contributions, as there generally are overlapping acoustic density distributions between targets and non-targets. It is difficult to establish an optimal threshold level, and any optimum would vary with the acoustic densities of both targets and non-targets. The Welwitchia strategy ensures a constant approach, removing any subjectivity involved in determining a variable threshold level (Dr. Fridtjof Nansen), but suffers the disadvantage of an inherent variable error that often could have been reduced. Conversely, the Dr. Fridtjof Nansen approach tempts to minimize the error, at the cost of added subjectivity. It should be noted, however, that it is possible to apply different thresholds during post-processing, examining their implications on the estimates and likely error ranges.

Statistical sampling error. - Any method based on some type of sampling technique will have inherent sampling errors caused by inference from samples of the population to the population level (e.g. Manly et al., 1997). Although sometimes overlooked, statistical sampling errors may constitute a fairly important error in acoustic fish abundance estimates. There is little firm documentation on statistical sampling errors in Cape horse mackerel surveys in the northern Benguela, but sample coefficients (CV) of about 25% have been suggested (Bauleth-D'Almeida, 2001).

Conversion errors

Vessel avoidance. - Avoidance from surveying vessels is well documented, both from vessel-radiated noise (e.g. Olsen et al., 1983; Anon., 1995; Vabø et al., 2002) and light (Levenez et al., 1990). While vessel avoidance in some situations does not represent any source of error at all (Fernandes et al., 2000), significant bias to acoustic abundance estimates has been broadly documented (Olsen et al., 1983; Fréon et al., 1993; Vabø et al., 2002). Vessel avoidance in Cape horse mackerel from the R/V Welwitchia was studied in northern Namibia in June 1999 (Anon., 2004). The Cape horse mackerel (16.5 to 20.5 cm total length) were aggregated in a dense, continuous layer extending from about 30 to 130 m depth, over an area of several square nautical miles. Figure 8 shows an echogram recorded over a 12 min. period, showing the first passage of the vessel. The Cape horse mackerel showed strong avoidance reactions to the vessel, as fish disappeared from the acoustic beam at depths shallower than 75 m at the time of the passage. The deepest extent of the layer changed from about 130 m to 150 m (bottom) at the time of passage. While the disappearance of fish in the upper part of the layer could be due to either horizontal (out-of-beam) or vertical (within-beam) avoidance, or a combination of both, the greater lower depth limit of the fish layer could only be due to fish diving at least 20 m. The high-density region at about 80 m depth at the time of passage strongly suggests that fish from the upper part of the layer dived in front of the vessel (Olsen et al., 1983), suggesting a fair component of vertical avoidance. In this example, the effective reduction in recorded s_A from the initial situation to the passage of the vessel corresponded to 41%. Interestingly, the lowest return (61% reduction) was recorded after the passage of the vessel, and echo levels remained low for several minutes after passage, suggesting some out-of-beam avoidance as well (Anon., 2004).

While the example shown here would suggest an avoidance error of about −40% in this case, the highest reduction rate recorded during the experiments was as high as 80% (B.E. Axelsen, unpubl. data), demonstrating that vessel avoidance could be an important factor during acoustic surveys of Cape horse mackerel in the northern Benguela. However, these particular examples, viewed in isolation, cannot in any way be taken as representative for the general survey situation, where many factors such as depth, and fish packing density, size/age and biological conditions combined determine the overall effect, if any (Fernandes et al., 2000; Vabø et al., 2002). Vessel avoidance effects are virtually always negative, and due to the variable nature of these effects, any corrections should be conducted with great caution.

Fishing gear avoidance. - Fishing gear avoidance from trawls and other sampling gear is primarily a problem during bottom trawl surveys (swept area estimation) (e.g. Morgan et al., 1997), where catch rates of demersal species such as hake and monk (*Lophius* spp) in Namibia are treated as indices of fish density. Fishing gear avoidance may, however, represent a source of error to acoustic abundance estimation as well. Trawl sampling during acoustic surveying serves both purposes of ground-truthing of acoustic targets and of estimating size structures in the populations, and consequently fishing gear avoidance can affect acoustic estimates both through target identification errors and conversion errors (Bauleth-D'Almeida et al., 2001). Horse mackerel are fast swimming and can sustain high swimming speeds in the net opening for long periods (Wardle et al., 1996). Since larger fish have greater endurances than smaller fish, a higher proportion of bigger fish tend to escape the sampling trawl. The resulting negatively skewed size distributions induce negative bias to the biomass estimates (Bauleth-D'Almeida et al., 2001). Fishing gear avoidance errors during acoustic surveys have been estimated to contribute with about 5%, negatively, of the estimates (Bauleth-D'Almeida et al., 2001). These errors could be reduced by improving the performance of the sampling gear, and by incorporating commercial catch data in the estimates.

Target strength. - Conversion from acoustic densities to fish abundance requires knowledge of the acoustic backscattering properties of the target species, specifically the mean dorsal aspect target strength (e.g. MacLennan and Simmons, 1992):

$$\text{TS} = 10 \log\left(\frac{\sigma}{4\pi}\right) \text{ (dB)} \qquad (1)$$

or in the form (Foote, 1980):

$$\text{TS} = 20 \log L + b_{20} \text{ (dB)}, \qquad (2)$$

where TS is the acoustic target strength (dB), σ the corresponding acoustic backscattering crossection in the linear domain (m^2), and L is the total fish length (cm). This relation between fish size and acoustic density at the survey frequency represents a key factor in acoustic abundance estimation. The TS of horse mackerel has been studied on the basis of survey data, volumetric considerations of the swimbladder, back-calculation from integration values using independent estimates of target densities (the comparison method), and by means of *in situ* techniques. An overview of published target strengths relevant for horse mackerel is given in Table 1. The range of TS relations reported here represents a factor of about four, and evidently no species-specific TS expression for Cape horse mackerel is conclusively supported by the literature. The TS to fish size relation corresponding to b_{20}= -72 dB in equation (2) was originally derived for clupeoids (Foote et al., 1986; Foote 1987), and is currently applied in surveys of horse mackerel in Namibia and Angola (*T. t. capensis* and *T. trecae*, respectively). The differences in the TS values in Table 1 could be due to a number of factors. Inevitably, some variation stems from experimental/model errors, but the acoustic target strength of fish is a highly variable parameter, and natural variation could explain a great deal of the reported variability.

The gas-filled swimbladder constitutes as much as 90-95% of the acoustic backscatter at 38 kHz due to the high sound speed contrast between the seawater and the swimbladder (e.g. Foote, 1980), and swimbladder volume and shape therefore significantly affect the TS (Ona, 1990). In the high frequency range (HF), like 38 kHz, which is a widely used carrier frequency in fisheries applications, the swimbladder scattering is highly directive. The angular orientation of the fish relative to the incident acoustic wave (approximately vertical) is therefore one of the most important factors determining the dorsal aspect TS. Vertical migrations and vessel avoidance therefore significantly affect the dorsal aspect TS. Cape horse mackerel have a closed swimbladder (*physoclist*) that enables them to produce (*rete mirabilae*) and resorb (*oval*) gas, and hence to compensate for swimbladder volume compression during vertical movements, which can be described according to:

$$v_z = v_{z=0} \left(1 + \frac{z}{10}\right)^{-1}, \qquad (3)$$

where v is the swimbladder volume and z is the depth in m. The extent to which horse mackerel compensate for swimbladder compression is, however, not known. In terms of TS conversions, it is generally assumed that physoclists maintain neutral buoyancy throughout their depth range (Foote, 1987), but the validity of this assumption has been questioned for Cape horse mackerel (Axelsen *et al.*, 2003). The inflation (Blaxter and Tytler, 1978) and deflation (Tytler and Blaxter, 1973) processes take five hours or longer, somewhat contrasting the rapid vertical excursions characteristic for horse mackerel (Figures 4 and 7) (Pillar and Barange, 1998). Horse mackerel are fast-swimming (Wardle *et al.*, 1996) and have large pectoral fins that provide an upward lift while swimming, and are therefore not dependent on neutral buoyancy to maintain their preferred depths. Negative buoyancy is advantageous for rapid vertical predator avoidance (Blaxter, 1985), and since predators like dolphins utilize echolocation to locate prey (Au and Benoit-Bird,

Table 1. Ranked overview of published target strengths expressed in terms of the b_{20}-constant relevant for horse mackerel.

b_{20}	Method	Species or group	Reference
-77.5	*In situ* tracking data	*T. trachurus capensis*	Axelsen *et al.* (2003)
-74.9	*In situ* tracking data	*T. trachurus capensis*	Axelsen *et al.* (2003)
-73.4	Comparison method	*T. trachurus*	Misund *et al.* (1997)
-71.9	Various *in situ* methods	Clupeoids	Foote (1987)
-68.9	*In situ* survey data	*T. symmetricus murphyi*	Lillo *et al.* (1996)
-68.2	*In situ* survey data	*T. symmetricus murphyi*	Gutiérrez and MacLennan (1998)
-67.5	Various *in situ* methods	Physoclists	Foote (1987)
-66.8	*In situ* survey data	*T. trachurus capensis*	Barange *et al.* (1996)
-66.8	*In situ* survey data	*T. trachurus capensis*	Svellingen and Ona (1999)
-66.7	Swimbladder volume	*T. symmetricus murphyi*	Torres *et al.* (1984)
-65.2	*In situ* survey data	*T. trachurus capensis*	Svellingen and Ona (1999)

2003), uncompensated swimbladder compression causing reduced TS in the HF range could be beneficial to Cape horse mackerel in terms of reduced conspicuousness towards acoustically mediated predators (Hammond et al., 1994). TS conversion errors first of all involve systematic errors, i.e. bias, but variations in TS between surveys represent variable noise, and estimation of TS to size relations and quantification of natural variations in TS in Cape horse mackerel have therefore been identified as a prioritised research area for the BENEFIT region (Anon., 2003b, 2004).

Error estimation

The previous sections have identified the main sources of error in acoustic abundance estimation of Cape horse mackerel, in light of the aggregation and behavioural dynamics in the horizontal and vertical planes. Identification of the different types of error is important in order to find ways to improve the estimates. Quantification of the inherent error ranges is essential for quantification of the survey accuracy and precision, providing valuable information for effective, precautionary fisheries management. Unfortunately, estimating error ranges is generally difficult.

Two BENEFIT workshops on survey errors held in Cape Town, South-Africa, in 2000 (Anon., 2003b) and 2003 (Anon., 2004) examined the types of survey errors and their associated error ranges for the commercially important pelagic and demersal fish species. The workshops aimed at identifying and, as far as possible, quantifying survey errors, and involved some of the most experienced fisheries scientists and survey practitioners in the region. With basis in this work, errors identified for acoustic surveys of Cape horse mackerel in Namibia are summarized in Table 2, including assumed error range (min-max), type of error (variable, systematic, random) and expected main direction (positive, negative). Although a rather tentative synopsis, this information represents a very important step towards providing accurate and precise estimates of Cape horse mackerel abundance with associated precision estimates.

CONCLUDING REMARKS

This chapter has focused on the behaviour of the Cape horse mackerel in the northern Benguela in relation to problems with acoustic abundance estimation. We have shown that the behaviour and distribution of the Cape horse mackerel in the Namibian waters is very dynamic, varying diurnally, seasonally and annually. The acoustic abundance estimation of the Cape horse mackerel stock therefore represents a challenge to scientists trying to understand the dynamics of the stock and to provide reliable indices of abundance.

Table 2. Error estimates in acoustic horse mackerel surveys ranked by expected error range (Anon. 2003b, 2004). Several of the error sources are either poorly documented or subject to large inherent variation, and the table should therefore be interpreted as a list of rough estimates (Type A: availability errors, B: acquisition errors, C: conversion errors).

Source	Type	Min (%)	Mean (%)	Max (%)	Range (%)	Type	Dependencies	Comments	Reference
Target strength*	C	-200	0*	200	400	Variable	Depth, day/night, angular orientation	b_{20} =-72 best guess, large variations	Axelsen et al. (2003)
Vessel avoidance	C	-80	-10	0	80	Variable	Depth, light, fish size and densities	Mean very uncertain, large variations	Anon. (2003b, 2004)
Statistical sampling error	B	-25	0	25	50	Random	Transect spacing, horizontal resolution	Survey CV typically 25 %	Anon. (2003b)
Target ID	B	-10	0	10	20	Random	Degree of mixing, non-target densities	Sensitive to plankton density	Anon. (2003b)
Equivalent beam factor	B	-10	0	10	20	Systematic	Echosounder system	Max ± 1db	Simmonds et al. (1992)
Weather effect	B	-15	-10	-5	10	Variable	Time of year, latitude	Signal attenuation due to bubbles	Dalen and Løvik (1981)
Incomplete coverage	A	-15	-10	-5	10	Variable	Distribution pattern	Inshore distribution of juveniles	Anon. (2003b)
Blind zone detection**	A	-10	-5	0	10	Variable	Availability	Diurnal variability (DVM)	Ona and Mitson (1996)
Fishing gear avoidance	C	-10	-5	0	10	Variable	Fish size, gear, light, tow duration	Affects estimated mean fish size	Bauleth-D'Almeida et al. (2001)
Sv Thresholding***	B	-5 (-20)	0 (-15)	5 (-10)	10	Variable	Plankton/ non-target densities	Plankton inclusion/target removal	Anon. (2003b)
Calibration	B	-5	0	5	10	Random	Calibration conditions	Max ± 0.2 db, all acoustic surveys	Foote et al. (1987)
Signal attenuation	C							Negligible	

*see Table 1; **including both upper blind zone and bottom dead zone,
***Values for R/V Dr. Fridtjof Nansen, values for R/V Welwitschia in brackets.

Notwithstanding all these challenges, it is important to emphasize the importance of the acoustic method, both as a tool to understand the behaviour of the fish as well as to obtain indices of abundance. The Cape horse mackerel is a species that is relatively easy to identify using the acoustic method, which combined with the large sample volumes covered by echosounders, provides a very strong assessment tool. All methods have their weaknesses, and some of the pitfalls of the acoustic method have been discussed here. It may be argued, however, that the methodological challenges involved in surveying Cape horse mackerel in Namibia are reasonably well understood in comparison with other species and ecosystems.

One of the greatest challenges for the future is to build on to the existing understanding of the dynamics of the Cape horse mackerel stock in relation to its environment in order to quantify the implications of fish behaviour and ecology to acoustic surveying. Knowledge of the changing horizontal distribution and dynamics should be a guideline for survey design and stock coverage, while vertical dynamics are important in terms of availability, target identification and conversion from acoustic densities to fish abundance during surveys. Ultimately, such ecological information should be integrated in the acoustic survey, the data interpretation and the stock assessment in a quantitative manner, in line with the ambition of the international research community to adopt an ecosystem approach to fisheries research and management.

References

Agenbag, J.J. (1980): General distribution of pelagic fish off South West Africa as deduced from aerial fish spotting (1971-1974 and 1977) and as influenced by hydrology. *Fisheries Bulletin of South Africa* 13: 55-67.

Agenbag, J.J., and Shannon, L.V. (1988): A suggested physical explanation for the existence of a biological boundary at 24°30'S in the Benguela system. *South African Journal of Marine Science* 6: 119-132.

Andronov, V.N. (1983): Feeding of Cape horse mackerel *Trachurus trachurus capensis* and Cape hake *Merluccius capensis* off Namibia in January 1982. *Collection of Scientific Papers. International Commission for the Southeast Atlantic* 10(I): 1-6.

Andronov, V.N. (1985): Feeding of Cape horse mackerel *Trachurus trachurus capensis* Castelnau in the Namibian area. *Collection of Scientific Papers. International Commission for the Southeast Atlantic* 12(I): 1-16.

Anonymous (1993): Cruise Reports "Dr. Fridtjof Nansen". Surveys of the fish resources of Namibia, Cruise Report No 2/1993, part II. Surveys of the pelagic stocks 26 May - 19 June 1993. Ministry of Fisheries and Marine Resources, Swakopmund, Namibia and Institute of Marine Research, Bergen, Norway. 45 pp.

Anonymous (1995) Underwater Noise of Research Vessels - Review and Recommendations. Ed.: R.B. Mitson. ICES Cooperative Research Reports 209. 47 pp.

Anonymous (1998): Cruise Reports "Dr. Fridtjof Nansen". BENEFIT Cruise Report 4/1998. Horse mackerel survey methodology 07-19 October 1998. BENEFIT Secretariat, Swakopmund, Namibia and Institute of Marine Research, Bergen, Norway. 48 pp.

Anonymous (1999): Cruise report of the R/V "Welwitschia". Horse mackerel and pre-recruit survey of the northern Benguela 6-28 February 1999. Ministry of Fisheries and Marine Resources, Swakopmund, Namibia. 36 pp.

Anonymous (2001a): Cruise Reports "Dr. Fridtjof Nansen". BENEFIT Cruise Report 1/2001. Recruitment and reproduction study on horse mackerel and anchovy. BENEFIT Secretariat, Swakopmund, Namibia and Institute of Marine Research, Bergen, Norway. 48 pp.

Anonymous (2001b): Cruise Reports "Dr. Fridtjof Nansen". BENEFIT Cruise Report 3/2001. Diel vertical migration in horse mackerel. BENEFIT Secretariat, Swakopmund, Namibia and Institute of Marine Research, Bergen, Norway. 24 pp.

Anonymous (2003a): Cruise report of the R/V "Welwitschia". Horse mackerel survey of the northern Benguela (16°´30´S - 25°00´S) 4-27 February 2003. Ministry of Fisheries and Marine Resources, Swakopmund, Namibia. 36 pp.

Anonymous (2003b): BENEFIT report on Survey Errors Workshop, Cape Town, South-Africa 4–7 December 2000. BENEFIT Secretariat, Swakopmund, Namibia. 45 pp.

Anonymous (2004): BENEFIT report on Survey Errors Workshop, Cape Town, South-Africa 14-15 November 2003. BENEFIT Secretariat, Swakopmund, Namibia. 31 pp.

Assorov, V.V., Dubrovin, B.I., Polischuk, I.A., Prits. S.E., and Prokofjeva, G.V. (1988): Seasonal distribution of the Cape horse mackerel (*Trachurus trachurus capensis*) population. *Collection of scientific papers. International Commission for the Southeast Atlantic* 15(I): 47-53.

Au, V.W.L., and Benoit-Bird, K.J. (2003): Automatic gain control in the echolocation system of dolphins. *Nature* 423(6942): 861-863.

Axelsen, B.E., Anker-Nilssen T., Fossum P., Kvamme C., and Nøttestad, L. (2001): Pretty patterns, but a simple strategy: predator-prey interactions between juvenile herring and Atlantic puffins observed with multi-beam sonar. *Canadian Journal of Zoology* 79: 1586-1596.

Axelsen, B.E., Bauleth-D'Almeida G., and Kanandjembo, A. (2003): In situ measurements of the acoustic target strength of Cape horse mackerel *Trachurus trachurus capensis* off Namibia. *South African Journal of Marine Science* 25: 239-251.

Barange, M., Hampton, I., and Soule, M. (1996): Empirical determination of *in situ* target strengths of three loosely aggregated fish species. *ICES Journal of Marine Science* 53(2): 225-232.

Barange, M., Pillar, S.C., and Hampton, I. (1998): Distribution patterns, stock size and life-history strategies of Cape horse mackerel *Trachurus trachurus capensis*, based on bottom trawl and acoustic surveys. *South African Journal of Marine Science* 19: 433-447.

Bauleth-D'Almeida, G. (2001): A review of the hydro-acoustic surveys conducted since 1990 to determine the abundance of horse mackerel *Trachurus capensis* in the Namibian waters. In: The Research and Management of Horse Mackerel in Namibian Waters. Proceedings of international workshop, Swakopmund, Namibia, 26-30 March 2001 (J-O. Krakstad, F. Botes, and A. Kanandjembo, eds.), Ministry of Fisheries and Marine Resources, Swakopmund, Namibia.

Bauleth-D'Almeida, G., Krakstad, J-O., and Kanandjembo, A. (2001): Comparison of horse mackerel length frequencies obtained from research vessels and commercial midwater trawlers: Implications for biomass estimation. *South African Journal of Marine Science* 23: 265-274.

Blaxter, J.H.S. (1985): The herring: a successful species? *Canadian Journal of Fisheries and Aquatic Sciences* 42(1): 21-30.

Blaxter, J.H.S., and Tytler, P. (1978): Physiology and function of the swimbladder. *Advances in comparative Physiology and Biochemistry* 7: 311-367.

Boyd, A.J., and Cruickshank, R.A. (1983): An environmental basin model for West Coast pelagic fish distribution. *South African Journal of Marine Science* 79(4): 150-151.

Boyer, D.C., and Hampton, I. (2001): An overview of the living marine resources of Namibia. *South African Journal of Marine Science* 23: 5-35.

Boyer, D.C., Boyer, H.J., Fossen, I., and Kreiner, A. (2001): Changes in abundance of the northern Benguela sardine stock during the decade 1990-2000, with comments on the relative importance of fishing and the environment. *South African Journal of Marine Science* 23: 67-84.

Butterworth, D.S., Hughes, G., and Strumpfer, F. (1990): VPA with ad-hoc tuning – Implantation for disaggregated fleet data, variance-estimation, and application to the Namibian stock of Cape horse mackerel *Trachurus trachurus capensis*.
South African Journal of Marine Science 9: 327-357.

Chapman, P., and Shannon, L.V. (1985): The Benguela Ecosystem 2. Chemistry and related processes. In: *Oceanography and Marine Biology. An Annual Review* 23 (M. Barnes, ed), Aberdeen, University Press: 183-251.

Crawford, R.J.M. (1999): Seabird responses to long term changes of prey resources off southern Africa. In: Proceedings of the 22nd International Ornithological Congress, Durban, August 1998 (N. J. Adams, and R. H. Slotow, eds.), Johannesburg. Birdlife South Africa: 688-705.

Cruickshank, R.A. (1983): Distribution of pelagic fish shoals determined by acoustic surveys in 1981-1982 and its relationship to environmental factors. *Collection of Scientific Papers. International Commission for the Southeast Atlantic* 10(II): 75-97.

Dalen, J., and Løvik, A. (1981): The influence of wind-induced bubbles on echo-integration surveys. *Journal of the Acoustic Society of America* 69: 1653-1659.

David, J.H.M. (1987): Diet of the South African fur seal and the assessment of competition with fisheries in southern Africa. In: *The Benguela and comparable ecosystems* (A.I.L. Paine, J.A. Guland, K. H. Brink, eds.), *South African Journal of Marine Science* 8: 693-713.

David, J.H.M. (1989): Seals. In: *Oceans of life off southern Africa* (A.I.L. Paine and R.J.M. Crawford, eds.), Vlaberg publishers. Cape Town.

Dingle, R.V., and Nelson, G. (1993): Seabottom temperature, salinity and dissolved oxygen on the continental margin off south-western Africa. *South African Journal of Marine Science* 13: 33-49.

Fernandes, P.G., Brierley, A.S., Simmonds, E.J., Millard, N.W., McPhail,

S.D., Armstrong, F., Stevenson, P., and Squires, M. (2000): Fish do not avoid research vessels. *Nature* 404(6773): 35-36.

Foote, K.G. (1978): Effects of fish behaviour on echo energy: the need for measurements of orientation distribution. *Journal du Conseil International pour L'exploration de la Mér* 39: 193-201.

Foote, K.G. (1980): Importance of the swimbladder in acoustic scattering by fish: a comparison of gadoid and mackerel target strengths. *Journal of the Acoustic Society of America* 67: 2084-2089.

Foote, K.G. (1987): Fish target strengths for use in echo integrator surveys. *Journal of the Acoustic Society of America* 82(3): 981-987.

Foote, K.G., Aglen, A., and Nakken, O. (1986): Measurements of fish target strength with a split-beam echosounder. *Journal of the Acoustic Society of America* 80(2): 612-621.

Foote, K.G., Knudsen, H.P., Vestnes, G., MacLennan, D.N., and Simmonds, E.J. (1987): Calibration of acoustic instruments for fish density estimation: a practical guide. ICES Cooperative Research Reports 144, 69 pp.

Fossen, I., Boyer, D.C., and Plarre, H. (2001): Changes in some key biological parameters of the northern Benguela sardine stock. *South African Journal of Marine Science* 23: 111-121.

Fréon, P., Gerlotto, F., and Misund, O.A. (1993): Consequences of fish behaviour for stock assessment. *ICES Marine Science Symposium* 196: 190-195.

Gutiérrez, M., and MacLennan, D.N. (1998): Resultados preliminares de las mediciones de Fuerza de Blanco in situ de las principales especies pelágicas. *Informe. Instituto del Mar del Perú.* Callao [Inf. Inst. Mar Perú] 135: 16-19.

Hammond, P.S., Hall, A.J., and Prime, J.H. (1994): The diet of grey seals in the inner and outer Hebrides. *Journal of Applied Ecology* 31(4): 737-746.

Harden Jones, F.R., and Scholes, P. (1985): Gas secretion and resorption in the swimbladder of the cod *Gadus morhua. Journal of comparative Physiology* 155: 319-331.

Hecht, T. (1990): On the life history of Cape horse mackerel (*Trachurus trachurus capensis*) of the south-east coast of South Africa. *South African Journal of Marine Science* 9: 317-326.

Jacobson, L.D., and MacCall, A.D. (1995): Stock recruitement models for the Pacific sardine (*Sardinops sagax*). *Canadian Journal of Fisheries and Aquatic Sciences* 52: 566-577.

Kerstan, M., and Leslie, R.W. (1994): Horse mackerel on the Agulhas Bank - summary of current knowledge. *South African Journal of Marine Science* 90: 173-178.

Konchina, Yu.V. (1986): Distribution and feeding of South African horse mackerel and hake in Namibian waters. *Collection of Scientific Papers. International Commission for the Southeast Atlantic* 13(II): 7-18.

Korneliussen, R.J. (2000): Measurement and removal of echo integration noise. *ICES Journal of Marine Science* 57(4): 1204-1217.

Korneliussen, R.J., and Ona, E. (2002): An operational system for processing and visualizing multi-frequency acoustic data. *ICES Journal of Marine Science* 59: 293-313.

Krakstad, J-O., and Kanandjembo, A. (2001): A summary of the fishery and biology of Cape horse mackerel in Namibian waters. In: The Research and Management of Horse Mackerel in Namibian Waters. Proceedings of international workshop, Swakopmund, Namibia, 26-30 March 2001. (J-O. Krakstad, F. Botes, and A. Kanandjembo, eds.), Ministry of Fisheries and Marine Resources, Swakopmund, Namibia.

Levenez, J.-J., Gerlotto, F., and Petit, D. (1990): Reaction of tropical coastal pelagic species to artificial lighting and implications for the assessment of abundance by echo integration. Developments In: Fisheries Acoustics: Symposium, Seattle, USA, 22-26 June 1987. *Rapports et procès-verbaux des réunions du conseil international pour l'exploration de la mér* 189: 128-134.

Lillo, S., Cordova, J., and Paillaman, A. (1996): Target strength measurements of hake and jack mackerel. *ICES Journal of Marine Science* 53(2): 267-271.

Lovell, A. (2000): Population structure of three commercially important species in the Gulf of Guinea. Thesis submitted for the degree of Doctor of Philosophy. Biological Sciences, University of Warwick, Coventry CV4 7AL, UK. Submitted August 2000. 121 pp.

MacLennan, D.N., and Simmonds, E.J. (1992): *Fisheries Acoustics*. Fish and Fisheries, Series 5. Chapman and Hall, New York.

Manly, B.F.M. (1997): *Randomization, bootstrap and Monte Carlo methods in biology*. 2nd edition. Chapman and Hall, London, UK. 376 pp.

Medwin, H., and Clay, C.S. (1998): *Fundamentals of Acoustic Oceanography*. Academic Press, San Diego. 712 pp.

Misund, O.A., Beltestad, A., and Castillo, J. (1997): Distribution and acoustic abundance of horse mackerel and mackerel in the northern North Sea, October 1996. ICES Doc. WG on the assessment of anchovy, horse mackerel, mackerel and sardine, Copenhagen 9/9 - 18/9, 1997.

Morgan, M.J., DeBlois, E.M., and Rose, G.A. (1997): An observation on the reaction of Atlantic cod (*Gadhus morhua*) in a spawning shoal to bottom trawling. *Canadian Journal of Fisheries and Aquatic Sciences* 54(Suppl 1): 217-223.

Naish, K.A., Hecht, T., and Payne, A.I.L. (1991): Growth of Cape horse mackerel *Trachurus trachurus capensis* off South Africa. *South African Journal of Marine Science* 10. 29-35.

Namwandi, T.M. (2002): Distribution pattern of Cape horse mackerel *Trachurus trachurus capensis* (Castelnau) in the Benguela system. M. Phil. thesis, University of Bergen, Norway. 62 pp.

Nøttestad, L, and Axelsen, B.E. (1999): Herring schooling manoeuvres in response to killer whale attack. *Canadian Journal of Zoology* 77(10):1540-1546.

Olsen, K., Angell, J. and Løvik, A. (1983): Quantitative estimations of the influence of fish behaviour on acoustically determined fish abundance. *FAO Fisheries Report* 300: 139-149.

Ona, E. (1990): Physiological factors causing natural variations in acoustic target strength of fish. *Journal of the Marine Biology Association of the U.K.* 70: 107-127.

Ona, E., and Mitson, R.B. (1996): Acoustic sampling and signal processing near the seabed: the deadzone revisited. *ICES Journal of Marine Science* 53: 677-690.

O'Toole, M.J. (1977): Investigation into some important fish larvae in the South-East Atlantic. Ph.D. thesis, University of Cape Town. 299 pp.

Pearre, S. Jr. (2003): Eat and run? The hunger/ satiation hypothesis in vertical migration: history, evidence and consequences. *Biological Reviews* 78: 1-79.

Pillar, S.C., and Barange, M. (1998): Feeding habits, daily ration and vertical migration of the Cape horse mackerel south of South Africa.

South African Journal of Marine Science 19: 263-274.

Pillar, S.C., and Wilkinson, I.S. (1995): The diet of Cape hake *Merluccius capensis* on the south coast of South Africa. *South African Journal of Marine Science* 15: 225-239.

Pitcher, T.J., and Wyche, C.J. (1983): Predator avoidance behaviour of sand-eel schools: why schools seldom split. In: *Predator and prey in fishes* (D.L.G. Noakes, B.G. Lindquist, G.S. Helfman, J.A. Ward, eds.), Junk, The Hague. p. 193-204.

Sekigushi, K., Klages, N.T.W., and Best, P.B. (1992): Comparative analysis of the diets of smaller odontocete cetaceans along the coast of southern Africa. *South African Journal of Marine Science* 12: 843-861.

Shannon, L.V. (1985): The Benguela ecosystem. 1. Evolution of the Benguela, physical features and processes. In: *Oceanography and Marine Biology. An Annual Review* 23. University Press. (M. Barnes, ed.), Aberdeen, Scotland. p. 105-182.

Simmonds, E.J., Williamson, NJ., Gerlotto, F., and Aglen, A. (1992): Acoustic survey design and analysis procedure: a comprehensive review of current practice. *ICES Cooperative Research Reports* 187: 127 pp.

Svellingen, I., and Ona, E. (1999): A summary of target strength observations on fishes from the shelf off West Africa. In: Proceedings from the 137th Meeting of the Acoustical Society of America and The Second Convention of the European Acoustics Association. Berlin 14-19 March 1999. File: 2PAO_2.pdf (available on CD only).

Torres, G.A., Guzman, F.O., and Castillo, P.I. (1984): The swimbladder as a resonant organ and its influence in sonic intensity. *Investigacion Pesquera* 31: 81-88.

Twatwa, N.M., and van der Lingen, C. (2000): Feeding selectivity of juvenile Cape horse mackerel (Trachurus trachurus capensis). 10th Southern African Marine Science Symposium (SAMSS 2000): Land, Sea and People in the New Millennium - Abstracts. p. 1.

Tytler, P., and Blaxter, J.H.S. (1973): Adaptation by cod and saithe to pressure changes. *Netherlands Journal of Sea Research* 7: 31-45.

Vabø, R., Olsen, K., and Huse, I. (2002): The effect of vessel avoidance of wintering Norwegian spring spawning herring. *Fisheries Research* 58(1): 59-77.

Venter, J.D. (1976): Voeding van die Suid-Afrikaanse maasbanker *Trachurus trachurus* Linnaeus. M.Sc. thesis, Randse Afrikaans Universiteit. 174 pp.

Wada, T., and Jacobson, L.D. (1998): Regimes and stock recruitment relationships in Japanese sardine (*Sardinops melanostictus*) 1951-1995. *Canadian Journal of Fisheries and Aquatic Sciences* 55: 2455-2463.

Wardle, C.S., Soofiani, N.M., O'Neil, F.G., Glass, C.W., and Johnstone, A.D.F. (1996): Measurements of aerobic metabolism of a school of horse mackerel at different swimming speeds. *Journal of Fish Biology* 49(5): 854-862.

Zenkin, V.S., and Komarov, Y.A. (1981): Genetic differentiation in the Cape horse mackerel (*Trachurus trachurus capensis* Castelnau) population. *Collection of Scientific Papers. International Commission for the Southeast Atlantic* 8(II) 291-298.

8 A BRIEF OVERVIEW OF CURRENT BIOECONOMIC STUDIES OF NAMIBIAN FISHERIES

*Ussif Rashid Sumaila and Stein I. Steinshamn**

Abstract

This contribution provides an overview of bioeconomic studies of Namibian fisheries available in the literature, and then lists interesting research ideas that are waiting to be explored. We found that there are very few papers in the literature that study the bioeconomics of Namibian fisheries. In all, three types of papers are present in the literature, namely, those that look at the bioeconomics of hake stocks; a few that develop bioeconomic models of Namibian sardine; and one paper that explores the economics of Namibian fisheries in an ecosystem context. Clearly, there is a severe lack of bioeconomic studies to support the optimal, sustainable management of Namibia's fisheries. The overview also revealed the need for bioeconomic models to support the transboundary management of the shared stocks in the Benguela Current Large Marine Ecosystem.

INTRODUCTION

The objective of this chapter is to briefly give an overview of the few bioeconomic models that have been developed to study the fisheries of Namibia. Bioeconomic models have been used extensively by fisheries economists since 1954 when Gordon published his seminal paper (Gordon, 1954) to help fisheries managers decide the amount of fishing effort (and therefore total allowable catches) to exert on fish stocks in order to maximize economic rent. Gordon's work was followed by works by Scott (see, Christy and Scott, 1965 and Hannesson, 1993), to mention but a few, that helped to spread the application of bioeconomic models and the results derived therefrom in the management of fisheries. Most of the models have been applied to analyze fisheries located in the economically more developed countries of the world. Hannesson (1983), Steinshamn (1992); and Sumaila (1995) are examples of

---

* Sumaila thanks the Research Council of Norway, the Sea Around Us Project and the Pew Charitable Trusts for their support.

applications of bioeconomic models to study the fisheries in the Barents Sea; Clark and Kirkwood (1979) is an application of bioeconomics to the Gulf of Carpentaria prawn fishery.

Bioeconomic models that study fisheries in the less developed countries of the world, such as Namibia, are rare. However, a number of other economic papers are more common. Two examples are (i) work on natural resource accounting (Lange and Motinga, 1997; Manning and Lange, 1998; Lange, 2000), and (ii) work on the valuation of recreational fisheries (Kirchner et al., 2000; Zeybrandt and Barnes, 2001; Sumaila, 2002).

In the next section we present bioeconomic models of Namibian hake. This is followed by an overview of bioeconomic models of Namibian sardine. Key elements of an ecosystem-economic model of the Benguela current ecosystem, which supports the Namibian Exclusive Economic Zone (EEZ), among others, are then presented. Finally, a discussion of potential research areas is given.

THE HAKE BIOECONOMIC MODELS

This section of the chapter summarizes work published in Sumaila (2000, 2001). These two papers addressed the question of which of the two main vessel types used to exploit hake, that is, wetfish and freezer trawlers, should harvest what proportion of Namibian hake, given Namibia's fisheries policy objectives? Sumaila (2000) addresses this question after the total allowable catch (TAC) for hake has already been decided (the 'after-TAC' analysis). In other words, the determination of the TAC for hake is exogenous to the analysis. The second paper, on the other hand, develops a full-fledged bioeconomic model in which both the biology of hake, and economics of the fishery are endogenous to the model (the 'before-TAC' analysis).

Why are these studies interesting? First, the fishing sector is an important part of the economy of Namibia, with the hake fishery being one of the main contributors to the sector. It has been estimated (by the Ministry of Fisheries and Marine Resources of Namibia, MFMR) that hake contributed about 7.4 per cent of Namibia's estimated exports in 1994, and contributed about N$ 951 million to Namibia's GDP in 1997 (MFMR, 1997).[1] It should be noted that these figures include only the direct contribution to GDP; additional contributions from secondary industries and the multiplier effects of spending hake-related incomes are not included. Secondly, before Namibia became independent in 1990, freezer trawlers have been the dominant vessels employed in the harvesting of the hake stock in the Namibian Exclusive

---

[1] One US$ was equivalent to about N$7.07 in March 2004.

Economic Zone (EEZ). Namibia saw its rich fishery resources as one of the vehicles available to it for the badly needed economic development of its people. Hence, the country took an important fishery policy decision that called for the restructuring of the trawler fleet in favour of wetfish trawlers. Third, because of the difference between the two vessel groups both in economic and biological terms, determining the optimal proportion of the hake TAC that should be landed by the two vessel types is not trivial.

In 1992 wetfish trawlers landed only about 5 000 tonnes out of a total reported hake landing of 87 498 tonnes. The announced government policy with respect to the allocation of the TAC between the two fleet groups is that 20%, 40% and 60% of it should be allocated to the wetfish trawlers in 1993, 1994 and 1995, respectively. The ultimate aim is to maintain an allocation of 60:40 in favour of the wetfish trawlers into the future. Performance against stated objectives has been quite good up to 1994: in 1993, 19.9% of the TAC was allocated to wetfish trawlers. The corresponding allocation for 1994 was 48.9%, well over the target of 40%. However, the target of 60% could not be achieved from 1995 to 1998, mainly because there was no increase in the TAC in these years. One reason for the decline of the wetfish trawler share of the TAC was that the biomass of Cape hake, which forms the basis of the wetfish catch, was decreasing during this period (Dave Boyer, pers. comm.). Progress in this direction is again being made given the recent upswing in the total allowable catches approved by the Minister of Fisheries and Marine Resources (see Oelofsen, 1999).

The main reason for committing to the above policy is to encourage onshore processing, and thereby reap benefits such as increased employment for workers from the north of the country. This move would make companies process their catch on land in Namibia to encourage the development of value-added fish products by the country. In this way, the country hopes to ensure participation in the fishery by Namibians, with the attendant positive effect on their economic welfare.

Interesting questions to ask here are, is this new policy economically rational; if not, are there any reasons other than economic that may justify this move? For example, is it the case that the gains in employment due to the restructuring can compensate for the resulting economic loss, if any?

The principle underlying the work from which we report here is *economic efficiency*, in other words, it is assumed that the primary objective is to harvest and process the stock in the most economically efficient manner. Notwithstanding the Namibianization policy of the Ministry of Fisheries and Marine Resources, this assumption appears to be reasonable in the case of Namibian fisheries, which unlike the fisheries of most developing countries, are mainly industrial.[2] Thus, most of the complications that usually arise due to the community-based nature of certain fisheries are simply not present here. There is, therefore, an excellent opportunity for pursuing and, indeed, achieving economically optimal management of the resources of the Namibian EEZ. Moreover, the results computed are meant to serve as benchmarks for determining the trade-offs between different government policies: we should, for instance, be able to discover what is being sacrificed in economic terms due to a government policy that is geared towards increasing Namibian employment in the industry (see Armstrong *et al.*, this volume), as against one based purely on economic efficiency criteria.

In more concrete terms, the study sought to: (i) test the government TAC allocation policy target of 60:40 for the wetfish and freezer trawlers, respectively, to see if it is optimal in an economic sense,[3] (ii) determine what discounted economic benefit would accrue to society at large under the optimal allocation regime, (iii) find out the optimal number of both wetfish and freezer trawlers needed to achieve these objectives, and (iv) look at the employment-generating capacities of the wetfish and freezer trawlers. The trade-offs between the economic gains and the employment-generating capacities of the two classes of vessels are also discussed.

WETFISH VERSUS FREEZER TRAWLERS: THE AFTER TAC ANALYSIS

In the after-TAC analysis, the objective was to determine the proportion of an already decided TAC that should be landed by the wetfish and freezer trawlers, respectively. In addition, the specific questions listed under the introductory section are addressed using the analytical framework outlined below.

Method

A typical freezer trawler is usually larger than a typical wetfish trawler. It fishes in deeper waters, probably catching larger and more valuable fish. In addition, it can stay offshore for longer periods than the wetfish trawler. The freezer trawler is equipped fully for catching, freezing and packaging at sea.

---

[2] Although small coastal communities caught fish in coastal lagoons during pre-colonial times, the only indigenous fishing tradition amongst the peoples of the interior was freshwater fishing in the streams and rivers of the north.

[3] By an optimal quota allocation we mean the allocation that would maximise the social planner's, that is, the Namibian government's overall economic benefit from the resource.

Therefore, all the processes needed, from actual harvesting to packaging in readiness for export, are undertaken offshore.[4]

The following assumptions underlie the analysis in Sumaila (2000):

- Annual TACs are assumed to be optimally and exogenously determined by the MFMR. Hence, this study does not seek to give advice on what the optimal TAC for hake should be, but rather on what percentage of the decided TAC should be harvested by the wetfish and freezer trawlers, respectively.
- It is assumed here that there are no interactions between the two agents at the market-place. This assumption is reasonable because the agents sell their landings at competitive markets where prices are exogenously determined.
- It is further assumed that there are no significant natural interactions between the hake species and others. This implies that externality, due to say, predator-prey relations, is ignored. Given the lack of adequate studies on interspecies interactions between the species living in the Namibian EEZ, this assumption is considered to be a pragmatic one, which will be relaxed as more biological information becomes available.
- The model is deterministic in the sense that all parameters of the model are assumed to be known with certainty. Also, future TACs are assumed to be known. Clearly, these are strong assumptions. In the case of future TACs, for instance, we know that yearly allocations are based on both scientific knowledge concerning the biomass of hake and policy-related considerations, both of which vary from year to year. In general, the MFMR has a policy of limiting the variability of TAC to an increase of no more than 5% and a decrease of no more than 10% in any one year. A future task would be to introduce uncertainty into the model.[5]

Input data, constraint and objective functions. - The assumption of no interaction at the market-place necessarily implies that both wetfish and frozen fish are supplied at given prices, implying that the price the fishermen receive for their produce is inelastic to the quantities of fish they supply to the market. It

---

[4] One major difference between wet and freezer trawlers is that to a large extent they target different species. The wetfish trawlers target shallow-water hake (*Merluccius capensis*) while the freezer trawlers catch deep-water hake (*Merluccius paradoxus*). It is worth noting that the incorporation of this information could have affected the results of this analysis. Similarly, the cannibalistic nature of hake, which is not explicitly incorporated in this paper, needs to be included in future studies.

[5] In the meantime, the model is designed to be flexible enough to allow quick sensitivity analysis, making it possible to vary important parameters as new information flows in.

should be noted that the main market for Namibian hake is Spain. This is a large international market supplied by many other sources, of which Namibia is only one of many suppliers. For Namibia, or any of the other suppliers, to be able to influence the market price, one of two things will have to happen. The supplier will have to withdraw from the market a large proportion (if not all) of its output, or else there has to be a sudden large increase in its supply to the market, both of which are unlikely to happen under normal conditions.

Modelling the cost of landing hake. - In general, two types of costs can be identified depending on whether one is talking about the costs directly incurred by the agents in the model, that is, *private costs* or costs incurred by society as a whole, that is, *social costs*.[6] Usually these two are not identical because of distortions in market prices and/or costs. As the focus of this study is on the benefits to society as a whole, the study applied social costs.

The production and profit functions. - In both theoretical and applied fishery economics, it is common to assume that the production (harvest) function depends on the stock size or biomass, the vessel efficiency (that is, the catchability coefficient) and the number of vessels taken out to fish in a given period. The underlying idea is that, other things being equal, the ability to harvest fish at any point in time is proportional to the biomass available in the habitat. This is particularly so in the case of non-schooling species such as hake.[7] The profit accruing to a given vessel group in any given year is the total revenue from fishing less the total cost of fishing in that year.

Stock dynamics and constraints. - The stock constraint in this model comes in the form of the TAC fixed annually by the government. The players are free to maximise their profits from the fishery so long as their combined harvest does not exceed the TAC. Given the assumption that TACs are optimally determined to ensure the long-term survival of the stock, they implicitly ensure that the underlying stock dynamics and constraints are respected all the time.

The social planner's objective. - It is assumed that the objective of the social planner, that is, the Ministry of Fisheries and Marine Resources of Namibia,

---

[6] It is the government that is concerned about these costs; private agents would usually be concerned with only their private costs.

[7] It should be noted that even though not a schooling species such as sardines, it also aggregates into less dense shoals.

is to choose a sequence of TAC shares, $\varsigma_{i,t}$ (t=1,2, .. ,T), to obtain the highest possible discounted profit from the TAC, using social costs and prices.

Results and Discussion
The data required on the Namibian hake fishery to numerically implement the model presented above were collected and/or calculated (Sumaila, 2000), and the algebraic modelling language AMPL (Fourer et al., 1993) was used as computational aid. The key results obtained are presented below (see Sumaila, 2000 for details).

The total present value of economic rent from the resource given the government target of 60:40 allocation of the annual TAC to w (wetfish trawlers) and f (freezer trawlers) is N$10.42 *billion*. On the other hand, the economically efficient allocation turns out to be 100:0 to w and f, respectively. This allocation results in a total present value of economic rent of N$11.69 *billion*. Thus, the economic loss due to the implementation of the current government target rather than the economically optimal share is N$1.27 *billion*, about 11% of what is achievable.

In terms of employment generation, more allocation to the wetfish trawlers is a good thing, as this class of vessels generates more than six times the employment generated by the freezer trawlers for the same TAC allocation. It is possible to generate a total of up to 7 800 positions of various kinds annually from the activities in the hake fishery if the optimal solution is implemented (Sumaila, 2000).

The fleet size necessary to land the optimal allocation of the TAC is 53 wetfish trawlers of size class 1400-2000 HP or their equivalent. In the case of the declared government policy of 60:40 allocation, the necessary fleet sizes of both wetfish (size class 1400-2000 HP) and freezer (size class 1500-1999 HP) trawlers are 32 and 13, respectively.

Based on the results outlined above, one may come to the conclusion that the freezer trawlers should be banned from the exploitation of hake altogether: both economic efficiency and employment generation criteria support this change. There are, however, other issues to be taken into consideration. First, we should be interested in benefiting from certain intrinsic advantages of harvesting hake with freezer trawlers. An advantage of the freezer trawlers (but a disadvantage of wetfish trawlers) which needs to be taken into account is the fact that generally freezer trawlers fish in deeper waters than their wetfish counterparts, thereby ensuring a better spread of fishing activity than would be possible if only wetfish trawlers were employed. Such a spread is positive for the biological well-being of the habitat and the fish contained therein, especially because many believe that the freezer trawlers are by and large targeting a different species of hake. Another point to note with respect to the finding of this paper is that turning the hake fishery into a wetfish-trawler-only fishery may be ill-advised, because building a large land-based industry in a situation where the industry can swing widely due to environmental and other human factors can be very risky. The task of isolating the risk factors and the potential benefits of using freezer trawlers is left to future research. Another important issue, not taken up in this chapter but addressed elsewhere (see Manning, 1998), is the question of rent capture, its distribution and Namibian participation and ownership (Armstrong et al., this volume).

WETFISH VERSUS FREEZER TRAWLERS: THE BEFORE-TAC ANALYSIS

The analysis reported in this section is more ambitious, in that the determination of the TAC is assumed to be endogenous to the analysis. In other words, both the biology of hake and the economics of the fishery are explicitly taken into account in deciding the total harvest in a given year.

From the biological perspective, an age-structured model of the Beverton and Holt type is employed. This basic model is extended to allow for the inclusion of demographic diversity concerns. To incorporate economic behaviour, game theory is applied. Combining biology and economics, Sumaila (2001) develops a game theoretic modelling framework for the assessment of the trade-off between economic efficiency gains and demographic diversity conservation in a fishery. The paper introduces a *demographic diversity index*, and develops an application of the method on Namibian hake fisheries.

Biodiversity refers to the variety of life forms: the different plants, animals and microorganisms, the genes they contain, and the ecosystems they form. The concept emphasizes the interrelated nature of the living world and its processes. Biodiversity is important because human beings rely on biological systems and processes for their sustenance, health and well-being. Biotic resources[8] also serve recreation and tourism, and underpin the ecosystems, which provide us with many services. In addition, biodiversity has important social and cultural values. Thus, due to the many values humans derive from biological diversity, studies of this type are necessary to help decision-makers make sound decisions in situations where biodiversity conservation is at stake.

An underlying premise of the paper is that one can divide the management goal of a fishery into two broad categories, namely, conserving the biological diversity of the resource, and ensuring economic efficiency in the use of the resource. It is easy to see that in most situations these goals conflict with each other. To maximize biodiversity fully implies zero harvesting of fish, in most instances. Also, to pursue the absolute maximization of economic benefits will most likely imply the depletion of biodiversity. The challenge for Sumaila (2001) was to develop a modelling framework that would allow the exploration of the nature of the trade-offs between economic benefits on the one hand, and biodiversity conservation, on the other.

The work from which we report here contributes to the literature by focusing on demographic diversity concerns in fisheries, and how this can be incorporated into a bioeconomic model. The paper provides a framework that can be used to value some aspects of biodiversity, and help determine the opportunity cost of achieving different levels of biological diversity. Furthermore, it extends the current literature by developing a 'rare' computational game-theoretic model, which explicitly incorporates demographic diversity concerns.

Method

Modelling demographic diversity. - The paper specifically studies the economics of conserving demographic diversity within a population of fish. Diversity measures generally take into account two factors, namely, *species richness* (that is, number of species) and *evenness*, sometimes known as equitability (Magurran, 1988). The present paper looks at evenness within a fish population. It is concerned with maintaining diversity between, say the male and female parts of a population; the juvenile and mature parts of a population; or even the proportion of the different age groups of fish in the population. The index of demographic diversity in this paper is therefore the proportion of male to female, juvenile to mature; the different age groups in a population, or any such measures.[8] The benchmark index is the ratio of say, the standing juvenile biomass to the standing adult biomass when there is no harvesting. This ratio is given a value of 1 or 100%. The benchmark index captures the demographic diversity of the population, and therefore the "perfect" diversity level. The index for any other scenario, I, is then given by the following equation:

$$I = 1 - \frac{|y - x|}{x}$$

where x denotes the ratio of the juvenile to the mature parts of the biomass when there is no harvesting, and y represents the ratio when there is harvesting. It should be noted that the smaller the value of I the less demographic diversity there is in the population, since that signifies either too many or too few juveniles in relation to the 'no harvest' situation.

---

[8] In the original paper, the index is labelled genetic biodiversity index. A reviewer of this contribution made the point that this index is actually an index of demographic diversity. We agree with the reviewer.

Modelling management scenarios. - It is assumed that the regulator of a given fishery, say a government ministry, has devolved or delegated the authority to manage the fishery to the fishers. Hence, they decide how much of the fish to harvest in each period. In such a decentralized setting, the participants can pursue their fishery goals in a cooperative or non-cooperative game situation.[9] Hence, to address the issues raised by this paper, cooperative and non-cooperative game theoretic models are developed.[10] Two versions of the cooperative model are presented: one in which a demographic diversity conservation objective is explicitly incorporated, and one in which it is not. In the case of the latter, we look at cooperative 'with' and 'without' side payments.

The cooperative model. - Consider a fishery with two groups of participants, say w and f.[11] Let the net private economic benefits to these groups be B_w and B_f, respectively. These benefits depend positively on the quantity of fish harvested by the two users, that is, H_w and H_f, which in turn depends on the stock size, N, and the amount of fishing effort, e, that each player takes out fishing. On the other hand, B_w depends negatively on the amount of fish harvested by f, and B_f, depends negatively on the harvest of w.

Discounted joint economic benefits and diversity indices for different values of a preference parameter $0 \leq \omega \leq 1$ for a given player are computed for different runs of the model under various management scenarios. The parameter ω plays an important role in differentiating between the outcomes for the different versions of the cooperative model. In the cooperative with an explicit demographic diversity conservation objective, the optimal ω is the one that produces the *highest diversity index*, 1. In the cooperative with side payments management scenario, the optimal ω is the one *that maximizes joint discounted economic benefits*, even if that implies pensioning out one of the players. Finally, under the cooperative without side payments regime, the optimal ω is the one that *maximizes the product of the distance be-*

---

[9] By a game we mean any activity involving two or more participants, each of whom recognizes that the outcome for himself depends not only on his own actions, but also those of other participants.

[10] A non-cooperative game is one in which there is no "good" communication between the players in the game; no binding contracts can be entered into; and players take the actions of the others in the game as given, and then decide their own actions unilaterally. A cooperative game is the opposite of the non-cooperative (see Nash, 1953).

[11] For mathematical and computational convenience, we develop what is termed in game theoretic terminology as "two-person" games. In principle, however, the qualitative results of the paper should be valid also for n-person games (n>1).

tween the cooperative outcomes and the threat point outcomes of the players (see Munro, 1979).

The non-cooperative model. - This model attempts to capture a situation in which devolving the responsibility to manage the resource leads to non-cooperative behaviour. This is likely to happen when there is no communication between the participants; the parties can enter into no binding agreements, and each of them goes about exploiting the resource unilaterally (see Sumaila, 1999, and the references therein). There are a number of good reasons why it is important to develop this model. First, it is well known that, most of the time, interaction between agents exploiting common property resources usually degenerates into something close to the kind of outcomes predicted by non-cooperative models. This is probably one of the causes of the many collapses of fish stocks around the world over time.[12] Second, it is important to keep reminding stakeholders, regulators and the general public of the potential losses from non-cooperative behaviour, with the hope that this will eventually help bring about more cooperative management regimes in the world's shared fisheries, and other natural resources.

Under non-cooperation, it is assumed that the objective of each player is to maximize *own private* benefits, B_p, $p=w,f$, from the use of the resource. The problem of the non-cooperating agents is therefore to choose harvest or effort levels in each fishing period so as to achieve own objective, without due regard to the consequences of their action on other participants.

There are two key differences between the cooperative with an explicit diversity objective and the non-cooperative model. First, the users in the non-cooperative setting do not incorporate diversity concerns into their reaction functions – they care only about the private benefit that accrues to them. Second, users race for the fish, as they unilaterally decide how much to take. It should be noted that the latter is also the main difference between the non-cooperative and the other versions of the cooperative model.

Results and Discussion

Again data on the Namibian hake fishery are applied to run the model. The general results of the study show that in the case of the cooperative with an explicit diversity conservation objective, the optimal ω (for the wetfish fleet) value is 0.2, as this is the ω value that gives the highest diversity index. At $\omega = 0.2$, the diversity index is 98% of the ideal "no harvest" scenario. When it

---

[12] Examples of collapses are Norwegian spring spawning herring; Atlantic cod fisheries off Newfoundland (Walters and Maguire, 1996); and Peruvian anchovies (Idyll, 1973). It should be noted that there are other competing reasons given for these collapses (El Niño, environmental regime shifts, etc.).

comes to the cooperative with side payments scenario, the optimal ω is 1, as this gives the highest possible joint discounted economic rent. On the other hand, the optimal ω for the cooperative without side payments turned out to be 0.6. This gives the Nash solution for a cooperative without side payments management scenario.

A key result of the paper is that the highest diversity index is achieved under the cooperative with an explicit demographic diversity conservation objective scenario. With an index of 98%, this version of the model conserves demographic diversity best. On the other hand, the non-cooperative model delivers the worst demographic diversity conservation, with only 56% of the 'no harvest' demographic diversity preserved. In between these two extremes lie the outcome of 65% in the case of cooperative with side payments, and 78% in the case of the cooperative without side payments.

Turning to the economic results, the study shows that the cooperative with side payments management regime produces the highest joint discounted economic rent, thus achieving what Munro (1979) calls the *optimum optimorum* (N$10.23 billion). The next best economic result is achieved under the cooperative without side payments scenario (N$7.14 billion). This is followed by the non-cooperative outcome (N$5.14 billion), and the cooperative with an explicit demographic diversity objective produces an economic result of N$4.94 billion, respectively.

These economic results are reflected in the harvest levels under the different scenarios. The only exception here is in the case of the non-cooperative scenario, where even though the harvest level is slightly more than that in the cooperative without side payments scenario, and about 50% higher than in the case of the explicit diversity objective scenario, the economic rent is much lower relative to that in the case of the former, and nearly the same in the case of the latter.

A number of interpretations, observations and deductions can be made from the results presented above. First, it is clear that the opportunity cost of achieving "near perfect" level of demograghic diversity conservation is high. This turned out to be over 50% of the potential discounted economic rent in our model. Second, a cooperative without side payments management scenario appears to do reasonably well on both economic efficiency and demographic diversity conservation criteria. This scenario produces an economic result, which is 70% of the potential discounted economic rent, and about 78% of the 'no harvest' diversity level. Hence, it is possible to attain reasonably high diversity level, and reasonably high economic rent at the same time. Third, non-cooperative behaviour is (as has been shown again and again by previous studies) bad on all counts – it delivers only 56% of the potential diversity level, and just over 60% of the potential economic rent. Fourth, the bargaining powers of the players, w and f, vary with the type of management scenario. For instance, in the scenario with an explicit demographic diversity objective, the freezer trawlers do most of the harvesting and therefore make the most economic rent. On the other hand, the wetfish trawlers do most of the harvesting in the cooperative without side payments scenario.

THE SARDINE BIOECONOMIC MODELS

Sardine is a pelagic schooling species mainly caught by purse seiners. It is the most important of the pelagic species in Namibia. It is a short-lived species with a life span of four to six years. Juvenile sardines measure up to 12 cm, immature from 12 cm to about 20 cm, whereas adult sardines are 20-26 cm. About 50 per cent maturity occurs at 21 cm.

There are two distinct stocks along the west coast of southern Africa, the South-African sardine and the Namibian sardine. The South-African and the Namibian sardine are biologically separated by the ecological barrier in the Benguela system not far from Lüderitz. Namibian sardine is shared with Angola, and this causes problems, as the fisheries management policy of Angola is different from the Namibian.

Commercial exploitation

The Namibian sardine was an unexploited stock until 1947 when the Walvis Bay Canning Company started an experimental fishery, first for fishmeal and fish oil, and already in 1949 for canning. The catch in 1948 was a modest 1000 tonnes. Only five years later, in 1953, six companies altogether were in operation, and the total catch had increased to 262 000 tonnes. After that the catch increased steadily, peaked at 1.4 million tonnes in 1968 and dropped steadily to a modest 100 000 tonnes a decade later. Unofficial sources have it, however, that the catch in 1968 was more like 2 million than 1.4 million. The catch dropped to 1.1 million in 1969, 565 000 in 1970 and 328 000 in 1971. After 1977 the stock more or less collapsed. In the eighties and nineties the stock size has varied between virtually zero and some 100 000 tonnes. This illustrates the need for bioeconomic models that can serve as basis for sound economic and biological management of the stock.

Models

Although there exist numerous biological studies of sardine, and quite a few of Namibian sardine, bioeconomic studies of sardine are rare (Johnston and Sutinen, 1996; de Anda-Montanez and Seijo, 1999); not to mention Namibian sardine. The only bioeconomic studies of Namibian sardine, to our knowledge, are the studies of Namibian sardine by Sumaila and Vasconcellos (2000) and Sandal and Steinshamn (1999, 2001a). Boyer and Hampton

(2001) give a nice overview over marine resources in Namibia and the socio-economic value of the Namibian fishing industry, and Cochrane et al. (1998) study South-African sardine. The latter draw some conclusions that may be relevant for Namibia as well.

Cochrane et al. (1998) discuss the lessons learned in the application of management procedures and their precursors in the pelagic fishery in South Africa (anchovy and sardine). The high variability in abundance of the two stocks, the trend in their relative abundance, the substantial uncertainties in information, strong pressure to meet socio-economic goals and the conflicting objectives that arose between the directed anchovy and directed sardine fishery are identified as major problems in the implementation of procedures and management of the resources. However, the use of management procedures is considered to have led to greatly improved communication with the industry and to substantial input by them into the management process in South Africa.

Sandal and Steinshamn (1999, 2001a) present applications of a single-species deterministic bio-economic model. An outline of the modelling technique can be found in Sandal and Steinshamn (2001b). One of the main aspects of this modelling technique is that it is a feedback model. This means that at any point in time, the optimal control (harvest) is a function of the state of the system (the stock size). It does not matter how the present situation has come about, it is the situation itself that determines the action. This is important as no decisions are made in advance; they are all made based on the most recent information about the system. The reason why it is possible to find the feedback solution of the dynamic optimization problem is that the model is non-linear. Non-linearities may enter through downward sloping demand or increasing marginal costs. Otherwise (with fixed prices or costs) the optimal solution is a very simple feedback, namely the so-called bang-bang solution defined as zero harvest when the stock is below the optimal steady state and maximum harvest when the stock is above the optimal steady state. This is obviously not a realistic alternative in practice.

The biological submodel is an aggregated model in the sense that it is not a year-class model. There are two types of aggregated models, namely the surplus production type (Schaefer model) or the recruitment type (Ricker model). In the applications, the surplus production model was chosen as this turned out to fit the data best, that is, it gave better statistical results.

The economic submodel is a net revenue function defined as gross revenue minus costs. The gross revenue is defined as price multiplied by harvest. The price determination (demand function) for sardine is particularly interesting as one part of the harvest goes to canning (human consumption) and another part goes to fishmeal. The size of the part going to canning depends on the total size of the harvest. For small harvests almost everything goes to canning. When the harvest exceeds a certain level, some of it goes to fishmeal. For an even higher level of the harvest, everything exceeding this level goes to fishmeal. To find this threshold level, however, may be problematic as hardly anything at all has gone to fishmeal over the last two decades.

The characteristic of the demand function mentioned above results in an optimal harvest function that is slightly stepwise, as shown in Figure 1.

The cost function is the most difficult one to estimate due to lack of appropriate data. The good thing about the modelling approach, however, is that it is possible to try a wide variety of non-linear cost functions.

Optimal harvest from a deterministic version of the model, as described by Sandal and Steinshamn (2001a), is illustrated in Figure 1. A stochastic version of the same model is described in the Steinshamn-Lund-Sandal chapter of this book.

Results and discussion

The main conclusion to be drawn from this section is that, according to the model, the sardine stock has been well below the harvest moratorium level almost continuously for two and a half decades even when we use discount rates up to 20 per cent. In practice this means that a harvest moratorium on sardine ought to have been instituted already in 1977.

At any point in time the model had been implemented, the stock would

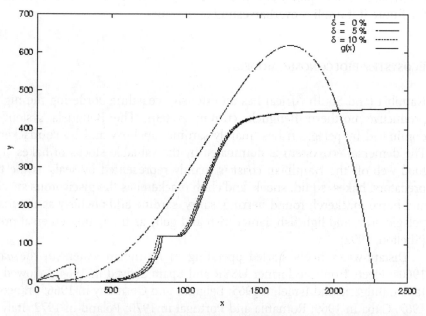

Figure 1. Optimal harvest paths as functions of the stock size with 0%, 10% and 20% discounting, and the estimated surplus production function. The uppermost harvest curve represents the highest discounting. Units are 1000 tonnes.

start growing towards the optimal steady state level of just over two million tonnes, which could accommodate a permanent annual harvest just over 400 000 tonnes. This is, however, in a deterministic setting. We all know that reality is much more volatile, and there is no guarantee that stocks grow according to models. At best, this is the case on average in the long run.

The good thing about the feedback model is that sporadic uncertainty is automatically taken care of. Therefore there is good reason to believe that after this has been applied for a while the stock will at least vary around two million tonnes and the harvest around 400 000 tonnes. If, on the other hand, there is systematic uncertainty, in the sense that the stochastic process depends on the stock level, a completely different approach is needed and stochastic dynamic programming must be used. This is the topic of another chapter in this book (Steinshamn et al., this volume).

The bottom line therefore is that even though a simple deterministic bioeconomic model would probably yield much better results both economically and biologically, than the actual harvest pattern has done, the best choice would be to apply a proper stochastic bioeconomic model. Stock management according to such a model would yield both higher economic returns and more employment in the long run than the actual management.

An argument that is sometimes brought forward is that a higher harvest is needed in order to avoid labour unemployment in the short run. Unfortunately, this kind of policy always has to be paid for by lower employment in the future if it implies overharvesting of the stock.

ECOSYSTEM BIOECONOMIC MODELS

Namibia (and South Africa) has an extensive coastline bordering the highly productive northern Benguela current system. The Benguela system is dominated by pelagic fishes, mainly sardine, anchovy and horse mackerel. The demersal ecosystem is dominated by the valuable stocks of hakes. The food web off the Namibian coast is mainly represented by seals as the top predators, hakes, squid, snoek, and chub mackerel as the piscivorous species and horse mackerel, round herring, saury, sardine and anchovy as the main pelagic prey, and lightfish, lanternfish and goby as the main demersal preys (Shelton, 1992).

Distant-water fleets started operating in Namibian waters in the early 1960s. Fleets from the former USSR and Spain arrived in 1964; followed by Japan, Bulgaria and Israel in 1965; Belgium and Germany in 1966; France in 1967; Cuba in 1969; Romania and Portugal in 1970; Poland in 1972; Italy in 1974; Iraq in 1979; Taiwan in 1981; and South Korea in 1982 (FAO Yearbooks of Fishery Statistics for hake). With the announcement of the EEZ regime by the independent government in 1990, there was almost an instant drop to zero of the number of unlicensed foreign vessels fishing in the area (Sumaila and Vasconcellos, 2000).

Sumaila and Vasconcellos (2000) analyse the impact of the activities of distant water fleets on Namibia's marine ecosystem through the use of simulation modelling. Dynamic simulations were performed to capture the ecological impacts of the activities of DWFs in Namibian Exclusive Economic Zone (EEZ) during the pre-independence era. These impacts are then valued to give an indication of their economic effects. Scenarios of the Namibian ecosystem are developed using information on the catches of hake, horse mackerel and sardine during this period. Based on these scenarios, the following three questions were raised and addressed. Given the harvest of hake, horse mackerel and sardine in the years prior to independence, what are the impacts of DWFs on the biomass of the major species in the ecosystem? How do DWF activities impact the biomass and potential catches of the major commercial stocks off Namibia? How do the impacts on the catch levels affect economic values?

Methods

Simulations were based on an Ecopath model of the Benguela ecosystem off Namibia developed by Jarre-Teichmann and Christensen (1998) for the period 1971 - 1977. Ecopath is a mass-balance static model that describes the trophic relationships in an ecosystem in steady-state conditions (Christensen and Pauly, 1992) while Ecosim is the dynamic version of Ecopath (Walters *et al.*, 1997).

To carry out their economic analysis, Sumaila and Vasconcellos (2000) took the catches and fishing efforts generated by Ecosim under different DWF fishing scenarios, and applied appropriately determined unit prices for the fish landed; the cost of exploiting the fish; and the discount rate. This allowed the authors to compute the net discounted economic rent that is achievable under the 'with' and 'without' distant water fishing fleets scenarios, which in turn allowed them to determine the economic impacts of DWFs under these scenarios.

Results

The main results of the analysis are (i) for hake, the average standing biomass in the "with" scenario is only 51% of that in the "without" scenario. The equivalent numbers for sardine and horse mackerel are 68% and 60%, respectively. For the ecosystem as a whole, the effect of DWF activities is to reduce the potential standing biomass by about 16%; (ii) substantially higher catches of hake, sardine and horse mackerel are taken in the "with" than in the "without" DWF scenarios; and (iii) in the "with" scenario, the DWFs make on average Namibian dollar (N$) 1 011 million annually while the domestic fleet made only about N$112 million.

Distance water fleet activities have been cited as one of the reasons for the buildup of excessive effort and increasing stock exploitation in the years leading up to the extended fishery jurisdiction. Sumaila and Vasconcellos (2000) showed that exploitation of Namibian fisheries in the pre-independence era by the DWF is a good example of the exertion of excess effort and the consequent stock over-exploitation and economic waste that follows.

POTENTIALLY INTERESTING RESEARCH REMAINING

As can be seen from the short list of bioeconomic studies of Namibian fisheries, a lot remains to be done in this area. Even with respect to single-species-deterministic models, very few studies have been performed. The analysis in Sumaila (2000) can be fruitfully extended to include new biological knowledge about hake published since the publication of this paper. Another useful extension of this paper would be to look at the secondary economic benefits of the hake fishery.

The bioeconomic model for sardine (Steinshamn et al., this volume) is a stochastic feedback model, but a lot still remains to be done with respect to the quantification of the stochastic processes. A useful method for quantifying stochastic processes, which is also relevant for Namibian fisheries, is the one described by McDonald and Sandal (1999). It is important to note here that this study is not only about passive uncertainty but also active uncertainty, in the sense that the stochastic process itself may be a function of the stock. The focus is mainly on uncertainty and stochasticity in the biology, but uncertainty may also be present in the economic part of the model. Bioeconomic models with both types of uncertainties deserve more attention.

Generally, even purely biological multi-species models are rare. To our knowledge, there are no bioeconomic multi-species models of Namibian fisheries. A multi-species bioeconomic model may be a model that only contains the multi-species aspect in the biological submodel, and this is the most common approach. However, there may not only be multi-species interactions on the biological side but also on the economic side, for example, if the supply of one species to the market affects the price of other species as well. In this case, a different kind of modelling that takes this into account is needed.

Another topic that is important for Namibia is the economics and management of shared fish stocks with South Africa and Namibia. This calls for a game-theoretic approach to the modelling of the Benguela upwelling system. Only very limited modelling has been carried out so far in this area (see Armstrong and Sumaila, this volume).

Finally, there are indications that the price of hake in Spain is not a constant that is determined by a competitive market. Namibia is one of the main suppliers of hake to the European market. The country could therefore be in a position to exert some market power. Studies that look at the market and price dynamics of hake will be very useful in helping us get away from the current assumption of constant prices applied in studying the bioeconomics of hake. Typically this will imply a more conservative harvesting pattern.

References

Anon. (1994): *Namibia Brief. Focus on Fisheries and Research*. Namibia.

Armstrong, C.W. and Sumaila, U.R. (2004): The Namibian-South African hake fishery – costs of non-cooperative management. In: *Namibia's fisheries: ecological, economic and social aspects* (U.R. Sumaila, D. Boyer, M. Skogen, and S.I. Steinshamn, eds.), pp. 231-244. Eburon, Delft.

Armstrong, C.W., Sumaila, U.R., Erastus, A. and Msiska, O. (2004): Benefits and costs of the Nambianization policy. In: *Namibia's fisheries: ecological, economic and social aspects* (U.R. Sumaila, D. Boyer, M. Skogen, and S.I. Steinshamn, eds.), pp. 203-214. Eburon, Netherlands.

Boyer, D.C., and Hampton, I. (2001): An overview of the living marine resources of Namibia. *South African Journal of Marine Science* 23: 5 – 35.

Christensen, V. and Pauly, D. (1992): Ecopath II-a system for balancing steady-state ecosystem models and calculating network characteristics. *Ecological Modeling* 61: 169-185.

Christy, F.T. Jr and Scott, A..D. (1965): *The Common Wealth in Ocean Fisheries*, Johns Hopkins University Press, Baltimore, Md.

Clark, C.W. and Kirkwood, G.P. (1979): Bioeconomic model of the Gulf of Carpentaria prawn fishery. *Journal of the Fisheries Research Board of Canada* 36: 1304-1312.

Cochrane, K.L., Butterworth, D.S., de Oliveira, J.A.A., and Roel, B.A. (1998): Management procedures in a fishery based on highly variable stocks and with conflicting objectives: experiences in the South African pelagic fishery. *Reviews in Fish Biology and Fisheries* 8: 177 – 214.

de Anda-Montanez, A., and Seijo, J.C. (1999): Bioeconomics of the Pacific sardine fishery in the Gulf of California Mexico. *California Cooperative Oceanic Fisheries Investigation Report*, 40: 170 – 178.

Fourer. R., Gay, D.M. and Kernighan, B.W. (1993): *AMPL: A Modelling Language for Mathematical Programming*, Scientific Press, South San Francisco, CA.

Gordon, H.S. (1954): Economic theory of a common property-resource: the fishery. *Journal of Political Economy* 62: 124-142.

Hannesson, R. (1993): *Bioeconomic analysis of fisheries*, Fishing News Books, London.

Hannesson, R. (1983): Optimal harvesting of ecologically interdependent fish species. *Journal of Environmental*

Economics and Management 10: 329-345.

Idyll, C.P. (1973): The anchovy crisis. *Scientific American* 228: 23-29.

Jarre-Teichman, A. and Christensen, V. (1998): Comparative modelling of trophic flows in four large upwelling ecosystems: Global versus local effects. In: *Global versus local changes in upwelling systems* (M.H. Durand, P. Cury, R. Mendelssohn, C. Roy, A. Bakun and D. Pauly eds.), pp. 425-443. ORSTOM Editions, Paris.

Johnston, R.J., and Sutinen, J.G. (1996): Uncertain biomass shift and collapse: Implications for harvest policy in the fishery. *Land Economics* 72: 500 – 518.

Kirchner, C.H., Sakko, A.L.,and Barnes, J.I. (2000): The Economic value of the Namibian recreational rock-and-surf fishery. *South African Journal of Marine Science* 22: 17-26.

Lange, G. and Motinga, D.J. (1997): The contribution of resource rents from minerals and fisheries to sustainable economic development in Namibia, 1980 to 1995. Research Discussion Paper #19, Directorate of Environmental Affairs, Ministry of Environment and Tourism: Windhoek, Namibia.

Lange, G. (2000): Fisheries Accounting in Namibia, Institute for Economic Analysis. New York University. Presented at International Workshop on Environmental and Economic Accounting, 18-22 September 2000, Manila, Philippines.

Magurran, A.E. (1988): *Ecological Diversity and Its Measurement*. Croom-Helm, London.

Manning, P.R. (1998): Managing Namibia's fisheries: optimal resource use and national development objectives, Ph.D. thesis, London, LSE.

Manning, P. and G-M.. Lange. (1998): The contribution of resource rent from Namibian fisheries to economic development: an evaluation of policies favoring Namibian ownership of fishing companies. Paper presented at the Fifth Biennial Conference of the International Society for Ecological Economics. Santiago, Chile, 15-19 November, 1998.

McDonald, A.D. and Sandal, L.K. (1999): Estimating the parameters of stochastic differential equations using a criterion function based on the Kolmogorov-Smirnov statistic. *Journal of Statistical Computation and Simulation* 64: 235 – 250.

Ministry of Fisheries and Marine Resources (1997): Report of Activities and State of the Fisheries Sector, Windhoek.

Munro, G. R. (1979): The optimal management of transboundary renewable resources Canadian *Journal of Economics* 12(8): 355-376.

Nash, J. (1953): Two-Person Cooperative Games. *Econometrica* 21: 128-140.

Oelofsen, B.W. (1999): Fisheries management: the Namibian approach. *ICES Journal of Marine Science* 56: 999-1004.

Sandal, L.K., and Steinshamn, S.I. (1999): Adaptive management: the case of the Namibian pilchard fishery, Report no. 71/99 (Institute for Research in Economics and Business Administration, Bergen, Norway).

Sandal, L.K., and Steinshamn, S.I. (2001a): A bioeconomic model for Namibian pilchard, *South African Journal of Economic* 69: 299 – 318.

Sandal, L.K., and Steinshamn, S.I. (2001b): A simplified feedback approach to optimal resource management, *Natural Resource Modelling* 14: 419 – 432.

Shelton, P. A. (1992): Detecting and incorporating multispecies effects into fisheries management in the North-West and South-East Atlantic. In:

Benguela trophic functioning (A. I. L. Payne, K. H. Brink, K. H. Mann, and R. Hilborn eds.). *South African Journal of Marine Science* 12: 723-737.

Steinshamn, S.I. (1992): Economic Evaluation of Alternative Harvesting Strategies for Fish Stocks (Ph.D. thesis, Norwegian School of Economics and Business Administration, Bergen, Norway).

Steinshamn, S.I., Lund, A-C and Sandal, L.K. (2004): A stochastic feedback model for optimal management of Namibian sardines. In: *Namibia's fisheries: ecological, economic and social aspects* (U.R. Sumaila, D. Boyer, M. Skogen and S.I. Steinshamn, eds.), pp. 245-266. Eburon, Delft.

Sumaila, U.R. (1995): Irreversible capital investment in a two-stage bimatrix fishery game model. *Marine Resource Economics* 10: 263 - 283.

Sumaila, U.R. (1999): A review of game theoretic models of fishing. *Marine Policy* 23(1): 1-10.

Sumaila, U.R. (2000): Fish as vehicle for economic development in Namibia. *Forum for Development* 2: 295-315.

Sumaila, U.R. and Vasconcellos, M. (2000): Simulation of ecological and economic impacts of distant water fleets on Namibian fisheries. *Ecological Economics* 32: 457 – 464.

Sumaila, U.R. (2001): Biodiversity in a game theoretic model of the fishery. In: *Proceedings of the 10th International Conference of the International Institute of Fisheries Economics & Trade* (R.S. Johnston and A. L. Shriver eds.), Corvallis, Oregon, USA (CD-ROM): http://www.oregonstate.edu/dept/IIFET/2000/papers/sumaila2.pdf.

Sumaila, U.R. (2002): Recreational and commercial fishers in the Namibian silver cob fishery. In: *Recreational fisheries: ecological, economic and social evaluation* (T.J. Pitcher and C.E. Hollingworth, eds.), pp. 51-62. Blackwell Science, Oxford, UK.

Walters C.J. and Maguire, J.J. (1996): Lessons for stock assessment from the northern cod collapse. *Reviews in Fish Biology and Fisheries* 6: 125-137.

Walters, C., Christensen, V. and Pauly, D. (1997): Structuring dynamic models from trophic mass-balance assessment. *Reviews in Fish Biology and Fisheries* 7: 139 - 172.

Zeybrandt F. and Barnes, J. (2001): Economic characteristics of demand in Namibian marine recreational fisheries. *South African Journal of Marine Science* 23: 145-156.

9 ECONOMIC VALUE OF FISH STOCKS AND THE NATIONAL WEALTH OF NAMIBIA

*Glenn-Marie Lange**

Abstract

Since the establishment of the EEZ at independence, Namibia's fish stocks have contributed an important share of national income and exports, and constitute an important component of national wealth. Proper management of fish assets is very important for Namibia's economy. Government must balance the pressure from industry for higher TAC, and with it the danger of further collapse of the fish stocks, against the goal of rebuilding fish stocks by restraining fishing activity to prudent levels. An economic assessment of the value of the fish stock, the economic loss incurred through over-exploitation and depletion of the stock, and the potential value of the stock under different management regimes is an essential tool for government. This assessment can be provided, in part, by natural resource accounts for fisheries, which estimate the value of natural capital, such as fisheries, and account for the use of resources in economic activities. This chapter presents the fisheries accounts for Namibia and the contribution of fisheries to national wealth. The chapter provides a brief introduction to environmental accounting, describes the methodology and the data sources used to construct accounts for Namibia's major commercial fish stocks: hake (*Merluccius capensis* and *Merluccius paradoxus*), horse mackerel (*Trachurus capensis*) and sardine (*Sardinops sagax*), presents the physical and monetary accounts for each fishery and discusses the contribution of fisheries to national wealth.

INTRODUCTION

National income and economic well-being depend on a country's wealth—produced assets, natural capital, human and social capital. Economic

---

* This work was made possible with support from the United States Agency for International Development, the Swedish International Development Agency, and several workshops sponsored by the Beijer Institute, Stockholm.

sustainability[1] has been shown to require that per capita national wealth is non-decreasing over time (e.g. Dasgupta and Maler, 2000; Hartwick, 1977; Pearce and Atkinson, 1993; Solow, 1986). Although natural capital is a large component of wealth, it has not yet been systematically included in the national economic accounts of most countries.

With regard to fisheries, for example, the system of national accounts (SNA) has treated aquaculture and capture fisheries quite differently[2]. With aquaculture, the SNA records both production and changes in the fish stock so that the consequences of depletion or increases in stocks are accounted for. With capture fisheries, however, the SNA records only the income from fishing, but not changes in stocks. Consequently, the economic impact of the historical devastation of Namibia's fish stocks, described elsewhere in this volume, appeared in the national accounts of that time to be an economic success because only the economic value of the fishing activity was recorded, not the corresponding depletion of the fish stocks on which that activity was based. This omission in the SNA is being corrected by the gradual implementation of environmental accounts, which treat both produced and non-produced assets in the same way (United Nations, 1993b, 2003).

Since the establishment of the EEZ at independence, Namibia's fish stocks have contributed an important share of national income and exports, and constitute a significant component of national wealth. Proper management of fishery assets is essential for Namibia's economy. An economic assessment of the value of fish stocks, the economic loss incurred through depletion of stocks, and the potential value of stocks under different management regimes is an essential tool for government. This assessment can be provided, in part, by environmental accounts for fisheries.

Namibia has joined a regional initiative in southern Africa to construct fisheries accounts, as well as accounts for other major resources. (See Lange et al., 2003 for a discussion of this initiative and its results.) The monetary fisheries accounts provide a useful assessment of the impact of the changing management strategy and also an indicator of what Namibia stands to lose if the fish stocks are depleted. The fisheries accounts also help to provide a more complete picture of Namibia's national wealth, which is necessary for sustainable macroeconomic development. Section two of this chapter begins with a brief overview of environmental accounts then describes the method-

---

[1] Strictly speaking, sustainability only requires non-declining wealth; sustainability that is equitable toward future generations requires non-declining per capita wealth.

[2] This differential treatment has a conceptual basis in the SNA definition of assets (United Nations, 1993a), but also a practical basis: the knowledge about capture fisheries is limited and very uncertain, and capture fisheries are not economically significant in many countries.

ology and data used to construct accounts for Namibia's major commercial fisheries: hake, sardine and horse mackerel. Section three presents the physical and monetary fisheries accounts for Namibia. This is followed in section four by a discussion of Namibia's national wealth.

METHODOLOGY AND DATA

Overview of Environmental Accounts

Environmental accounts have evolved since the 1970s, initially through the efforts of individual countries and practitioners, and later through the concerted effort of major international organizations: United Nations Statistics Division, Eurostat, OECD, World Bank, and national statistical offices. The United Nations published an interim handbook on environmental accounting in 1993 called the *Handbook of Integrated Economic and Environmental Accounting*; a revised and expanded version of this handbook is now available (UN, 1993b, 2003).

Environmental accounts consist of stocks and flows of environmental goods and services. They provide a set of aggregate indicators to monitor environmental-economic performance at the sectoral and macroeconomic level, as well as a detailed set of statistics to guide resource managers toward policy decisions that will improve environmental-economic performance in the future. The accounts have four major components:

- asset accounts record stocks and changes in stocks of natural resources such as fisheries, forests and minerals;
- flow or production accounts provide information at the industry level about use of materials and energy and emissions of pollution, in effect an extended input-output table for the environment;
- environmental protection and resource management expenditure accounts identify government expenditures for resource management as well as the taxes and other levies on resource use;
- environmentally adjusted macroeconomic aggregates include indicators of sustainability such as environmentally adjusted Gross Domestic Product (GDP), Net Domestic Product (NDP), national saving and national wealth.

Fisheries accounts are based on the UN's System of Integrated Environmental and Economic Accounts (SEEA) and a more specialised manual for fisheries accounts (FAO and UN, in press). A number of countries have constructed accounts for fish or are planning to do so in the near future (Table 1). Relative to other resources, fisheries accounts are not constructed by many countries. In part, that may be because fisheries are not economically significant in many countries. In addition, the compilation of fisheries accounts presents a combination of challenges greater than many other resources. For example, fish, like minerals, cannot be directly observed the way forest resources can, but in addition, multi-species fisheries are affected by complex predator-prey interactions, inter-annual variations can be quite large, and fish may migrate out of a country's exclusive economic zone (EEZ).

Structure of physical and monetary accounts for Namibia[3]

Namibia's accounts have been constructed for the three main commercial fisheries: hake, sardine, and horse mackerel. Namibia also has other commercial fisheries, artisanal freshwater fisheries that form an important component of rural livelihoods in northern Namibia, and a recreational fishing industry that is important for tourism. There are not yet sufficient data to include these resources in the accounts.

Table 1. Countries compiling accounts for fisheries.

	Physical	Monetary
Regular compilation by statistical offices		
Norway	X	Pilot studies
Iceland	X	Pilot studies
Namibia	X	X
New Zealand	X	Planned for future
Canada*	X	X
United Kingdom*	X	X
Occasional studies		
Australia	X	
Philippines		X
Korea		X

* Countries planning to introduce fisheries accounts.

The list includes only countries for which accounts were constructed within government offices and does not include one-time academic or other studies.

---

[3] For more detailed discussion of methodology and data sources, see Lange et al. (2003) and Lange (2003).

Physical stock accounts for fish are constructed for opening stocks, changes that occur during the accounting period (usually one year), and the closing stock. Changes that occur during the year consist of catch, recruitment, natural mortality, and other volume changes. Other volume changes can include factors such as the migration of a fish stock out of the country's EEZ due to environmental events. In practice, there is not enough information to quantify recruitment, natural mortality and other volume changes, so the changes in the accounts collapse into two categories: catch and other volume changes.

Monetary accounts are constructed by estimating the value of the physical accounts. The value of fish stocks, like any other asset, is the net present value of the stream of resource rent it is expected to generate in the future. Constructing monetary accounts has two components: 1) defining how rent is to be calculated and 2) making projections about the future rent a fishery is likely to generate. Both these components raise unique challenges for fisheries.

Measuring Resource Rent

Resource rent is the value of the resource *in situ*, defined as the value of production minus the marginal exploitation costs. Fisheries managed under an individually transferable quota (ITQ) system such as Iceland or New Zealand may develop a market for quotas that, under the right circumstances, accurately reflect the rent. When markets are lacking, as in Namibia, rent is measured with the residual approach in which resource rent, RR, is calculated in each year as total revenue, TR, minus intermediate consumption, IC, compensation of employees, CE, consumption of fixed capital, CFC, and normal profit, NP. Normal profit is calculated as the product of Fixed capital stock, K, used in fishing and the rate of return on capital stock, i:

$$RR_{j,t} = TR_{j,t} - (IC_{j,t} + CE_{j,t} + CFC_{j,t} + NP_{j,t}) \qquad (1)$$

$$NP_{j,t} = i \times K_{j,t} \qquad (2)$$

for each fishery, j, where j = 1,2,3 for hake, sardines, horse mackerel.

Note that in actual implementation of the residual approach, average cost is used rather than marginal cost because data about marginal cost are not generally available. This practice introduces an upward bias into the measure of rent when average cost is lower than marginal cost, which is normally assumed to be the case (e.g. see Vincent, 1997 for an estimate of marginal and average costs). The relationship between marginal and average cost is an empirical issue that must be determined for each fishery.

In Namibia, data from the national accounts have been used to calculate rent because they are the only source that covers all species and all years since independence in 1990[4]. Revenues are obtained for each species from highly detailed information about catch and the value of catch gathered by the Ministry of Fisheries and Marine Resources (MFMR); the data are reasonably good. Cost data, on the other hand, are based on average production costs for each species estimated with a model developed by the MFMR (pers.comm., N. Kali, CBS, 1998). Estimated costs may differ from actual production costs in any given year. Quotas are awarded to each company for a single species only, and high by-catch fees discourage excessive by-catch (see other chapters in this book for further discussion of Namibia's fisheries policy).

The only figure required by equations 1 and 2 that is not provided by the national accounts is the rate of return, or opportunity cost, on fixed capital. There is little long-term borrowing in the fishing industry that might indicate an appropriate cost of capital for that sector. The Ministry of Fisheries recommended a 20 to 30 per cent return on fixed capital because unpredictable environmental disturbances make fishing a very high-risk activity. In this report, a 20% return is used, which is considerably higher than the return used in other countries.

In calculating resource rent, fishing and fish processing industries were combined. This was done for two reasons. First, much of the fish is processed offshore on factory trawlers whose continuous-process operation makes the separation of fishing from fish processing somewhat arbitrary. Second, there is a high degree of vertical integration in the industry, which makes the separation very difficult. The combination of a primary industry and its immediately downstream processing industry is common practice in accounting for forestry (Eurostat, 2000).

Projecting Future Resource Rent

The value of each fish stock is the net present value of the rent it will generate in the future. The present value calculations require projections of future prices, technology, costs of production, fish stock levels, and resource exploitation paths. Future stock levels depend partly on fisheries policies and partly

---

[4] There is a survey of fishing companies, which would provide a better source of data, but reliable data from the survey are not yet available for a sufficient number of years. There is no ITQ system which might provide a reasonable measure of resource rent. Namibia has a system of individual *non*-tradable quotas. Trading of fishing quotas is discouraged and failure to utilise one's assigned fishing quota can result in loss of rights of exploitation. Some limited trading occurs, but most is unofficial and unrecorded. For these reasons, the quota trading price is not expected to reflect the resource rent very well, nor is it easy to obtain reliable information about the trading price.

on environmental conditions and their impact on fish stocks, which are difficult to forecast. The economics of fishing also depends on fisheries policy: a more efficient fishery generates higher rent and is of greater economic value.

In some relatively well-understood fisheries, a bioeconomic model can be used to assess the likely future stocks, costs of fishing, and rent under different management regimes. Such a model was used, for example, to assess the value of Iceland's fisheries resources (Danielsson, 2000). In addition to uncertainty about future prices, it is exceedingly difficult to determine whether Namibia's fish stocks will, in the long term, remain at current levels, increase to previous higher levels, or collapse further. Each of these possibilities has different implications for future rent and the present value of the asset. Assuming prices and costs are constant, if fish stocks remain constant, then rent and asset value will remain constant. If there is a recovery from depletion and fish stocks increase, the rent will increase over time and the present value of the asset is much higher, than under the constant-stocks assumption. If, on the other hand, fish stocks decline, then the asset value will be much lower.

Despite the MFMR's goal of restoring fish stocks to previous higher levels, there is little evidence that this objective will be achieved in the near future. For the calculation of monetary fisheries accounts it has been assumed that the stocks have stabilized at current levels and will generate the same rent in the future. While the fluctuation of fish stocks and rent over the past ten years shows that this is an unrealistic assumption on a year-to-year basis, this assumption is used for lack of any other information at this time. Under this assumption, the net present value of the stock, V, is simply the rent, RR, divided by the social discount rate, r, for each fish stock, j:

$$V_{j,t} = \frac{RR_{j,t}}{r} \qquad (3)$$

A social discount rate of 10 per cent is used[5]. While this rate may appear rather high, it is used for comparability with other public sector policy analysis. This rate is commonly used by a number of governments in southern Africa for project evaluation.

For Namibia's mineral accounts, which extend back to 1980, the value is estimated using a 5-year moving average of rent. Because the time series for

---

[5] The social discount rate is typically lower than the private discount rate or the private rate of return on capital used in calculating resource rent for a number of reasons (explained by Hanley and Spash, 1993, pp. 127-151).

Namibia's fisheries is still so short, the moving average has not yet been introduced. It will be introduced when 15 years' data are available.

As with many economic variables, in order to assess trends over time, current values must be converted to constant value measures. Two different approaches have been proposed for constructing a constant price measure of natural capital: one is production oriented and the other is income oriented. For its minerals and natural forest accounts, the Australian Bureau of Statistics takes the production approach and treats the annual unit rent as the price of the asset *in situ*. Constant price asset accounts are then obtained by applying the prices for the benchmark year to physical accounts throughout the times series (Johnson, ABS, pers. comm.). An alternative, income-based approach, under consideration by Statistics Canada (Gravel, pers. comm.), deflates current-price unit rent using the GDP deflator to represent the changing purchasing power of rent over time, similar to deflating financial assets or wages. Informal discussions with other economists and national accountants indicate more support to the income-oriented approach, so that is the method applied here. See Lange (2002) for further discussion of this issue and its implications for estimates of constant value natural capital.

PHYSICAL AND MONETARY ACCOUNTS FOR FISH

The physical accounts for the 11-year period, 1990 to 2000, show that net change has been positive only for hake, which ended the decade 30% higher than in 1990 (Table 2). If the lower bound of the confidence interval was used for both years, the stock growth was much lower. Indeed, the confidence interval was 20% of the stock size or greater in all years. Sardine was less than half its volume at the beginning of the decade and horse mackerel, which improved during the late 1990s, fell below the 1990 level in 2000. (Confidence intervals were not estimated for these fisheries.) The tremendous amount of inter-annual variation in stock indicates how difficult it is to manage Namibia's fisheries.

Sardine generated the most rent at the beginning of the decade, but was eventually surpassed by hake (Table 3). This is not surprising since Namibia had an established sardine fishery prior to independence but only achieved control over the other fisheries over the past decade. The rent per tonne for hake has been steadily rising, reflecting both improvements in the industry and also the devaluation of the Namibian dollar over time, which has a major impact on earnings because most Namibian hake is sold to the lucrative European market. Sardine has shown the greatest volatility of rent over the decade. Rent fell to near zero in 1996 when almost no sardine (only two thousand tons) was caught and the industry suffered considerable losses.

Table 2. Physical accounts for hake, sardines, and horse mackerel in Namibia, 1990-2000 (thousands of tons).

		1	2	3	4	5
		Opening Stock	Catch	Other volume changes	Net annual change (col. 3- col. 2)	Closing stock (col. 1+ col. 4)
Hake	1990	906	55	100	45	951
	1991	951	56	176	120	1072
	1992	1072	87	127	40	1112
	1993	1112	108	90	-18	1094
	1994	1094	112	108	-4	1090
	1995	1090	130	158	28	1118
	1996	1118	129	170	41	1159
	1997	1159	110	145	35	1194
	1998	1194	141	136	-5	1188
	1999	1188	161	159	-2	1186
	2000	1186	160	143	-17	1170
Sardine	1990	500	89	249	160	660
	1991	660	68	49	-19	641
	1992	641	82	-128	-210	431
	1993	431	116	-100	-216	215
	1994	215	115	25	-90	125
	1995	125	95	-25	-120	5
	1996	5	2	147	145	150
	1997	150	32	182	150	300
	1998	300	65	40	-25	275
	1999	275	42	-8	-50	225
	2000	225	27	-108	-135	90
Horse Mackerel	1990	1450	409	309	-100	1350
	1991	1350	434	1184	750	2100
	1992	2100	426	126	-300	1800
	1993	1800	479	179	-300	1500
	1994	1500	360	260	-100	1400
	1995	1400	314	114	-200	1200
	1996	1200	319	119	-200	1000
	1997	1000	306	1106	800	1800
	1998	1800	258	258	0	1800
	1999	1800	288	238	-50	1750
	2000	1750	320	-180	-500	1250

Source: based on National Marine Research and Information Centre (2001).

Table 3. Resource rent for sardine, hake, and horse mackerel, 1990-1998 (millions of Namibian dollars)[1].

	Sardine	Hake	Horse mackerel	Total rent
1990	117	27	9	153
1991	65	30	30	125
1992	135	36	20	192
1993	201	68	37	306
1994	229	159	40	429
1995	201	209	39	449
1996	*	192	51	243
1997	95	261	49	406
1998	150	640	91	881

* Less than 1.0
[1] Assumes a 20% rate of return to fixed capital.
Source: based on Lange (2003).

Horse mackerel, though harvested in higher volumes than either of the others, generates the least rent; its unit rent has been positive, but an order of magnitude lower than hake and sardines.

Table 4 shows the monetary accounts in current prices and in constant 1990 prices. Over the past decade, there has been a remarkable 180 per cent increase in the real value of fish stocks from N$1,526 million to N$4,276 million in 1998, even though there was a decline in physical stocks of sardine and horse mackerel over that period. This increase in value is attributable to the increase in the hake stock as well as an increase in the international price of fish. The emergence of hake as the most valuable fish stock represents a success for government policy, which targeted the development of the hake fishery, controlled almost entirely by foreigners prior to independence.

FISHERIES AND NAMIBIA'S NATIONAL WEALTH

As described in the introduction, sustainable development requires that the value of assets be non-decreasing over time; sustainable development that is also equitable with regard to future generations requires that *per capita* wealth does not decline over time. A country's total wealth amounts to a national asset portfolio that can be analysed in terms of its diversity, distribution of ownership, and volatility. Diversity is important because, in general, the more diverse an economy is, the more resilient it will be to economic disturbances. Volatility is also important in planning for the future—lower volatility contributes to more stable economic development. The distribution of the ownership of assets—between public and private sector, the concentration among different groups in society, and between domestic and foreign owners—can have significant economic implications and can influence the sustainable management of resources.

In using national wealth to monitor economic sustainability, it is crucial to include all assets, or at least the most important ones. Namibia's asset accounts include produced and natural capital but exclude human and social capital because there is not yet a method for measuring it. The natural capital accounts include minerals and fisheries, but omit three major resources due to measurement problems: land, wildlife and water. While the impact of water and wildlife is expected to be rather small, the omission of land is significant. Given the large share of land under traditional tenure and state

Table 4. Monetary accounts for hake, sardine, and horse mackerel in Namibia, 1990-1998 (millions of Namibian dollars)[1].

	Sardine	Hake	Horse Mackerel	Total
1990	1,168	268	90	1,526
1991	646	304	301	1,250
1992	1,348	365	204	1,916
1993	2,008	683	365	3,056
1994	2,292	1,591	402	4,285
1995	2,011	2,089	389	4,489
1996	3	1,918	509	2,431
1997	950	2,615	493	4,057
1998	1,500	6,402	911	8,813
Constant 1990 prices				
1990	1,168	268	90	1,526
1991	617	290	287	1,194
1992	1,175	318	178	1,671
1993	1,613	549	293	2,455
1994	1,590	1,103	279	2,972
1995	1,308	1,359	253	2,921
1996	2	1,111	295	1,407
1997	508	1,399	264	2,171
1998	728	3,106	442	4,276

[1] Values were estimated for the closing stock using the present discounted value method assuming a 10% social discount rate and a 20% cost of fixed capital. Constant price values were calculated using the GDP deflator. Figures may not sum to total because of rounding.
Source: based on Lange (2003).

ownership (roughly 56 per cent), it is not possible to estimate a market value for land at this time. It is most likely that the economic value of land has been relatively stable over the past 20 years[6] with the result that growth in wealth, as reported here, is overestimated.

Namibia's total national wealth has increased 16 per cent in real terms from N$29 billion to N$34 billion over the period 1980 to 1998 (Table 5).

The period from 1990 onward is especially important for Namibia. Prior to independence in 1990, the exploitation of Namibia's resources was con-

Table 5. National wealth of Namibia (millions of Namibian dollars in constant 1990 prices) 1980 to 1998.

	Produced Assets	Minerals	Fisheries	Total wealth
1980	22,485	6,729		29,214
1981	23,201	4,863		28,063
1982	23,622	3,815		27,438
1983	23,717	3,237		26,954
1984	23,670	2,740		26,410
1985	23,629	3,013		26,641
1986	23,530	3,933		27,464
1987	23,558	4,108		27,666
1988	23,599	4,009		27,608
1989	23,731	3,824		27,556
1990	23,989	3,227	1,526	28,743
1991	23,987	2,787	1,194	27,969
1992	24,381	2,243	1,671	28,295
1993	24,814	1,440	2,455	28,709
1994	25,374	1,070	2,972	29,416
1995	25,907	864	2,921	29,692
1996	26,505	1,121	1,407	29,034
1997	27,002	1,377	2,171	30,549
1998	28,350	1,262	4,276	33,888
Percent change, 1980 to 1998	26%	-81%		16%

Source: based on Lange (in press).

---

[6] Because the physical volume of land does not change over time, only changes in land productivity would cause the value of land to change. Namibia has experienced both land degradation (reducing value) and land improvements (increasing value), so, for lack of other information, it seems reasonable to assume that the value of land has remained fairly constant.

trolled by South Africa, with relatively little concern for Namibia's own national development. As a result, by the time Namibia had achieved independence its resources were vastly depleted—major fish stocks were less than 25 per cent of their former level and onshore diamonds were largely exhausted, forcing the industry to move to more costly offshore diamond mining. However, Namibia's wealth increased significantly with the establishment of its 200 mile Exclusive Economic Zone at independence, which brought many of its fisheries under national control for the first time, adding this asset to the national wealth[7]. Without fisheries, Namibia's wealth would have remained virtually unchanged from 1980 and 1998.

As in most countries, produced capital dominates Namibia's national wealth. Nevertheless, there is a fair degree of diversity among sources of wealth. Natural capital accounted for 17% of total wealth in 1998, of which fisheries accounted for 13% and minerals for 4%, which has declined from 23% in 1980. But the value of natural capital is quite volatile, leaving the country's economy vulnerable to changes in natural conditions and international markets.

So far, only trends in total wealth have been considered. However, Namibia's population is still growing and per capita wealth for Namibia shows a disturbing trend: per capita wealth has declined by 33 per cent from 1980 to 1998 (Figure 1). Not surprisingly, Namibia's real per capita GDP has stagnated over time, falling at an annual rate of –0.025% (CBS, 2001).

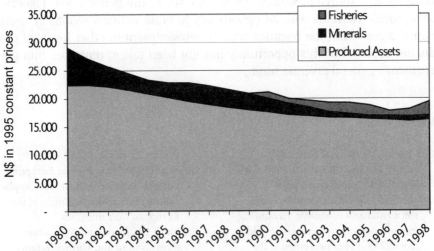

Figure 1. Per capita national wealth of Namibia, 1980 to 1998. Source: based on Lange (2002a).

---

[7] Resources are only considered assets when they are under the ownership and economic control of a country. Prior to 1990, Namibia only controlled its 12 mile territorial zone.

Conclusions

In a relatively short period of time since independence, Namibia has achieved some success in managing its fisheries. It has established a regime for sustainable fisheries management based on a system of TAC that is relatively well enforced. It has also vastly increased the economic contribution of fisheries to the Namibian economy while avoiding the subsidization of the industry seen in so many other countries, and Namibia's fisheries have become an increasingly important asset, accounting for 13% of national wealth. However, it is unclear whether this trend will continue. The uncertainties surrounding estimation of stocks and establishing an appropriate TAC, combined with pressure from the fishing industry, make sustainable management a difficult challenge. The confidence interval for the stock of hake was 20% of the stock size or greater, which was larger than the TAC. Stocks of both sardine and horse mackerel ended the first decade of independence lower than they started, and hake was not much higher. The Ministry's goal of restoring fisheries to previous highs, last seen in the 1960s, seems unlikely to occur. Despite the addition of fish to Namibia's national asset portfolio, per capita national wealth has declined by 33 per cent over the past two decades and real per capita income has stagnated. In addition, the economy is vulnerable to fluctuations in the value of natural capital. These factors suggest that it is especially important to protect Namibia's fisheries as a source of wealth and income for future generations. Fisheries also provide Namibia with an opportunity to build national wealth by designating a portion of the resource rent for reinvestment in other forms of productive capital, but this opportunity has not been taken; rent goes into government's general revenue fund.

References

CBS (Central Bureau of Statistics) (2001): National Accounts 2000. CBS: Windhoek, Namibia. 33 pp.

Danielsson, A. (2000): Integrated environmental and economic accounting for commercial exploitation of wild fish stocks. Paper presented at the Tenth Biennial IIFET Conference, Corvallis, Oregon, USA, 10-14 July. 17 pp.

Dasgupta, P. and K-G. Maler (2000): Net national product, wealth, and social well-being. *Environment and Development Economics*, 5: 69-94.

Eurostat (2000): Valuation of European Forests: Result of IEEAF Test Applications. Luxembourg: Office of the European Communities.

FAO and UN (Food and Agriculture Organisation and United Nations Statistical Division) (in press): *Handbook on Economic and Environmental Accounting of Fisheries*. UN, New York.

Hanley, N. and Spash, C. (1993): *Cost-Benefit Analysis and the Environment*.

Edward Elgar Publishing, Cheltenham, UK. 288 pp.

Hartwick, J.M. (1977): Intergenerational equity and the investing of rents from exhaustible resources. In: *American Economic Review* 67 (5): 972-974.

Kirchner, C., Sakko, A. and Barnes, J. (1999): An economic value of the Namibian recreational shore angling fishery. *South African Journal of Marine Science* 22: 17-25.

Lange, G. (2002): Alternative measures of the value of natural capital in constant prices. Paper presented at the workshop, Putting Theory to Work: The Measurement of Genuine Wealth, 25-26 May 2002, Stanford University.

Lange, G. (2003): Fisheries accounts: management of a recovering fishery. In: *Environmental Accounting in Action: Case studies from Southern Africa* (G. Lange, R. Hassan, and K. Hamilton), Edward Elgar Publishers, Cheltenham.

Lange, G. (in press): National wealth and economic development: the case of Botswana and Namibia. *Environmental and Resource Economics* 28 pp.

Lange, G., Hassan, R. and Hamilton, K. (2003): *Environmental Accounting in Action: Case studies from Southern Africa*. Edward Elgar Publishers, Cheltenham.

National Marine Information and Research Centre, Ministry of Fisheries and Marine Resources (2001): Unpublished data on fish stocks and fish landings. Swakopmund, Namibia.

Pearce, D. and Atkinson, G. (1993): Capital theory and the measurement of sustainable development: an indicator of 'weak' sustainability. *Ecological Economics* 8: 103-108.

Solow, R. (1974): Intergenerational equity and exhaustible resources. *Review of Economic Studies* 41: 29-45.

Solow, R. (1986): On the intergenerational allocation of natural resources. *Scandinavian Journal of Economics* 88: 141-149.

United Nations (1993a): System of National Accounts. UN, New York 711 pp.

United Nations (1993b): Interim Handbook of Integrated Environmental and Economic Accounts. UN, New York. 235 pp.

United Nations (2003): Handbook of Integrated Environmental and Economic Accounts. Available through the UN website: www.un.org.

Vincent, J. (1997): Resource depletion and economic sustainability in Malaysia. *Environment and Development Economics* 2: 19-37.

Zeybrandt, F. and Barnes, J. (2002): Economic characteristics of demand in Namibia's recreational marine shore fishery. *South African Journal of Marine Science*, 29 pp.

10 BENEFITS AND COSTS OF THE NAMIBIANISATION POLICY

*Claire W. Armstrong, Ussif Rashid Sumaila, Anna Erastus and Orton Msiska**

Abstract

When Namibia gained independence in 1990, the country faced a multitude of decisions regarding how to manage its relatively vast fish resources, and how to facilitate and ensure a greater degree of Namibian participation. The government opted for policies based on economic incentives in the form of quota taxation combined with tax reductions depending on the degree of Namibian involvement in the fisheries ventures. In this chapter, the innovative new policy called Namibianisation is analysed, specifically focusing on the benefits and costs that may have resulted from its implementation. The main benefits of this policy are increasing Namibian ownership and employment, while the losses consist of reduced quota tax revenues. Hailed generally as a success, the Namibianisation policy may offer lessons worth following for those countries whose fisheries resources are still exploited by foreign fishing fleets. The analysis provides some basis for further improvement in the development and implementation of the policy.

INTRODUCTION

After Namibia gained independence in 1990, the country faced serious decisions with regards to how the national marine resources should be managed to offer permanent respite to Namibians. This was deemed crucial because

---

* Several staff members of the National Marine Information and Research Centre are thanked for assisting with data acquisition. These are Mr T. Iilende on hake, Ms A. Kanandjembo on horse mackerel, Mrs H. Boyer on small pelagics, Ms H. Skrypzeck on Tuna and Mr F. Haufiku on crabs at the National Marine Information and Research Centre and Mr A.P. Ithindi at Ministry of Fisheries and Marine Resources headquarters. Sumaila acknowledges the support of the SAUP at the Fisheries Centre and the Research Council of Norway.

of the history of abuse inherited from the old apartheid system. The issue of sustainability in a biological sense of resource management was imperative, but perhaps even more so was the need for an equitable distribution of the benefits from the resources, even though many white Namibians also benefited from the policy. Namibia faced the difficult task of correcting years of unjust resource allocation and policies, where large groups of the populace had been excluded from enjoying the benefits of the country's rich natural resources. At the same time there was an urgent need to modernize the fishing industry to make it capable of meeting the demands of the global market for fish, where most of Namibia's fish products are traded.

There seemed to be at least two possible ways of correcting past injustices; 1) command and control, in the sense of appropriation of previous rights and the redistribution of these rights to the previously disadvantaged, or 2) incentive-based systems, where just reallocations are brought about through economic motivation. The first mode of injustice correction is familiar from other parts of the African continent, but the Namibian government chose to use the second mode in the fisheries sector, applying what has come to be known simply as the Namibianisation policy. This policy gives the stakeholders in the fishery economic incentives to encourage them to increase Namibian participation in fisheries in the form both of ownership and of employment, with particular regard to the previously disadvantaged groups of the population. The incentive is in the form of a tax reduction connected to the quota price the firms have to pay for fishing quota rights, given the level of Namibian employment, ownership and other societal issues applied by the quota holder. Hence, by using the carrot rather than the stick, the Namibian government hoped that social injustices could be remedied. There is, however, a price connected with the success of this policy, in the form of lost tax income to state coffers. The positive and negative impacts of the Namibianisation policy are therefore the focus of this study.

For the rest of the chapter, we will first present the Namibian quota system, and show how the Namibianisation policy fits hand in glove with this management regime. Then we will present some indicators of benefits and costs arising from policy intervention.

THE NAMIBIAN QUOTA SYSTEM

Today, Namibia has a highly managed fisheries sector, where 90% of the catch is determined by total allowable catches (TACs), while the remainder is bycatch managed via effort limitations. The major fisheries are managed after a rigorous biological assessment and entry to the fisheries is 100% limited. Furthermore, Namibia is one of few countries in the world to have succeeded in capturing substantial economic rent from her fisheries by the imposition of quota fees (Namibia Brief, 1998; Manning, 2000).

Quota fees were imposed for the first time in 1990, after Namibia gained independence. This was possible partly because fortuitously historic rights, which have plagued fisheries management all over the world, did not need to impact upon future management, providing an opportunity for the Namibian fisheries policy to start afresh (Oelofsen, 1999). Thus, from 1994 anyone could apply or re-apply for the rights to fish in Namibian waters as all rights were reviewed that year.

In 1992 a new policy was introduced, namely, the Namibianisation policy, which was intended to correct some of the pervasive human injustices brought about by the apartheid system in the country prior to independence, as well as to secure greater Namibian involvement in the fisheries of the country. The main policy instrument used to attain Namibianisation was the introduction of rebates on the quota fees, which are determined by the degree of Namibian ownership, employment of Namibian crew, and whether fish was landed and/or processed in Namibia, as well as other less specific conditions.[1] The guiding principles underlying Namibianisation are outlined in the 1992 Sea Fisheries Act, such that rights to fish on specific species are allocated in accordance to whether:
- the applicant is a Namibian citizen;
- the applicant company's beneficial control is vested in Namibian citizens;
- the applicant has the ability to exercise the right of exploitation in a satisfactory manner;
- the applicant must have beneficial ownership of any vessel to be used.

Furthermore, the following special criteria advance the possibility of rights allocations to an applicant:
- the advancement of previously disadvantaged persons socially, economically or educationally;
- contribution to regional development in Namibia;
- cooperation with other countries (especially within SADC);
- the conservation of marine resources;
- the successful performance of the applicant under exploratory right;
- contribution of marine resources to food security;
- any other matter that may be prescribed.

There are no economic criteria for allocation, and no transferability of rights. Leasing is, however, possible for short periods such that if the rights are not utilised by the original rights holder in the period they are allocated, a new application by the holder will not be successful (Les Clark, Fisheries Mana-

---

[1] In addition, levies per tonne of fish were charged and paid to the Sea Fisheries Fund, for research and training.

gement Advisor, Forum Fisheries Agency, Solomon Islands, pers. comm.). The rights were originally allocated for 4, 7 or 10 year periods depending on a number of conditions listed below. In 2000 these periods were increased to 7, 10, 15 or 20 years, motivated by a desire to make conditions more investor friendly and to bring about more stability to the sector.

Seven-year exploitation rights are granted to: (i) applicants with less than 50% Namibian ownership of vessels or onshore processing plants in the fishery where rights are granted; and (ii) applicants with less than 51% Namibian ownership in ventures without significant onshore investments in the fishery where rights are granted.

Ten-year exploitation rights are granted to: (i) applicants with at least 50% Namibian ownership of vessels or onshore processing plants in the fishery where rights are granted; (ii) applicants with less than 51% Namibian ownership in onshore investments in the fishery where rights are granted.

Fifteen-year exploitation rights are granted to: (i) ventures at least 90% beneficially Namibian owned with significant investments in vessels or onshore processing plants (50% ownership in facilities in the fishery where rights are granted, is seen to be significant); (ii) Namibian rights holders with smaller shares in larger ventures; (iii) majority foreign-owned ventures with the capacity to make a major contribution to economic and overall development in Namibia (e.g. onshore employment of 500 Namibians is seen as a major contribution); and (iv) smaller joint or wholly foreign-owned ventures, which can make innovative contributions to the development of the fishing industry in Namibia, such as developing new products or export markets, and where a long-term right is necessary to secure the investment involved.

Twenty-year exploitation rights are granted for ventures that fulfil the requirements for fifteen-year rights, and employ at least 5,000 permanent employees in onshore processing facilities.

The original terms for 4, 7 and 10 years are equivalent to the existing 7, 10 and 15 year terms. Shorter-term fishing rights could originally also be granted, e.g. in the early stage of a new fishery. Shorter-term fisheries rights could be upgraded to a longer-term right when the operations fulfilled the necessary requirements. If a rights owner who is given a four-year right, fails to increase Namibian ownership or otherwise invest to satisfy the 7-year requirement by the end of the third year, a new application would not be favourably considered.

Hake fees demanded from vessels with access rights in year 2000 were (i) N$880 per tonne paid by foreign freezer vessels; (ii) N$680 per tonne paid by Namibia-based freezer vessels (N$200 rebate); (iii) N$480 per tonne demanded from fully Namibian-owned freezer vessel (further N$200 rebate); and (iv) a further rebate of N$200 is given if the catch is processed on shore.

If allocated quota is not caught, the rights holder must pay N$800 per tonne for the uncaught part of the quota regardless of fee rebates, unless more than 20% of TAC is left uncaught, when fees are waived (Scheepers, 1998). Failure of a rights holder to harvest his or her full quota allocation in a given allocation period is an invitation to a quota cut in the future.

In order that the benefits from Namibia's fisheries directly impact the previously disadvantaged groups of the population, the Ministry of Fisheries and Marine Resources (MFMR) has enacted a so-called empowerment policy (Erastus, 2002), where it has attempted to allocate fish quota to Namibian newcomer applicants. In 1994-1995 the MFMR allocated 25% of the TAC to newcomers, which otherwise would have been allocated to existing companies. The new entrants are a varied group of business people, fishers and people in prominent positions. It has been observed that some of the newcomer firms have subsequently leased their quotas to more established firms. In the absence of many financing options, this may have been done to build up capital for investment in their own harvesting potential (Erastus, 2002) but this essentially means that some attributes of individual transferable quota systems may have unintentionally crept into the system (Manning, 2000).

Benefits from Namibianisation

The gains from Namibianisation are many and varied, and not always easy to quantify. In the ensuing discussion, we will limit ourselves to an analysis of trends in Namibian employment and ownership development since the introduction of the Namibianisation policy. It is, however, clear that an analysis of policy gains would have to take into account income tax revenues from the employment enhancement that Namibianisation presumably facilitates. There are also more non-tangible issues such as what ownership amongst previously disadvantaged Namibians does to the self-confidence of both those who directly benefit and those who do not, as well as the potential future generational gains that cannot be clearly visualised today. Hence, this analysis of benefits is only a first step towards studying the topic fully.

Figure 1 plots the percentage of Namibian ownership from 1993 to 1998. We see from this figure that the percentage of Namibian-owned licensed fishing vessels has increased from about 60% in 1993 to 85% in 1998.

From Figure 2, which graphs the percentage Namibian employment on land and sea from 1991 to 1998, we see that the share of Namibian employment in the fishing industry increased from just under 55% in 1991 to more than 75% in 1998. We observe that the share of Namibian employment declined in the years 1994-1996, before increasing again. In absolute numbers Namibian employment increased by 6,000 in this time period, doubling the number of Namibians employed in the fishing industry (Erastus, 2002).

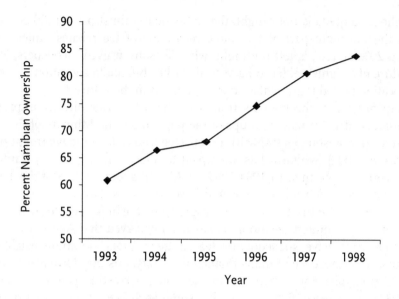

Figure 1. The percentage of Namibian ownership[2] of licensed fishing vessels, 1993-1998.

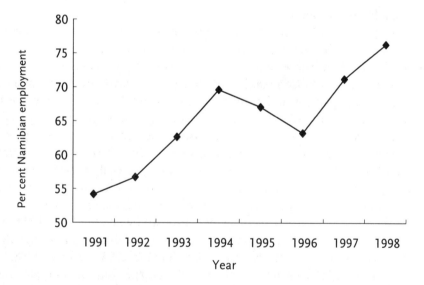

Figure 2. Percentage Namibian employment on land and sea in the Namibian fishing industry, 1991-1998 (Erastus, 2002).

---

[2] Namibian ownership is defined as 51% Namibian beneficial ownership and 80% Namibian crew (Erastus, 2002).

Table 1 shows that employment in some parts of the Namibian fishing fleet consisted of a large proportion of Namibians even as early as 1994, while others have increased the proportion of Namibians employed significantly since 1994. The exception to this is the midwater fleet, which in 1999 employed only 9% Namibians out of a total of more than 1,500 employees. Many of the remaining foreign-held posts are most likely to be those requiring higher skills. It is therefore important that complementary policies such as skill-enhancement schemes are implemented in order to further Namibianise these positions.

So far, it seems that the newcomer-empowerment policy functions in such a way that substantial newcomer quota is being leased to established companies, thereby at best limiting the benefits to the channelling of money towards the building of harvesting capacity, i.e. companies or private persons use the income from leasing out their quotas to, over time, enable the financing of fishing vessels. It is, however, clear that even if the empowerment policy fails to succeed in this sense, there will nonetheless be greater national gains, as compared with the allocation of quota to existing companies. This is because allocations to newcomers have been made with the requirement that fish is processed on land, generating greater multiplier effects in the economy, especially in employment.[3] Hence over time, the benefits from the empowerment policy can be expected to increase.

Table 1. Percentage Namibian employment (total person-years) in some sectors of Namibian fisheries, in the years 1994-1999 (Central Bureau of Statistics, 2001; MFMR, 1997).

	1994	1995	1996	1997	1998	1999
Pelagic	89 (493)	94 (445)	95 (476)	97 (427)	96 (562)	97 (395)
Demersal	65 (1,967)	72 (2,427)	73 (3,016)	81 (2,649)	84 (2,212)	79 (2,506)
Midwater	1 (2,664)	1 (2,409)	5 (2,141)	8 (2,100)	6 (1,606)	9 (1,531)
Linefish	93 (268)	95 (342)	99 (294)	100 (277)	100 (316)	84 (310)
Crab	57 (190)	46 (147)	53 (118)	64 (73)	65 (101)	71 (76)
Rock lobster	99 (550)	99 (674)	99 (541)	100 (525)	100 (429)	99 (439)
Deep water	N/A	N/A	N/A	70 (261)	78 (139)	72 (156)
Tuna	31 (1,112)	33 (1,356)	65 (940)	63 (957)	76 (1,218)	65 (1,205)

---

[3] In the hake fishery, quota allocated to newcomers is for wet fish harvest, with delivery to onshore processing, rather than onboard processing (Erastus, 2002).

The cost of Namibianisation

In a straightforward sense, the cost of Namibianisation is assumed to be the sum of fee rebates given to Namibian rights holders, as this is the tax revenue which the state foregoes in order to achieve its Namibianisation goals.

Tables 2 and 3 present the total allowable catches (TACs) and the potential quota fees for major Namibian fish species caught from 1993 to 1998, respectively. The potential quota fees are calculated by multiplying the TAC by the quota 'full' fee (that is, the no-rebate quota fee) in the respective years. The actual quota fees collected and the cost of Namibianisation are also reported in Table 3. This cost is defined as the difference between the potential quota fees and the actual fees collected.

Table 2. TAC in 1000 tonnes (MFMR, 1998, 1999).

	1993	1994	1995	1996	1997	1998
Hake	120	150	150	170	120	165
Sardine	115	125	40	20	25	65
Horse mackerel	450	500	400	400	350	375
Tuna	3.5	4.1	2.6	1.8	1.3	-
Crab	4.9	4.9	3	2.5	2	2

Table 3. Full quota fees (N$/tonne), quota fees collected, the potential total quota fees, and the cost of Namibianisation (N$1000).

Year	1993	1994	1995	1996	1997	1998	Unit fees
Hake	96,000	120,000	120,000	136,000	96,000	132,000	800
Sardine	17,250	18,750	6,000	3,000	3,750	9,750	150
Horse mackerel	28,148	31,275	25,020	25,020	21,893	23,456	63
Tuna	1,419	1,639	1,053	719	526	-	400
Crab	2,940	2,940	1,800	1,500	1,200	1,200	600
Quota fees collected	97,800	108,466	90,500	46,543	72,239	67,695	
Potential quota fees	145,757	174,604	153,873	166,239	123,368	166,406	
Cost of Namibianisation	47,957	66,138	63,373	119,696	51,129	98,711	
Cost as percentage of quota fees	33%	38%	41%	72%	41%	59%	

We see from Table 3 that since fee rebates were implemented in 1993, the cost of Namibianisation gradually increased as the incentives of the policy of Namibianisation came into effect, resulting in more Namibian employment and ownership, and hence less fees being collected. There was a big decrease in fees collected in 1997. However, this was due more to reductions in the allowable catch than the Namibianisation policy.

The cost due to Namibianisation is the difference between the potential quota fees and the actual collected quota fees. We note that this difference is growing, a sign that the Namibianisation policy is performing well. However, it is also clear that there is a substantial loss to state coffers; for instance, the loss in 1996 is 72% of the potential quota fee revenues. It must, however, be noted that when total uncaught harvests have exceeded 20% of TAC, as was the case in 1996, uncaught quota fees have been waived, making the loss figures biased upwards. Using TACs to calculate the full quota may also be somewhat questionable, as uncaught quota may not have anything to do with the quota system as such.

Determining the net gains of Namibianisation

Given that benefits and costs are in different units (dollars versus employment of previously disadvantaged Namibians), it is difficult to calculate net benefits. We will avoid this problem by comparing (i) the change in Namibian proportion of total fishing sector employment, and (ii) the change in the proportion of Namibian ownership of licensed fishing vessels with the change in the proportion of the potential quota fees that is not captured due to the Namibianisation policy.

In absolute terms the average annual foregone quota fee income is N$ 74.5 million in the time period 1993-1998. In the same time period the increase in number of Namibians employed was 6000, or an average annual increase of 1000 Namibians employed. Disregarding Namibian ownership increases, the cost to state coffers per unit increase in employment is N$ 74.5 thousand *per annum*. If labour is scarce, this seems a rather steep price. However, in the case that the alternative for those being engaged is unemployment, the price may not be overly high. And if this is an investment that gives permanent employment increases, as well as ownership changes, Namibians can see this in a positive light. A study that differentiates between the ownership and employment fee rebates would enhance this discussion, but data are unfortunately not available to the authors.

From the previous two sections, we calculated that the share of Namibian employment in the fishing industry increased from 62.7% in 1993 to 76.3% in 1998. That is a gain of about 22 percentage points in 6 years or about 3.6 percentage points per annum. The gains in the proportion of Namibian ownership of licensed fishing vessels increased from 60.8% to 83.8% in the same period, implying a per annum gain of 6.3%. Finally, a per annum increase in the percentage cost of Namibianisation to total fishing fees collected in the same period is 4.4 percentage points. This is obtained by dividing the cost of Namibianisation by the total fees collected (see Table 2), and calculating the percentage annual increase in this variable. These numbers indicate that there is just under a 1 and 1.5 percentage point gain in employment and Namibian ownership, respectively, for every percentage point loss in fees collected.

DISCUSSION AND CONCLUDING REMARKS

In summary, our results show that there has been a greater degree of Namibian ownership and employment in the country's fisheries since the Namibianisation policy came into existence. However, it is virtually impossible to be certain that the achieved gains are solely a result of Namibianisation. The vagaries of natural resource harvesting clearly also affect employment, and other policies increasing Namibian skills may also have benefited Namibian employment and ownership during the period under discussion.

Regardless, the achievements of the recent years are laudable since they have helped the country to meet one of its cardinal objectives, namely, to open up the fishing sector to previously disadvantaged Namibians. The study has also shown that this achievement, as would be expected, came at a cost in terms of tax revenues lost that otherwise would have accrued to the national coffers.

All the listed positives above not withstanding, there are strong critiques of the Namibianisation policy, and fisheries management in general (see Manning, 2000). The main points raised are: (i) much more rent could have been captured by Namibia than is currently the case; (ii) the little rent currently captured goes to a few rich operators, rather than to Namibians as a whole; (iii) actual ownership is largely still foreign, as there is evidence to show that many Namibians act as straw men for foreign owners.

As always there are counter arguments to the criticisms listed above (Oelofsen, 1999, and Les Clark, pers. comm.). The points made by these and other people are: (i) Namibia has succeeded in capturing a good percentage of the potential rent from its fisheries compared with other fisheries in the world; (ii) Namibian and black ownership have enjoyed an increase since independence; (iii) shareholders of Namibian fishing companies are mostly Namibian pension and insurance funds, as well as Namibian businessmen; and (iv) Namibian ownership creates greater income/business tax revenues.

Within the limitations of this analysis we claim that the Namibianisation policy has served Namibia well, and there may be some lessons to be learnt by other developing countries whose fisheries resources are still being largely exploited by foreign fleets.[4]

Issues that have received little attention are how the nationalisation of the Namibian resources may have increased pressure on the managers as regards increasing harvesting in the short term, with possible detrimental long-term conservation effects. Furthermore, a larger analysis remains to be undertaken determining *who* has benefited from Namibianisation. Manning (2000) does an early analysis to this effect, but a further broadly qualitative analysis similar to that of Isaacs and Hersoug (2002) and Isaacs (in press) in South Africa would shed more light on the issue. A central question is whether revenues accruing to state coffers could be reallocated to the needy in the community in a better way than that achieved through the Namibianisation policy. And would *not* implementing Namibianisation have been at all possible from a political point of view? That is, would a policy of purely foreign quota rental have been acceptable in light of the apartheid policy preceding independence? Furthermore, a question that policy makers must ask themselves is whether the benefits from Namibianisation in the form of ownership and employment are now at a point of strongly declining marginal increase, and whether the benefits that are achieved are permanently entrenched. That is, if further rebates only give small increases in employment and ownership, could the rebates be substantially reduced without affecting the existing Namibian employment and ownership? If the costs of hiring foreign labour are greater than that of Namibian labour, the fee rebate incentives for employment may well be reduced or eliminated, without loss of employment, and with a gain of fees. Likewise if the new Namibian owners are *de facto* agents in the fisheries, eliminating fee rebates will not reduce ownership. From the foregoing discussion, it would appear obvious that the longevity of the Namibianisation policy also needs to be investigated. These are important questions that the study raises but nonetheless are outside the scope of this chapter.

References

Atta-Mills, J., Alder, J. and Sumaila, U.R. (2004): The decline of a regional fishing nation: the case of Ghana in West Africa. *Natural Resources Forum* 28:13-21.

Central Bureau of Statistics (2001): Preliminary National Accounts 2000, March, Namibia.

Erastus, A.N. (2002): The Development of the Namibianisation Policy in the

---

[4] See Atta-Mills et al. (2004) on fisheries in West Africa and European fishing fleets.

Hake Subsector, 1994-1999. NEPRU Working Paper No. 82. Windhoek, Namibia.

Isaacs, M. (in press): Understanding the social processes and politics of implementing a new fisheries policy, the Marine Living Resource Act 18 of 1998, in South Africa. PhD thesis submitted to the University of Western Cape, 2003.

Isaacs, M. and Hersoug, B. (2002): According to need, greed or politics – redistribution of fishery rights within South Africa's new fisheries policy. In: *Fishing in a Sea of Sharks* (B. Hersoug, ed.), Eburon, Delft.

Manning, P. (2000): Review of the Distributive Aspects of Namibia's Fisheries Policy. NEPRU Research Report No. 21.

MFMR (1997): Report of the Activities and State of the Fisheries Sector. Ministry of Fisheries and Marine Resources, Windhoek, Namibia.

MFMR (1998): Fisheries Statistics. Summary Statistical Report. Ministry of Fisheries and Marine Resources, Windhoek, Namibia.

MFMR (1999): Fisheries statistics 1999. Summary stastistical report. Ministry of Fisheries and Marine Resources, Windhoek, Namibia.

Namibia Brief (1998): No. 20, January. The Namibia Foundation, Windhoek, Namibia.

Oelofsen, B.W. (1999): Fisheries management: the Namibian Approach. *ICES Journal of Marine Science* 56: 999-1004.

Scheepers, E. (1998): Quota Fees. Namibia Brief, No. 20, The Namibia Foundation, Windhoek.

11 ECONOMIC VALUATION OF THE RECREATIONAL SHORE FISHERY: A COMPARISON OF TECHNIQUES

Jonathan Barnes, Fredrik Zeybrandt, Carola Kirchner, Alison Sakko and James MacGregor[*]

Abstract

This chapter describes work to determine the economic values associated with the recreational marine line fishery in Namibia. This highly esteemed fishery involves angling from the shore with bait for bottom-feeding fish, mostly kob, steenbras and galjoen. Anglers come from South Africa (46%), inland Namibia (38%) and coastal Namibia (16%). In 1997 and 1998 several field surveys and valuation approaches were applied to the fishery. A roving creel survey was used to determine angler numbers and catches. Responses from two questionnaire surveys involving 240 and 372 anglers, respectively, were analysed to estimate angler expenditures, consumer surpluses, and the price elasticity of demand. Both the travel cost and the contingent valuation methods were used. Some 8,300 anglers spent a total of 173,000 days angling, and each angler spent some N$3,400 in this activity. Aggregate direct expenditures by anglers were between N$23 million and N$31 million. Gross value added associated with this was between N$11 million and N$15 million. This represents 2% to 4% of the whole fisheries sector, which itself makes up 4% of the economy. The aggregate consumer surplus enjoyed by anglers was N$24 million, of which 30% to 50% accrued to foreigners. Demand for angling is price inelastic, making it possible to capture rents from the fishery. Results from the different valuation exercises showed considerable convergent validation. All methods are best employed together, but each alone can provide useful values for policy analysis.

INTRODUCTION

The Benguela system, described in detail elsewhere in this volume, is characterised by cold but nutrient-rich upwellings, relatively low species diversity,

---

[*] This study was supported through funding from the Swedish Government (Sida), the United States Agency for International Development (USAID, through the World

and high production. It forms the basis for a highly esteemed recreational fishery. Anglers mostly fish from the shore, in the surf, using bait. Most frequently landed are kob (mostly silver kob, *Argyrosomus inodorus*, Sciaenidae, but also dusky kob, *A. coronus*), west coast steenbras (*Lithognathus aureti*, Sparidae), galjoen (*Dichistius capensis*, Coracinidae) and blacktail (*Diplodus sargus*, Sparidae). To a lesser extent, sharks are targeted, including the copper shark (*Carcharhinus brachyurus*, Carcharhinidae), the spotted gulley shark (*Triakis megalopterus*, Carcharhinidae) and the smoothhound (*Mustelus mustelus*, Carcharhinidae).

Access to shore angling on the Namibian coast is restricted to about one quarter of the coastline: 90% of angling is in the West Coast Recreation Area (WCRA), some 260 km, stretching from Sandwich Harbour, south of Walvis Bay to the Ugab River in the north, but additional small sites exist at Torra Bay and Terrace Bay to the north, and Lüderitz in the south. Anglers originate from coastal Namibia, inland Namibia, and South Africa, of whom very small numbers fish for subsistence. Recently, in 2001, angling licences were introduced, and the daily bag limit of 30 fish (or 30 kg of fillet) was reduced to 10 fish (or 10 kg of fillet).

The recreational line-fish resource is shared with a commercial line fishery, which operates inshore, from Walvis Bay, in some twelve vessels. These vessels target the same species, but also seasonally seek the pelagic snoek (*Thysites atun*, Gempylidae). The stocks are perceived to be declining (Kirchner, 1998; Holtzhausen, 1999; Holtzhausen et al., 2001). There is a need for economic data on the fishery, to inform sound policy development, planning and management. This chapter reviews and compares work done by ourselves, notably Kirchner et al. (2000) and Zeybrandt and Barnes (2001), on the economic valuation of the recreational shore fishery. The review is focused on comparing results from the different methodological approaches.

Wildlife Fund (US) LIFE Programme, under terms of Agreement no. 623-02510A-00-3135-00), the Overseas Development Institute (with funding from the British Department of International Development, DFID) and the Namibian Government. The opinions expressed do not necessarily reflect those of these agencies. We are very grateful to the researchers and anglers, who gave of their time. H. Holtzhausen, S. Voges, B. Louw, V. Kazapua, S. Pahl, R. Schommarz, and L. Polster assisted with enumeration and interviews. In the Ministry of Environment and Tourism, Permanent Secretary, T. Erkana, P. Tarr, C. Brown, H. Fourie, and R. Braby provided support, and assisted with logistics. G. Köhlin assisted in securing funding and with essential advice. H. Suich assisted with editing.

Methods

Economic values

The values (measured in Namibia dollars (N$))[1] can be placed in the context of "total economic value" for natural resources. Total economic value consists of *use values*, which embrace direct and indirect use values, and *non-use values*, which embrace option, bequest and existence values. Pearce and Turner (1990) describe these components. All of our measures of gross output, value added, and consumer surplus given reflect *direct use value*.

In Namibia, a primary macro-economic measure of direct use value is the gross national income (GNI). This can be estimated either as the total value of consumption of all final products in the economy, or as the total value added by all productive activities in the economy. Value added in an enterprise is defined as the return to internal factors of production (labour and capital), and is the gross output less expenditures on external factors (intermediate expenditures). Net national income (NNI) is gross national income less capital asset depreciation.

Central to the recreational fishery is the activity of angling. The total direct expenditures made by the fishers in angling make up the gross output of the fishery. The value added for the fishery is a proportion of the output. We had no measures of this proportion for angling tourism, but were able to extract estimates from the broader nature-based tourism sector in Namibia. Empirical data collected during the 1990s (Ashley, 1995; unpublished data, Environmental Economics Unit, Directorate of Environmental Affairs, Ministry of Environment and Tourism) showed that gross value added was 48% of gross output, and net value added was 41% of gross output. We applied these proportions to calculate gross and net national income for the recreational fishery.

Instead of simply determining the "value" of the fishery, we can measure its "impact" on the economy. Here, we determine the values generated by primary direct expenditures, plus those resulting indirectly through induced linkages and further rounds of spending. Although Kirchner *et al.* (2000) did employ a crude national income multiplier to estimate impact, we do not include it here.

Price levels for outdoor recreational activities are often set lower than those the users are willing to pay. Any positive difference between the price paid by a user and his/her willingness to pay is the user's *consumer surplus*, and it forms part of the economic direct use value of the activity. We used the travel cost and contingent valuation methods, described below, to measure this component of value.

---

[1] At the time of the studies, N$1.00 was equal to ZAR1.00 or approximately US$0.20

Surveys

The first of three surveys was a roving creel survey, to determine relative angler numbers and catches (Kirchner et al., 2000). Sampling was conducted from October 1996 to September 1997. Sampling was stratified to capture differences between the high and low seasons, as well as to adequately cover six spatial zones. Data were analysed to estimate the mean daily number of anglers and mean daily catch, for all angler categories. Lüderitz, where angler numbers are very low, was left out of the study.

The second survey involved a targeted sample of 240 anglers, 80 from each of the three categories: coastal Namibians, inland Namibians, or foreigners (South Africans), who were interviewed, while they were fishing, to determine their daily expenditures. The sample was made within the West Coast Recreational Area by two researchers. Subsistence anglers were few in numbers, very localised, and were left out of the survey. Foreign visitors were asked to estimate costs of fuel, accommodation, bait, tackle, groceries, refreshments and entertainment, in addition to costs of any fishing equipment purchased in the last calendar year within Namibia. Anglers from inland Namibia were asked to estimate the same costs, excluding those for groceries. For coastal residents, the costs of fuel, bait, tackle and equipment purchased within the last calendar year were included in the analysis.

The third survey involved a sample of 626 anglers made at angling destinations, from Walvis Bay in the south to Terrace Bay in the north, to determine trip expenditures and willingness to pay for angling and conservation (Zeybrandt and Barnes, 2001). The survey took place between January and April 1998. Sampling was not systematic or random, but non-selective at sites, with the aim of getting the highest possible number of responses. Stratification of sampling between sites was undertaken, aimed at achieving representative spatial coverage.

The sample contained different proportions of angler categories (foreign visitors, inland Namibians and coastal Namibians) from those measured in the roving creel survey (Kirchner et al., 2000). This sample bias was corrected for by weighting the results for the three segments. The questionnaire used in the third survey was similar to that used by Barnes et al. (1999) and Barnes (1996) to survey broader tourism populations and wildlife viewing tourists. It was designed to elicit data, for both travel cost and contingent valuation analysis. In addition to general tourist characteristics and reasons for the visit, respondents were asked to state their travel costs, total costs, specific angling costs such as bait, tackle, rods and reels and the replacement cost of their vehicle/skiboat (if any). A team of five enumerators distributed questionnaires, assisted respondents when needed, and collected completed questionnaires. The questionnaire was in most cases handed out to respondents for their own completion, but some regular interviews were held. Refusal rate was very low. From 626 returned questionnaires, 372 were selected for use after cleaning.

Analysis

Expenditure analysis. Data from the second survey of 240 anglers were used to estimate mean daily expenditure and expenditure per fish caught, for the three categories of recreational angler. Data from the third survey of 626 anglers were also used to extract details of direct expenditures on the angling experience. Here, the questions had been designed to form the base for the development of travel cost and contingent valuation models. These analyses are explained in detail by Zeybrandt and Barnes (2001) and below.

Travel cost analysis. In travel cost analysis, anglers' costs of consuming the services of the environmental asset are used as a proxy for price. These consumption costs include travel costs, entry fees, on-site expenditures, and the annualised costs of outlay on capital equipment needed for consumption. The basic premise is that the user population is homogeneous in its willingness to pay, and that differences in the costs of consumption (due, for example, to different travel costs) result in different rates of visitation. The visitation rate is used as the quantity measure of the angling experience. The travel cost method is thus an *indirect* method of valuation. By varying the travel costs and visitation rates, it is possible to derive a demand curve that expresses the demand for trips to the recreational area (Kerr, 1986; Hanley and Spash, 1993). The consumer surplus for the activity can be calculated from the demand function.

The travel cost method has not been used much in the context of southern African tourism activities, because it depends for success on assumptions, which are commonly not applicable. We considered the Namibian angling population to be suitable for analysis using the travel cost approach, because the angling population is relatively homogeneous, visits are made exclusively for angling, and substitute sites are remote and somewhat different.

Depending on the degree of homogeneity of the sample population regarding travelling distance and social characteristics, an *individual* or *zonal* travel cost model can be used. Our data were best suited to a zonal model, with highly variable costs and variable frequencies of visitation. For zonal models, population figures are derived for the zones and numbers of visits per capita, per zone are calculated. A typical zonal visitation rate model is:

$$(VPC)_{zj} = f(TC_{zj}, S_z) \quad (1)$$

where $(VPC)_{zj}$ is visits per capita from zone z to site j, TC_{zj} is trip (including travel) costs from zone z to site j; and S_z is a vector for the social characteristics of the zone z. It is assumed that the visitors travelling from different zones have the same willingness to pay and the same social characteristics. The zonal model is somewhat sensitive to the selection of the zones used. This can affect the resulting consumer surplus estimates (Hanley and Spash, 1993).

Thirteen geographical zones were identified for our model. These were made up of South Africa's nine provinces, three Namibian coastal zones, and one Namibian inland zone. The populations and mean incomes for the South African zones were derived from data from the South African Centre for Statistical Services (CSS). The populations for zones in Namibia were derived by adding the populations for each city or town in the zone represented in the zone samples. No official estimates of local Namibian incomes were available, and these were derived from the questionnaire data.

The travel costs included the fuel cost of a return trip to the Namibian coast and the on-site expenditure. We considered that the fuel costs only, rather than full cost of the vehicle (including depreciation of the car, tyres etc.), was closest to the typical respondent's perception of vehicle costs. A contentious issue regarding travel cost models relates to the inclusion and estimation of opportunity costs for travel time. Hanley and Spash (1993) suggest inclusion of a question about enjoyment during travelling, and imputing opportunity costs only to those not enjoying the travel time. Because 95% of our respondents enjoyed the time travelling, we included time costs for only 5% of respondents in the basic model.

The cost of time for the South African zones was determined by deriving hourly income from mean zonal incomes, as acquired from the CSS. For Namibian zones, mean incomes from questionnaire responses were used. The travel cost was determined by multiplying the distance travelled to and from the coast with the Automobile Association of South Africa's (AARSA, 1998) estimate of cost per km for two-wheel-drive and four-wheel-drive vehicles. Time costs were computed assuming average travel speed of 70 km per hour.

The inclusion of on-site and other non-travel costs such as accommodation or entry fees, is also contentious. Whether these should be included depends on whether they can be deemed to affect rates of participation and, as with travel time, the degree of enjoyment derived from the consumption. We considered that, along with the cost of travel, these expenditures overwhelmingly *do* affect visitation rates, and therefore should be included in the analysis.

Many travel cost models (e.g. Navrud and Mungatana, 1994) include social characteristics such as gender, income, and other relevant variables to obtain better specification for the model. In our case, lack of data and problems with multicollinearity precluded this. Different functional forms were tested. The model that had the "best" fit was chosen for the following stages of the analysis, i.e. developing a second stage demand function (Kerr, 1986; Hanley and Spash, 1993), and calculating the consumer surplus.

Contingent valuation analysis. Data from the third survey of 626 anglers were also analysed using contingent valuation, to estimate consumer surpluses (Zeybrandt and Barnes, 2001). Unlike travel cost, which is based on revealed preferences, contingent valuation is a *direct* method and is based on stated preferences. In it, the respondent's willingness to pay (WTP) for an increased amount of a specific good, or her/his willingness to accept (WTA) to avoid a decrease of a good, are elicited through surveys. It is generally agreed that willingness to pay is preferable to willingness to accept (NOAA, 1993; Mitchell and Carson, 1989).

We used a variation of the contingent valuation method, which Barnes *et al.* (1999) and Zeybrandt and Barnes (2001) describe in some detail. Among general questions on their personal characteristics, origin, trip and trip preferences, respondents were asked how much their *travel* to and from their angling destination was costing, what their *total angling trip* was costing, how much of this they were personally spending within Namibia, and what their annual income was. They were informed that their answers were to assist with planning and could not affect actual prices.

A payment card was used to ask the respondents what they would be willing to pay for a *similar, return,* angling trip. They were first asked whether their current trip was value for money and then whether they would be willing to return on a similar trip. If they said "yes" (nearly all did), they were asked to identify the cost level (in relation to their present or actual cost) that would *prevent* them from returning. If they said "no", they were asked to identify the cost level (also in relation to their actual cost) that would *induce* them to return. These cost levels were taken as the maximum willingness to pay for a return trip. For each respondent, a positive difference between willingness to pay for return trips and actual trip cost was taken as an estimate of that individual's consumer surplus for the whole trip. For foreign anglers, the consumer surplus for the Namibian part of the trip was calculated proportionally, based on the ratio between expenditures for the *whole trip* and the *Namibian component of the whole trip.*

The cost of travel and the cost of the overall trip were common to all respondents, and most seemed able to make a good estimate of these. They were first asked for these two costs in that order, before being asked to value any other specific components of the trip such as accommodation. The order of questions was selected with care after the pilot survey, and was thought to reduce the potential for both *budget constraint* bias (Mitchell and Carson, 1989) and also *embedding* or *part-whole* bias (Navrud and Mungatana, 1994;

Kahneman and Knetsch, 1992). Focus on the *overall* trip cost for the willingness to pay question was also thought to reduce the tendency for these biases (Moran, 1994; Navrud and Mungatana, 1994).

Getting anglers to focus on *return* trips in their consideration of willingness to pay was thought to reduce confusion between actual and maximum estimates, which might arise if they were to focus on the *actual* trip. In as much as desire for return trips is likely to be less than that for *first time* trips, the estimates of actual demand and consumer surplus are likely to be conservative. We consider this of value in reducing any effects of *avidity* bias, as described by Thomson (1991).

Use of the actual angling experience as the reference point, and the use of the words "prevent" and "induce", was thought to reduce the possibility of *strategic, mis-specification, compliance, starting point, range, relational* and *positional* bias, all described by Mitchell and Carson (1989) and Zeybrandt and Barnes (2001). To avoid possible sponsor bias, respondents were informed that the study was an environmental evaluation of recreational angling. Care was taken with the order of questions, to minimise the possibility of embedding or part-whole bias. In order to corroborate the results from the payment card, we also used an open-ended question, where we asked the respondent to state his/her maximum willingness to pay for the return trip.

Price elasticity. We derived measures of price elasticity from the data and the demand functions developed using the travel cost and contingent valuation methods. First multiple and then simple regressions were run on the raw variables, to try to determine price, income, success and other elasticities. Secondly, the second-stage demand functions from the travel cost analysis were used to calculate price elasticities. Thirdly, the variable for willingness to pay, obtained in the contingent valuation study, was manipulated to develop a derived demand function, which was also used to calculate price elasticities. In this case, the range of willingness to pay was divided into 20 equal segments, and a frequency histogram depicting the distribution of responses along the range was drawn. Simple regression on the histogram data was carried out to obtain the *price* (willingness to pay) to *quantity* (number of respondents per price category) relationship.

Double log, lin-log, log-lin, linear and reciprocal functional forms were tested for both multiple and simple regression models. In multiple regressions, different combinations of explanatory variables were tested in an attempt to minimise multicollinearity effects. Only models displaying significance, overall and with respect to the coefficients, were retained. Point elasticities, at mean and median price values, were calculated for all other than double log functions.

Results

Data from the roving creel survey revealed that some 8 300 anglers spent some 173 000 days angling on the Namibian coast during the 12 months of the 1997/98 season. The average angler thus spent some 26 days fishing and spent some N$ 3 400 doing it. Some 690 000 fish were caught, the mean weight of the daily catch was 6.06 kilograms, and the mean number of fish caught per day was 3.98. Of anglers, 46% were foreign, 38% were inland Namibians, and 16% were from coastal Namibia; 94% of anglers were male, the mean age was 45 years, and the mean income was N$115 680 per annum.

Travel cost model

Five visitation rate models were tested with different functional forms. The lin-log function had the best explanatory power for each of the five models. This is consistent with earlier research, where the semi-log function has been widely used (Ziemer *et al.*, 1980; Strong, 1983). All independent variables were, as expected, negative (i.e., the lower the travel costs, the more frequently anglers visit the coast). Further, they were all significant at the 99% level of significance ($p<0,01$). The modelling was thus successful and consistent with theory. Attempts to include other variables, such as income, in models were unsuccessful, with very low levels of significance and multicollinearity problems.

The base case model we selected for recreational angling can be described by the following function:

$$VPC = 0,004232 - 0,00055 \ln P \qquad (2)$$

where *VPC* is the number of visits per capita and *P* is the trip cost. The travel cost method estimates, for trip expenditure and consumer surplus, differ markedly between angler categories. The mean consumer surplus per trip for foreign anglers was more than three times larger than that for the Namibians. Inland Namibians enjoyed a more than two times larger consumer surplus than did the coastal Namibians. However, seen as percentage of trip costs, the coastal Namibian anglers enjoyed the largest consumer surplus, while the foreign anglers had the smallest.

The inclusion or not of on-site and other non-travel costs (accommodation, food, entry fees, costs of capital items) in the model was tested in sensitivity analysis. The consumer surplus estimates were sensitive to their inclusion. This finding points to the need for care in determining which costs to include in travel cost analysis. As explained above, our base case model was based on full inclusion of these costs, since we consider that they affect visitation rates.

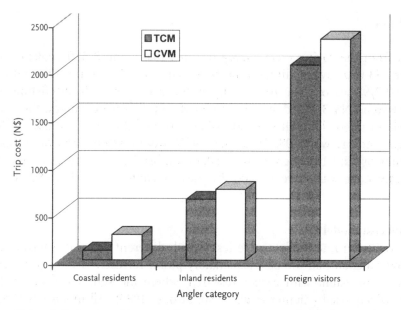

Figure 1. Estimates of mean angling trip costs for recreational shore angler categories, made using the travel cost (TCM) and contingent valuation (CVM) methods (Namibia, 1997/98).

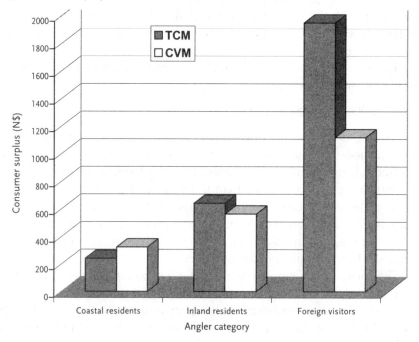

Figure 2. Estimates of mean consumer surplus for recreational shore angler categories, made using travel cost (TCM) and contingent valuation (CVM) methods (Namibia, 1997/98).

Contingent valuation

In the contingent valuation study, values from the payment card and open-ended questions were broadly comparable, and our findings confirm those of Kealy and Turner (1993), namely, that open-ended questions tend to give lower consumer surplus estimates than close-ended ones. The consumer surplus, in absolute terms, was greatest for foreigners. It was double that of the inland Namibians and more than triple that of the coastal Namibians. Expressed as percentage of expenditure, though, the coastal Namibians enjoyed a surplus of 121% compared with the foreigners' 48%.

Comparison of selected values from the travel cost and contingent valuation analyses is shown in Figures 1, 2 and 3. In Figure 1, the travel cost and contingent valuation estimates for trip costs are compared. In Figure 2 and Figure 3, the estimates for consumer surplus, and the consumer surplus expressed as a percentage of trip cost, are similarly compared. In all these comparisons there is remarkable consistency of pattern between the values. There is good consistency between techniques in estimation of expenditures, but in the estimation of consumer surplus, the travel cost method tends generally to yield higher values, particularly among foreign visitors.

Price elasticity of demand

Multiple regression models, constructed from the unaltered data, had extremely poor fit, were affected by multicollinearity, and were abandoned. Elasticity estimates were obtained, as explained above, from second-stage demand functions developed in the travel cost analysis, and derived price-

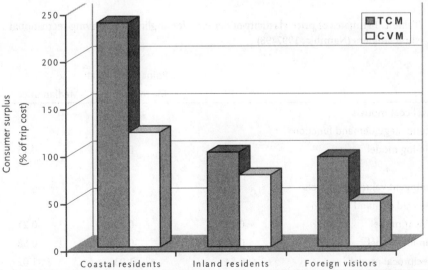

Figure 3. Estimates of consumer surplus, as percentage of trip cost, for recreational shore angler categories, made using travel cost (TCM) and contingent valuation (CVM) methods (Namibia, 1997/98).

quantity demand functions developed in the contingent valuation analysis. The lin-log form consistently provided good fit and significance. The second-stage lin-log travel cost demand function used is described as

$$Q = 18052.43 - 25.48 \ln P - 1186.61 \ln I - 837.02 \ln C \qquad (3)$$

where Q is the quantity of angling trips, P is trip cost, I is angler annual income, and C is angler consumer surplus. This model shows a negative response to rising price, as expected, but (not as expected) negative signs to the income and consumer surplus variables. The derived lin-log demand function constructed from the contingent valuation (willingness to pay) data is described as

$$Q = 266.09 - 29.43 \ln P_w \qquad (4)$$

where P_w is the willingness to pay for angling trips.

The results, shown in Table 1, suggest that demand of shore angling on the Namibian coast is price inelastic. The variation in values, depending on the model used, highlights the need for sensitivity analysis in such exercises. The simple regression models are mis-specified to the extent that other, possibly explanatory variables are omitted. Price elasticities derived from simple regressions were consistently higher than those from multiple regressions. True price elasticity is probably lower than indicated in Table 1, but comparison of results derived from the travel cost and contingent valuation models suggests broad consistency.

Table 1. Estimates of price elasticity of demand for angling trips among recreational shore-anglers (Namibia, 1997/98)

	R^2	Point elasticity at:	
		Mean price	Median price
Travel cost models			
Second-stage demand function			
Lin-log model*	1.00	-0.16	-0.15
Contingent valuation models			
Derived demand function			
Linear model**	0.73	-0.32	-0.21
Lin-log model**	0.93	-0.71	-0.58
Reciprocal model**	0.84	-1.03	-1.02

* multiple regression.
** simple regression.

Aggregate values

The aggregate angler numbers and mean values estimated for anglers were used to calculate aggregate economic values for the recreational shore fishery. The values for total direct expenditures, between N$23 million and N$31 million, are effectively measures of gross output for the recreational fishery. This gross output and the aggregated consumer surplus added together provide a gross measure of direct economic use value, between N$50 million and N$55 million. The part of this measure attributable to Namibia excludes the foreign consumer surplus (N$7 million to N$12 million). The value added to gross national income by the fishery is the proportion of gross output made up by gross value added (between N$11 million and N$15 million). Similarly, the value added to net national income is the proportion of gross output made up by net value added (between N$9 million and N$13 million).

Figure 4 shows a comparison between the aggregate expenditure and consumer surplus estimates, as derived from the three different techniques. There is general consistency in results, although the travel cost method tends to yield relatively lower value for direct expenditure, and relatively higher value for consumer surplus.

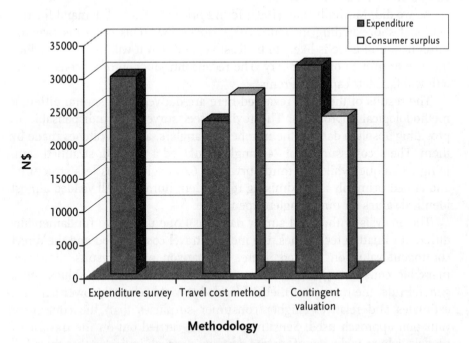

Figure 4. Estimates of aggregate direct expenditure and consumer surplus for the recreational shore fishery, as determined using expenditure analysis, travel cost method and contingent valuation (Namibia, 1997/98).

Conclusion

The studies reviewed have provided the authorities with considerable information to inform policy and management in the fishery. The recent introduction of angling licences and reduced bag limits reflect this. Comparison of results of Kirchner et al. (2000) and Zeybrandt and Barnes (2001) from Namibia with those of McGrath et al. (1997) and Brouwer et al. (1997) from South Africa, shows that Namibia's recreational shore fishery is very much smaller (between 2% and 5% of that in South Africa). There would seem to be potential for growth in the Namibian fishery but this needs to be planned with great care. More research is needed on bag limits, and the feasibility of promoting "catch and release" measures. Research is also needed to establish the most economically and biologically efficient allocation of line-fish stocks between the commercial fishery and recreational use. The gross value added of the recreational fishery (between N$11 million and N$15 million per annum) amounts to between 2,8% and 3,8% of the total gross value added in the whole Namibian fisheries sector, which was some N$391 million in 1996 (Central Bureau of Statistics, 1998). It thus has important value within the whole fisheries sector, which itself contributes some 4% of the Namibian national economy.

As we did, McGrath et al. (1997) found price elasticity of demand for recreational shore-angling in South Africa to be low. This confirms that the introduction of a fee is likely to be feasible, and that it will likely not reduce the size or growth of the fishery. The recent introduction of a licensing system will facilitate capture of rents.

The results of the work reviewed here are derived from several different methodological approaches. The roving creel survey was indispensable in providing absolute data on the numbers of anglers, and the catches made by them. The second survey, of 240 anglers entailed a targeted, stratified sampling technique, while the third survey, of 626 anglers, was less structured and aimed primarily at maximising respondent numbers. All yielded almost identical values for direct angler expenditure.

The analysis of the third survey data was done using two fundamentally different valuation techniques, the indirect travel cost method and the direct contingent valuation method. Here, comparison of the results shows remarkable consistency in pattern, and regular consistency in values. As a general rule, the travel cost method tended to yield relatively lower trip cost estimates and relatively higher consumer surpluses than the contingent valuation approach used. Sensitivity analysis, carried out on the travel cost models, where inclusion of on-site costs was varied, indicates that their full inclusion yields results closest to those of the contingent valuation. Use of both the travel cost and contingent valuation models to derive price elasticity estimates is possible. Generally, greater variation is evident between estimates from different functional forms, than between estimates from the two types of model. It can be concluded that our use of the two widely disparate methods to value the recreational fishery has shown significant convergent validation of the economic measures. For best results all methods should be employed together, but each separately can provide useful values for policy analysis.

REFERENCES

AARSA (1998): Vehicle ownership cost schedule. Automobile Association of South Africa, Johannesburg, South Africa. 6 pp.

Ashley, C. (1995): Tourism, communities, and the potential impacts on local communities and conservation. Research Discussion Paper No. 10, Directorate of Environmental Affairs, Ministry of Environment and Tourism, Windhoek, Namibia. 50 pp.

Barnes, J.I. (1996): Economic characteristics of the demand for wildlife viewing tourism in Botswana. *Development Southern Africa* 13: 377-397.

Barnes, J.I., Schier, C. and van Rooy, G. (1999): Tourists' willingness to pay for wildlife viewing and wildlife conservation in Namibia. *South African Journal of Wildlife Research* 29: 101-111.

Brouwer, S.L., Mann, B.Q., Lamberth, S.J., Sauer, W.H.H. and Erasmus, C. (1997): A survey of the South African shore-angling fishery. *South African Journal of Marine Science* 18: 165-177.

Central Bureau of Statistics (1998): Republic of Namibia: National Accounts 1982-1997. National Planning Commission, Windhoek, Namibia: 54 pp.

Hanley, N. and Spash, C.L. (1993): *Cost-Benefit Analysis and the Environment*. Edward Elgar Publishing, Aldershot, UK.

Holtzhausen, J.A. (1999): Population dynamics and life history of west-coast steenbras (*Lithognathus aureti* (Sparidae)), and management options for the sustainable exploitation of the steenbras resource in Namibian waters. PhD Thesis, University of Port Elizabeth, South Africa, 213 pp.

Holtzhausen, J.A., Kirchner, C.H. and Voges, S.F. (2001): Observations on the linefish resources of Namibia, 1990-2000, with special reference to west coast steenbras and silver kob. *South African Journal of Marine Science* 23: 135-144.

Kahneman, D. and Knetsch, J. L. (1992): Valuing public goods: the purchase of moral satisfaction. *Journal of Environmental Economics and Management* 22: 90-94.

Kealy, M.J. and Turner, R.W. (1993): A test of the equality of close-ended and open-ended contingent valuations. *American Journal of Agricultural Economics* 75: 321-331.

Kerr, G.N. (1986): *Introduction to Non-Market Valuation: Theory and Methods*. Centre for Resource Management, Lincoln College, Christchurch, Canterbury, New Zealand.

Kirchner, C.H. (1998): Population dynamics and stock assessment of the exploited silver kob (*Argyrosomus inodorus*) stock in Namibian waters. PhD Thesis, University of Port Elizabeth, Port Elizabeth, South Africa,

246 pp.

Kirchner, C.H., Sakko A.L. and Barnes, J.I. (2000): An economic valuation of the Namibian recreational shore-angling fishery. *South African Journal of Marine Science* 22: 17-25.

McGrath, M.D., Horner, C.C.M., Brouwer, S.L., Lamberth, S.J., Mann, B.Q., Sauer, W.H.H. and Erasmus, C. (1997): An economic valuation of the South African linefishery. *South African Journal of Marine Science* 18: 203-211.

Mitchell, R. and Carson, R. (1989): *Using Surveys to Value Public Goods, the Contingent Valuation Method*. Resources for the Future, Washington, DC, USA.

Moran, D. (1994): Contingent valuation and biodiversity: measuring the user surplus of Kenyan protected areas. *Biodiversity and Conservation* 3: 663-684.

Navrud, S. and Mungatana, E.D. (1994): Environmental valuation in developing countries: the recreational value of wildlife viewing. *Ecological Economics* 11: 135-151.

NOAA Panel (1993): Natural resource damage assessments under the Oil Pollution Act of 1990. U.S. Department of Commerce, National Oceanic and Atmospheric Administration, Federal Register 58(10): 4601-4614.

Pearce, D.W. and Turner, R.K. (1990): *Economics of Natural Resources and the Environment*. Harvester Wheatsleaf, London, UK.

Strong, E. (1983): A note on the functional form of travel cost models with unequal zonal populations. *Land Economics* 59: 342-349.

Thomson, C. J. (1991): Effects of the avidity bias on survey estimates of fishing effort and economic values. *American Fisheries Society Symposium* 12: 356-366.

Zeybrandt, F. and Barnes, J.I. (2001): Economic characteristics of demand in Namibia's marine recreational shore fishery. *South African Journal of Marine Science* 23: 145-156.

Ziemer, R., Musser, W. and Hill, C. (1980): Recreation demand equations: functional form and constructions. *American Journal of Agricultural Economics* 62: 136-141.

12 THE NAMIBIAN-SOUTH AFRICAN HAKE FISHERY: COSTS OF NON-COOPERATIVE MANAGEMENT

*Claire W. Armstrong and Ussif Rashid Sumaila**

Abstract

The Namibian Exclusive Economic Zone borders South African, Angolan and international waters. Several commercially important species migrate in and out of these different areas. Hake is one such stock which is shared between these countries since it hardly straddles international waters. Despite the literature showing that such shared stocks make cooperative management between the respective states sharing them substantially more profitable than the alternative, at the moment we do not see agreements of this kind between Namibia and its neighbours. In this paper, we apply bioeconomic theory to study the potential for loss of benefits to Namibia and South Africa from not cooperating in the management of the hake fishery. Angola is not included in the analysis because it catches less than 0.5% of the total annual landings of hake.

INTRODUCTION

The literature on the management of transboundary fisheries has demonstrated the large potential gains that may be achieved from cooperation between nations exploiting shared stocks[1] (see for instance, Munro, 1979; Armstrong and Flaaten, 1991). In countries with long traditions for industrial

---

* We thank Dave Boyer, Gordon Munro and two anonymous referees for comments. Sumaila thanks the Research Council of Norway, the Sea Around Us Project and the Pew Charitable Trusts for their support.
[1] The fishery studied in this paper fits into the definition of shared stocks, as we will be studying a stock that crosses the boundaries of two countries' Exclusive Economic Zones (EEZs) (Munro, 2000). Hence, we will not be discussing so-called highly migratory or straddling stocks, where for the former, tuna is the most prominent example (Duarte etal., 2000), and the latter covers all species that move between EEZs and adjacent high seas (Lindroos and Kaitala, 2000)

fisheries, bilateral agreements are common and historic – a case in point is Norway and Russia in the management of the Northeast Atlantic cod and other stocks in the Barents Sea. In countries where artisanal fisheries are the rule, these types of agreements are not often seen, most likely due to social and political concerns.

A recent Norway FAO Expert consultation on the management of shared fish stocks took place in Bergen, Norway (FAO, 2002; 2003). The main goal of this consultation was to provide information and analysis to foster international understanding of the management of shared fish stocks. Several case studies of the state of shared management of joint stocks of fish from different parts of the world were presented (FAO, 2003). FAO (2002) reports the conclusions of three Working Groups of the Consultation: (i) resolving allocation issues, (ii) achieving coordination of management plans and objectives and research problems, and (iii) ensuring implementation and enforcement of management agreements. These are issues that participants at the Consultation believed needed to be properly addressed to ensure that the world's joint fishery resources are managed sustainably, and in an economically sensible manner.

In the case of Namibia, which historically has been dominated by foreign fleets (Sumaila and Vasconcellos, 2000), cooperation was to a large degree directed towards South Africa, from which Namibia obtained independence in 1990. Despite or perhaps as a result of this, we do not observe specific fisheries agreements between the two countries. Though the political instability in Namibia's second neighbouring country, Angola, has not been a situation that encourages cooperative agreements,[2] the lack of agreements with South Africa shows that a more stable situation in Angola might not be expected to affect this.[3] The question therefore becomes, in a backward-

---

[2] Since 1994 some Namibian purse seiners have been licensed to fish in Angolan waters. This was originally negotiated due to declining catches of sardine in Namibian waters. In 1995 a large part of the sardine stock moved into the Angolan EEZ and half of the total catch of that year came from Angola. The stock collapsed and since then these vessels have been forced to target sardinella (*Sardinella aurita* and *S. maderensis*) and Angolan horse mackerel (*Trachurus trecae*), neither of which are considered shared stocks. (Dave Boyer, pers. comm.).

[3] In Anon (1999) the problem of stock depletion of transboundary stocks is connected to "lack of collaborative management of shared resources" (Table A1). Yet under suggestions for activities to remedy this, cooperation is limited only to industrial financing, resource assessment, information gathering, and the like. Actual cooperative fisheries management is not mentioned. Sydnes (2002) finds this lack of cooperation at the management level to be common amongst developing countries.

looking perspective, has there been a price paid for not having such cooperative agreements?[4] We will, in the following, concentrate on the hake fishery.

The hake fishery

Namibia's EEZ borders both Angolan and South African waters, and several different species including hake cross these borders. Hake is found in the Benguela Current Large Marine Ecosystem (BCLME), which covers the west coast of South Africa, the entire Namibian coast, and southern Angola depending on the position of the Angola-Benguela front, which moves seasonally between 14 and 17 degrees South.

Three of the most important fisheries in the EEZ are hake, horse mackerel and sardine (Sumaila, 2000). The most economically important ones are sardine, which is found mainly in Namibian, South African and occasionally Angolan waters, and hake, which migrates between South Africa and Namibia (see Namibian and South African hake catches in Figure 1). In the following, we will study the issues connected with this latter shared stock. However, there are actually two species of hake involved,[5] *Merluccius paradoxus* and *Merluccius capensis*, that mix in Namibian waters, increasing the complexity of this fishery. In practical management, this fact is not taken into account, as the hake is treated as a single stock fishery, which we will also do in the following. Hake accounts for around half of the value of Namibia's fish catch, while the strongly fluctuating sardine fishery has contributed much less (Manning, 2000).

The shallow-water Cape hake, *M. capensis*, dominates the Namibian catch, while South Africa catches mostly the deep-water hake, *M. paradoxus*. However, recent data show that the proportion of *M. paradoxus* in Namibia's catch continues to increase. The *M. capensis* is found further inshore than the *M. paradoxus*, and as the fish grow older they tend to migrate into deeper waters, where the older *M. capensis* mix with the younger *M. paradoxus*, leading to predation by the former upon the latter (Hutchings, 1995). However, both *M. paradoxus* and *M. capensis* are transboundary species (Boyer and Hampton, 2001), and the fact that they interact "cannibalistically" brings in further stock-management issues. Hence, the way and manner in which

---

[4] The interesting question of what price may be paid by posterity if such an agreement does not come about, is left for future research to determine. We would like to stress that this paper must be seen as a minor first step in a deeper future analysis of the fishery in question.

[5] In fact there are three hake species in Namibian waters, the third being the Benguela hake (*Merluccius polli*), which is mainly found in Angolan waters (Boyer and Hampton, 2001).

Namibia manages *M. capensis* can be expected to affect the South African harvest of *M. paradoxus*, and vice versa.

Figure 1 shows a threefold increase in the Namibian harvest of hake in the last decade. This is due to Namibian conservative harvesting policy immediately after independence following years of mismanagement. This policy resulted in very low post independence harvests, which were allowed to increase over the ten-year period presented. South African harvests, on the other hand, stayed relatively constant.

Model

To help manage a given shared fish stock such as hake of the Benguela current ecosystem, a number of equilibrium solutions can be computed from bioeconomic models (Sumaila, 1999):
- sole owner solution;
- open access equilibrium solution;
- non-cooperative solution, and;
- cooperative solution.

A sole owner solution is relevant if the countries sharing the resource were to agree to allow only one of them to exploit the resource under some form of arrangement. Or, if the countries could exploit the resource as if they were a sole owner in all respects, that is, they work in full harmony in terms of their objectives with respect to the use of the shared resource. This is clearly a difficult equilibrium to reach in practice but it can serve as a benchmark for the ideal economic solution that can be achieved. The open access equilibrium is the other end of the story – this solution is arrived at when there is no attempt to manage the exploitation of the resource in any way. Hence,

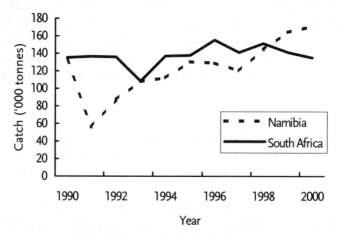

Figure 1. Hake catch by South Africa and Namibia from 1990 to 2000.

fishers of both countries pursue their fishing interest unilaterally. This equilibrium solution can also serve as a benchmark for the worst equilibrium that can be reached. A non-cooperative equilibrium will be used to depict the situation in which each country manages the fishery alone under the assumption that the fish is not a shared resource. Thus, the non-cooperative solution (Clark, 1980) lies between the open access and sole owner solution. This solution we believe captures the current management situation for hake. Finally, a cooperative solution is one in which the two countries work together to maximize their joint benefits from fishing after taking into account their different preferences (Munro, 1979).

We begin by developing the sole owner and open access solutions in the case of the shared hake stocks. We then compute and compare the solutions from these management scenarios with the outcome obtained given the current management regimes in the two countries. Finally, we develop and compute a simple cooperative solution in which the two countries differ only in the prices their fishers receive per unit weight of fish. It should be noted that the models presented are simple 'first cut' models that should lead to more detailed models in our future research. A detailed model will, for example, relate and discuss the different equilibrium solutions to the particularities of the fishery more rigorously.

The biological model

On the biological side of the model, a logistic growth function is applied. We assume, in similar fashion to that of biologists working on Namibian hake, that the two main hake stocks found in the ecosystem are similar enough ecologically to be considered as one and the same for modelling purposes. Hence, growth is described by the single species logistic function:

$$F(x) = rx(1 - \frac{x}{K}), \qquad (1)$$

where x is the stock biomass, r is the intrinsic growth rate and K is the carrying capacity of the environment. It is assumed that in equilibrium harvest equals growth, i.e. $h(t)=F(x)$, hence equation (1) gives the equilibrium harvest.

The sole owner model

The sole owner is assumed to maximise the following objective function:

$$PV = \int e^{-\delta t}(p - c(x))h(t)dt, \qquad (2)$$

where PV is present value, p, is the unit price of harvest $h(t)$, and $c(x)$ is the unit cost of harvesting, as a function of the stock biomass x, while δ is the discount rate. The variables x and h must satisfy the non-negativity constraints.

The solution to the above problem is (Clark, 1990):

$$F'(x) - \frac{c'(x)F(x)}{p-c(x)} = \delta \qquad (3)$$

The left side of equation (3) describes the stocks' own rate of economic rent creation, which under optimal conditions must equal society's rate of discount. This equation implicitly gives the optimal stock size, economic rent, etc.

The open access model

According to Gordon's theory of the open access fishery, fishing effort expands to a level E_∞ at which revenues exactly equal opportunity costs (Gordon, 1954). In terms of our logistic model, the equilibrium effort level is given by (Clark, 1990):

$$E_\infty = \frac{r}{q}(1 - \frac{c}{pqK}), \qquad (4)$$

given that $c(x)=c/qx$, where c is the unit cost of effort, q is the catchability coefficient and K is the carrying capacity of the stock. The corresponding stock level $x = x_\infty$ is:

$$x_\infty = \frac{c}{pq} \qquad (5)$$

Hence, in open access the optimal stock size is determined only by the cost, price and catchability data.

The cooperative model

In Munro's 1979 paper on transboundary fishery management, a bioeconomic model of two states managing a single transboundary fish species is presented. The two countries are modelled as having different costs, discount rates or consumer tastes. In this work we apply Munro's transboundary management model to the hake fishery in order to determine whether there is anything to gain from cooperative interaction between South Africa and Namibia in the management of this species. We will, however, limit this study by assuming that the key difference between the two countries is the prices they receive per unit weight of hake they land.

It is demonstrated in Clark (1980) and also in Levhari and Mirman (1980) that the outcome to the fishing nations of non-cooperation is, as would be expected, of unquestionable undesirability. In developing a cooperative management analysis, Munro (1979) combined the standard economic model of the fishery with cooperative game theory. In the following, we redefine this model to suit our game of two players harvesting the same fish stock while facing different prices in the marketplace. It is assumed that there is potential for a binding agreement under cooperation.

The objective functions that players 1 and 2 wish to maximise are assumed to be:

$$PV_1 = \int e^{-\delta t}(p_1 - c(x))\alpha(t)h(t)dt, \tag{6a}$$

$$PV_2 = \int e^{-\delta t}(p_2 - c(x))(1-\alpha(t))h(t)dt \tag{6b}$$

where $0 \leq \alpha(t) \leq 1$, is player 1's share of the harvest, p is the unit price of harvest $h(t)$, and $c(x)$ is the unit cost of harvesting, as a function of the stock biomass x, while δ is the discount rate.

Given the above objectives, a potential cooperative agreement can be characterised by:

$$\max PV = \beta PV_1 + (1-\beta)PV_2 \tag{7}$$

Where β, a constant that can take values from 0 to 1, is a bargaining parameter and a measure of the weight given to 1's management preferences.

By solving the above maximisation problem for each possible β ($0 \leq \beta \leq 1$),[6] we determine the Pareto efficient frontier in the space of realised payoffs. By choosing a cooperative game solution option, β can then be determined for this. Assuming that only prices differ for the two countries, the optimal biomass time path is given by the following modified golden rule equation (Munro, 1979):

$$F'(x^*) - \frac{[\alpha\beta + (1-\alpha)(1-\beta)]c'(x^*)F(x^*)}{\alpha\beta p_1 + (1-\alpha)(1-\beta)p_2 - [\alpha\beta + (1-\alpha)(1-\beta)]c(x^*)} = \delta \tag{8}$$

For the cooperative regime, we assume that there is a binding agreement between the two countries, that the harvest shares are constant, and that no side payments are allowed.

In solving (8) for x^*, we obtain the optimal stock level as a function of β, which can be applied to equations (6a and 6b), in order to obtain the two countries' optimal present values (again as functions of β). By varying the bargaining parameter β between 0 and 1, we find the payoff possibility fron-

---

[6] For $\beta = 0$ or $\beta = 1$, the optimal harvest program reduces to that of single owner (country 2 if $\beta = 0$; country 1 if $\beta = 1$).

tier. The optimal β can be determined by using the Nash bargaining solution, (Munro, 1979).[7]

Discrete versions of the models described above are run for the last decade, that is from 1990 to 2000.

Data

Namibia exports almost all of its hake harvests, while South Africa has exported approximately 45% of its hake harvests (Anon, undated)[8]. The reasons for this difference between the two countries are different technology, quality of production, as well as market opportunities. This results in different first-hand prices in the two countries. We have calculated the average first-hand value of hake in 1997 to be 3,852 and 5,140 N$/tonne for South Africa and Namibia respectively (data from Anon, 2000). The prices are applied to the cooperative solution studied. In the sole owner case we apply the higher price, as we are interested in the overall optimal solution, the optimum optimorum (Munro, 1979). Since many of the same companies participate in the hake fisheries both in Namibia and in South Africa, we assume that the costs are the same for the two countries. Costs per unit harvest are assumed to be approximately half of the price per unit harvest (Aina Ulenga, MFMR, pers. comm.). Following Backeberg (1997), a discount rate of 10% is used. With regards to biological data, we assume that the intrinsic growth rate r is 0.344 (Punt, 1994), and the carrying capacity K is 4.4 million tonnes, which is twice that reported for the hake stock outside South Africa in the beginning of the twentieth century (Hampton et al., 1999). The stock size prior to the analysis is assumed to be 0.9 million tonnes (Hampton et al., 1999).[9]

---

[7] For a review of bargaining schemes, see Kaitala (1985). In this paper we will only study the so-called dictator solutions, that is, β equals 0 or 1, as with the data given below this will be sufficient to show that a cooperative solution will not differ substantially from the sole owner solution.

[8] Recent price increases in the Spanish market have, however, increased this share to 65% (AllAfrica, 2002).

[9] This is an average of survey data from South Africa and Namibia, and an estimate from an age-structured production model (Hampton et al., 1999). The two numbers differed by about 100,000 tonnes. The survey estimate for 1990 was, however, uncharacteristically high as compared with the closest years, but nonetheless fit well with the data in the age-structured production model.

RESULTS

When studying the cooperative solution with different prices, we find that the outer limits as regards the payoff possibility frontier, that is β equals 1 and 0, result in optimal stock sizes that hardly differ (1,871,491 and 1,780,404 tonnes, respectively). Hence, weighting of management preferences to the advantage of either of the countries hardly affects the optimal harvesting strategies, this despite the South African price being only 75% of the Namibian price. Hence, even greater differences in the parameters involved are needed in order to obtain more significant effects of cooperative management.[10] In the following, we will derive results under sole ownership management and open access, and compare this with the computed rent from the actual harvest profile in the 1990s.

Using the data presented earlier, we calculated the economic rent, biomass and catch levels obtainable under the various equilibrium solutions presented above. The results show that under sole ownership the equilibrium stock size is about 1.9 million tonnes (see Figure 2); the equilibrium

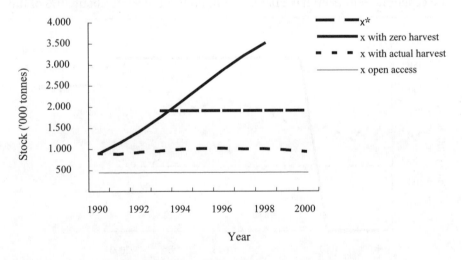

Figure 2. Stock size from 1990 to 2000 using the presented model.

---

[10] This lack of big difference may well be due to the fact that our single species biological model does not capture the natural interaction between juvenile and adult hake, and the interaction between the fishing gears employed to exploit the resource in the two countries.

harvest is about 370 thousand tonnes (see Figure 3), and the total economic rent over the 11-year period of 1990 to 2000 is about N$8.6 billion[11]. The equivalent numbers for the open access equilibrium are just under 0.5 million tonnes; 138 thousand tonnes; and zero N$.[12] These numbers depict the potential bioeconomic loss when hake is exploited under pure open access. The numbers show that while as much as N$8.6 billion of economic rent is dissipated under open access compared with what is obtainable under sole ownership, the stock size is also much lower, at about only a quarter of the biomass under sole ownership.

We know that hake is not exploited under pure open access, since Namibia and South Africa both have elaborate management arrangements for hake. However, the stocks are currently non-cooperatively managed because even though hake is a shared stock, it is not managed cooperatively. We allow the actual catch profiles of the two countries to help us depict the non-cooperative solution for this fishery. The justification for this is that the actual management has been one of non-cooperation between the two countries involved, hence both parties can be expected to have acted in expectation of the other maximizing own interests. Our calculations indicate that the economic rent from hake harvested in both countries is about 70% of the

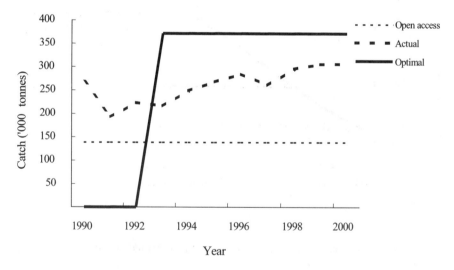

Figure 3. Catch in tonnes from 1990-2000.

---

[11] The economic rent is calculated for the 11-year period and discounted back to 1990. In the case of optimal harvests, the harvesting commences in 1993, when the build-up phase to optimal stock size is over. Hence economic rent in the optimal scenario is calculated for the years 1993-2000.

[12] We disregard rent from the "fished-down" phase.

N$8.6 billion computed when sole ownership is assumed. The current estimated biomass is just less than 1 million tonnes; hence, it is approximately half of the sole ownership equilibrium biomass. The results, however, show that just over two thirds of the potential rent from the hake stocks is actually realized under current management regimes. Thus, our estimate of the potential cost of non-cooperation is about N$3.6 billion over the 11-year period being studied. This implies an average annual loss of about N$327 million or 31% of the total achievable rent under joint management. The low estimated stock level compared with the optimal case is an indication of potentially greater losses in the future.

Conclusion

We have presented a simple model for the hake fisheries of Namibia and South Africa, and used it to estimate the potential cost of non-cooperative management of this shared stock. Because of the simple nature of the model, our results can only be taken as indicative of this cost. More detailed modelling of both the biological and economic aspects of the fishery is required before something definitive can be said about the potential gains from cooperative management.

With respect to biology, detailed modelling that takes into account the natural interactions between the different stocks of hake, between juvenile and adult hake through cannibalism, and between hake and other species in the ecosystem is required. On the economic side, we need to incorporate more on preferences and differences between Namibia and South Africa into the model to be able to make more categorical statements about the cost of non-cooperation. Also, how the two countries negotiate and decide on the sharing of benefits from cooperation is left to future work.

Having said the above, the current results demonstrate that the potential losses due to non-cooperation can be quite high. Thus, there is a need for the development of cooperative agreements between Namibia and South Africa for the management of the hake stocks of the Benguela current ecosystem, if the full benefit from the resource is to be extracted. It must be added that this potential loss notwithstanding, Namibia retains, relative to other countries, a significant amount of rent from its management of the hake fishery. Still, there are further gains to be made from cooperative management between Namibia and South Africa.

REFERENCES

AllAfrica (2002): Export Needs Cause Local Hake Shortage. *allAfrica.com*, Aug. 16; http://allafrica.com.

Anon (1999): Transboundary Diagnostic Analysis (TDA) UNDP Windhoek, Namibia. http://www.ioinst.org/bclme/resources/index.htm, http://www.ioinst.org/bclme/resources/table_a1.htm.

Anon (2000): *Fishing Industry Handbook South Africa, Namibia and Mocambique 2000*. Exbury Publications, South Africa.

Anon (undated): The Economic and Sectoral Study (ESS) of the South African fishing industry. http://www.envirofishafrica.co.za/ess/ESS2000WEBSITE/essindex.htm.

Armstrong, C. and Flaaten, O. (1991): The optimal management of a transboundary resource: The Arcto-Norwegian cod stock. In: *Essays on the Economics of Migratory Fish Stocks*. (R. Arnason and T. Bjørndal, eds.), pp. 137-152. Springer-Verlag, Berlin.

Backeberg, G. (1997): Water institutions, markets and decentralised management: prospects for innovative policy reforms in irrigated agriculture. Presidential address, 35th Annual Conference of the Agricultural Economics Association of South Africa, East London.

Boyer, D.C. and Hampton, I. (2001): An overview of the living marine resources of Namibia. *South African Journal of Marine Science* 23: 5-35.

Clark, C.W. (1980): Restricted access to common-property fishery resources: a game theoretic analysis. In: *Dynamic Optimisation and Mathematical Economics*. (P.Liu, ed.), Plenum Press, New York.

Clark, C.W. (1990): *Mathematical Bioeconomics. Optimal Management of Renewable Resources*. Wiley and Sons, Toronto, New York.

Duarte, C.C., Brasao, A. and Pontassilgo, P. (2000): The application of C-Games. *Marine Resource Economics* 15(1): 21-36.

FAO (2003): Papers presented at the Norway-FAO Expert Consultation on the Management of Shared Fish Stocks, Bergen, Norway, 7-10 October 2002. FAO Fisheries Report No. 695 Supplement.

FAO (2002): Report of the Norway-FAO Expert Consultation on the Management of Shared Fish Stocks, Bergen, Norway, 7-10 October 2002. FAO Fisheries Report No. 695.

Gordon, H.S. (1954): Economic theory of a common property-resource: the fishery. *Journal of Political Economy* 62: 124-142.

Hampton, I., Boyer, D.C., Penney, A.J., Pereira, A.F. and Sardinha, M. (1999): Integrated overview of fisheries of the Benguela Current region. Thematic report for the Benguela Current Large Marine Ecosystem Programme. Windhoek: United Nations Development Programme.

Hutchings, L. (1995): An overview of South African hake fisheries research. In: *The Benguela Current and Comparable Eastern Boundary Upwelling Ecosystems*. (M.J. O'Toole, ed.), Ministry of Fisheries and Marine Resources, Swakopmund, Namibia.

Kaitala, V.T. (1985): Game Theory Models in Fisheries Management: A Survey. In: *Dynamic Games and Applications in Economics*. (T. Basar, ed.), Springer Verlag, Berlin.

Levhari, D. and Mirman, L.J. (1980): The great fish war: An example using dynamic Cournot-Nash solution. *Bell*

Journal of Economics 11: 322-334.

Lindroos, M. and V. Kaitala (2000): Coalition Game of Herring. *Marine Resource Economics* 15(4): 321-340.

Manning, P. (2000): Review of the distributive aspects of Namibia's fisheries policy. NEPRU Research Report no. 21.

Munro, G.R. (1979): The optimal management of transboundary renewable resources. *Canadian Journal of Economics* 12: 355-376.

Munro, G.R. (2000): The UN fish stocks Agreement of 1995. *Marine Resource Economics*, 15(4): 265-280.

Punt, A.E. (1994): Assessment of the stocks of cape hakes off South Africa. *South African Journal of Marine Science* 14: 159-186.

Sumaila, U.R. (1999): A review of game theoretic models of fishing. *Marine Policy* 23(1): 1-10.

Sumaila, U.R. (2000): Fish as vehicle for economic development in Namibia, 2000. *Forum for Development Studies* 2: 295-315.

Sumaila, U.R. and Vasconcellos, M. (2000): Simulation of ecological and economic impacts of distant water fleets on Namibian fisheries. *Ecological Economics* 32: 457-464.

Sydnes, A.K. (2002): Regional fishery organisations in developing regions: adapting to changes in international fisheries law. *Marine Policy* 26: 373-381.

13 A STOCHASTIC FEEDBACK MODEL FOR OPTIMAL MANAGEMENT OF NAMIBIAN SARDINE

*Stein Ivar Steinshamn, Arne-Christian Lund and Leif K. Sandal**

Abstract

A stochastic feedback model is applied in order to derive optimal harvesting strategies for the Namibian sardine that takes into account socio-economic considerations in addition to biological safety and sustainability. There are certain features pertaining to the sardine fishery that need special attention. These features are the technology of the fishery, uncertainty and stochasticity and market conditions. One of the main results from this study is that more uncertainty does not necessarily imply more conservative management and lower harvest; under certain circumstances the opposite may be the case. Based on what we think is the most realistic version of the model, the main result is that in the present situation, that is for stock levels less than about 500 000 tonnes, a harvest moratorium should be instituted. The optimal harvest path is then a piecewise curve due to the demand function. The optimal long-term steady state is a stock at around two million tonnes, and the economic optimal harvest is around 450 000 tonnes at this stock level. The overall conclusion, then, is that in the present situation severe measures have to be taken in order to get the sardine stock out of the trap where it is at present. If the right measures are taken, the sardine stock may give a valuable contribution to the Namibian economy in the future, otherwise it may disappear for a long time. However, 2002 is the first year since 1981 with a zero TAC, indicating that drastic action is now beginning to take place.

INTRODUCTION

In this chapter a stochastic model is applied in order to derive optimal harvesting strategies for the Namibian sardine that takes into account socio-economic considerations in addition to biological safety and sustainability. The deterministic version of the model is described by Sandal and Stein-

---

* The authors are very grateful to Dave Boyer and Les Clark for valuable suggestions and input to the chapter.

shamn (1997a, 2001a) and the stochastic version by Sandal and Steinshamn (1997b). Application of the deterministic model to the Arcto-Norwegian cod and Atlanto-Scandian herring can be found in Arnason et al. (2000), to the cod off Newfoundland in Grafton et al. (2000) and to Namibian sardine in Sandal and Steinshamn (2001b). Application of the stochastic model to Southern Bluefin Tuna can be found in McDonald et al. (2002).

There are certain features pertaining to the sardine fishery that need special attention. These features are the technology of the fishery, uncertainty and stochasticity and market conditions. More about the sardine can be found in other chapters in this book, for example the chapters by Boyer and Oelofsen or Sumaila and Steinshamn.

The Namibian sardine fishery is characterized by purse seine technology. This means that there is relatively little relationship between stock size and harvesting costs. The costs of catching one school of fish do not depend on the total number of schools to any large extent. An important implication of this is that the optimal steady state may not be as sensitive to economic parameters as otherwise would have been the case.

The Namibian sardine is also characterized by a high degree of uncertainty and stochasticity in the biological functions. This is mainly due to the unpredictable Benguela Current system, and it calls for explicit inclusion of rather general stochastic processes in the models.

Despite the stochasticity and unpredictability it is possible to detect certain time trends in the productivity of the sardine stock. The biology of the stock seems to go through different regimes over time. This calls for independent analysis of different time periods, and it gives interesting and novel insight regarding the biological productivity of the sardine stock in different periods.

One of the main results from this study is that the consequence of introducing stochasticity is not monotone regarding optimal management. In other words, more uncertainty does not necessarily imply more conservative management and lower harvest, under certain circumstances the opposite may be the case.

DETERMINISTIC MODEL[1]

The general dynamic optimization can be described by a couple of equations, namely:

---

[1] This section is rather technical and can be skipped by those who are mainly interested in the results.

$$W = \int_0^\infty \Pi(x,y,t)dt \qquad (1)$$

subject to the dynamic constraint

$$\dot{x} = g(x) - y$$

and the appropriate transversality conditions. Here x is the fish stock, y denotes the yield rate (or harvest rate) from the stock and t denotes time. The function $g(x)$ is the biological surplus production function for the stock. Time is the basic variable here as the variables $x = x(t)$ and $y = y(t)$ are themselves functions of time. Dot-notation is used to represent time derivatives, that is:

$$\dot{x} \equiv \frac{dx}{dt}.$$

The function W is the objective that is to be maximized. Further, the function Π can represent private net revenue, that is producers' surplus, or some measure of social welfare, for example the sum of producers' surplus or consumers' surplus. In this chapter, Π will usually represent the producers' surplus for the harvesting sector, and it assumed that Π is concave. The processing sector is also important for the Namibian economy, but due to lack of economic data this sector has to be excluded from the analysis.

In the basic version of the model dealt with here the only sort of explicit time dependence will be discounting of the future. That is,

$$\Pi(x,y,t) = e^{-\delta t}\Pi(x,y)$$

where δ is the discount rate (and $e^{\delta t}$ is the discount factor). The method applied to solve this maximization problem is optimal control theory, see, for example, Kamien and Schwartz (1991) or Clark (1990) for application to fisheries.

In the following all variables are in current values. The Hamiltonian for the problem is given by

$$H(x,y,m) = \Pi(x,y) + m \cdot [g(x) - y] \qquad (2)$$

where m is the so-called costate variable. The first-order conditions for maximum are[2]

$$H_y = 0, \quad \dot{x} = g(x) - y$$
$$\dot{m} = \delta m - H_x, \quad \dot{H} = \delta m \dot{x}.$$

---

[2] Subscripts are used to denote partial derivatives.

The last of these equations follows from the three previous ones. The conventional approach in optimal control theory would be to find an alternative expression for m from $H_y = 0$, equate this to the expression we already have for m, and use this together with the equation for x to construct a system of two differential equations, one for x and one for y. A typical approach is to set these equal to zero and solve the system for the steady state.

In this chapter we are more interested in the optimal way to approach steady state than we are in the steady state itself. The reason for this is that faced with practical management problems one is usually far away from the optimal steady state, and it is therefore not sufficient to know the optimal policy only in, or very close to, the steady state. We therefore take the following alternative approach. The so-called maximum principle, $H_y = 0$, implies that $II_y - m = 0$. From this follows that the costate variable (which in optimum can be interpreted as the shadow value of the resource) can be written as a function

$$m = M(x, y) = II_y.$$

This is a known function when II and g are known, and it can therefore be used to eliminate m from the system. Substituting M for m in the Hamiltonian we get a new function that is equal to the Hamiltonian in value along an optimal trajectory, but different as a function. This new function can be defined as

$$P(x, y) = H(x, y, M(x, y)).$$

This function has the same interpretation as the Hamiltonian, namely as the rate of increase of total assets, and it will here be called total economic rent for short.

Here we are primarily looking for optimal policies to set total allowable catches as a function of the stock level, that is to find y as a function of x, $y = y(x)$. This inserted into $H = \delta m x$ from (2) yields a first-order differential equation that can be used to determine the feedback control. This equation can be written

$$\frac{dP}{dx} = \frac{\partial P}{\partial x} + \frac{\partial P}{\partial y}\frac{dy}{dx} = \delta \cdot M(x, y). \quad (3)$$

Equation 3 is the basic equation that will be used in the deterministic model. The term feedback is important here. This means that we have a rule to determine optimal harvest as a function only of the last observed stock estimate given that the parameters in the model have been properly estimated. The implication of this is that management becomes genuinely adaptive, and no forecasting is required.

BIOLOGICAL MODEL

Biological fisheries models can be divided in two broad categories: aggregated and disaggregated (year-class) models. Year-class models are primarily used for short term purposes, mainly stock assessment. These usually have a high degree of detail with respect to individual weight, sexual maturity, natural mortality, etc., between year-classes, that are useful for short-term prediction of stock abundance. These details are less useful for long-term predictions, however, as such biological characteristics tend to vary over time. Hence aggregated biological models are more useful for describing long-term variations of the stock, and derive long-term harvesting strategies, as they use averaged parameters.

Several functional forms of the surplus production model given by Equation (1) were tried, and the following gave the best statistical properties[3]

$$g(x) = \alpha x^3 + \beta x^4. \tag{4}$$

The function given by (4) entails depensation, but not critical depensation. Depensation means that there is a convex area of the function close to the origin implying that the production of the stock is low here. Critical depensation means that there is a convex area with negative growth close to the origin. In a deterministic model this implies that below a certain stock level the population will go extinct even without any harvest, see also Clark (1990). The apparent collapses of the stock in the mid-sixties, and again in the mid-seventies, as seen from Figure 1 may partly be attributed to such depensation.

The results of the estimation are given in Table 1 with biomass measured in 1000 tonnes.

Table 1. Statistical properties of the surplus production model estimated with data from 1969 to 2000 (1975 removed as an outlier).

Parameter	t-value	
α = 3.87E-7	6.7	R^2 = 0.60
β = -1.61E-10	-6.2	F = 45.3, DW = 1.21

ECONOMIC MODEL

The main components of the economic model are revenue and costs. As the

---

[3] The estimations were performed using the program NLREG.

objective of the model is to maximize the welfare of the Namibian people, it would be natural to include the harvesting as well as the processing sector; that is, to maximize the producers' surplus from both these sectors. Unfortunately it has not been possible to get sufficient data from the processing sector to include this in the optimization, so we concentrate entirely on the harvesting sector in the following. The so-called consumers' surplus is not included in the welfare concept, as most of the fish products based on sardine are exported abroad, mainly to South Africa.

Gross Revenue Function

The revenue function can be described by the equation

$$R(y) = p(y)y$$

where y is harvest as earlier and $p(y)$ is the inverse demand function. The inverse demand is the price as a function of output instead of the other way around. In general the demand is a function of several variables, in particular own price, price of close substitutes and income. As income and price of substitutes typically change over time, this would call for non-autonomous, or time-dependent, functions. Here we concentrate on demand as a function of own price, or rather the inverse, the price as a function of output or harvest. For simplicity we drop the word inverse in the following.

We do not attempt to estimate the demand as a function of multiple variables. This is partly because sufficiently good data do not exist and partly because it is not required for this type of model. What is needed here is to find a formulation of the revenue function that can describe reality reasonably well.

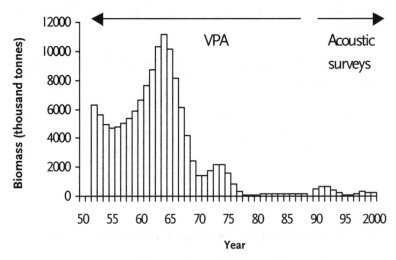

Figure 1. Estimated biomass of Namibian sardine. Source: VPA estimates are from Butterworth (1983). Acoustic estimates are annual means.

In this work the following function has been applied:

$$p(y) = \frac{ap_{max} + p_{min}y}{a+y} \qquad (5)$$

where p_{max} is the maximum price, p_{min} is the minimum price and a is a parameter to be determined. This function has some useful characteristics: When $y = 0$ the price is p_{max} and when the quantity approaches infinity ($y \to \infty$) the price approaches p_{min}. The maximum and minimum price can be pre-specified or estimated from data. Given that the parameter $a>0$, the demand function is always downward sloping

$(\frac{dp}{dy} < 0)$

and convex

$(\frac{d^2 p}{dy^2} > 0)$.

Furthermore, this function is also always elastic. The elasticity of demand as a function of p is given by

$$El = \frac{p(p_{max} - p_{min})}{(p_{max} - p)(p - p_{min})}.$$

It is relatively easy to show that this is always greater than one, and it is at its minimum when $p = \sqrt{p_{max} * p_{min}}$. Further, $El \to \infty$ both when $p \to p_{max}$ and $p \to p_{min}$. With this demand function, gross revenue will be an increasing but concave function in output.

Cost function
The cost function applied here is supposed to include the costs of the harvesting sector. The costs are in general supposed to be a function of both the stock size and the harvest. In the special case of the sardine fishery, these costs, however, are only weakly dependent upon the size of the fish stock. The reason for this is that the sardine is a schooling species and the fishery is mainly purse seining. It is usually almost as easy to locate and catch a school of fish when the stock is large as when the stock is small. Another way to say this is there is relatively little relationship between the stock size and catch per unit effort. Hence there is little reason to believe that costs will go down when the stock increases for the same quantity harvested. To the extent that there is any difference, it is the time it takes to locate the schools that differs when the total stock changes.

On the other hand, it is natural to assume that the costs are convex in the harvested level. One functional form that can be relevant for this case is

$$c(x, y) = k \frac{y^2}{x^b} \qquad (6)$$

where k and b are non-negative parameters, and b is small. This includes both the convexity with respect to harvest and the weak dependence upon the stock. If $b = 0$, there will be no relationship between costs and the stock level. On the other hand, if $b \geq 1$, there will be a strong relationship, which is typical for demersal fisheries. A $b > 0$ also provides an economic guarantee against extinction as the cost of harvesting goes to infinity when the stock approaches zero. Further the cost function has the following characteristics:

$$c_y > 0, \; c_{yy} > 0, \; c_x \leq 0.$$

This means that costs are increasing and convex in harvest and decreasing in the stock.

Net revenue function
The net revenue function is simply gross revenue minus costs:

$$\Pi(x, y) = R(y) - c(x, y) = p(y)y - c(x, y).$$

As we have seen, this function can in principle be completely general in both arguments. With the functions given above, we are guaranteed that the net revenue function is meaningful, and that the sufficiency conditions are fulfilled.

ESTIMATING THE ECONOMIC SUBMODEL

It is much harder to find relevant economic data for sardine than relevant biological data. Cost data are especially difficult to obtain as is the case in most fisheries. The two main uses of sardine are either canning or fishmeal, and the price for fish going to canning has always been higher than for fish going to fishmeal. This price difference has increased over the 90s. Whereas the price for canning has increased steadily the price for meal has been stable for long periods.

The volume going to meal has been decreasing; also in periods when total catch increased. Note, however, that in the entire period 1980 - 1999 the catch has not been above 120 000 tonnes. For larger quantities harvested a larger proportion will be used for meal.

Estimating the demand function

In order to determine a reasonable demand function for sardine, one has to take into account that the harvest is partly used for canning, and partly used for fishmeal. For small quantities most of the catch goes to canning whereas for larger quantities it is expected that most of the catch goes to fishmeal. In the following some simplifying, but reasonable, assumptions are made: for TACs less than 60 000 tonnes everything goes to canning, for the part of the TAC between 60 and 120 000 tonnes 80 per cent goes to canning and 20 per cent to fishmeal, and for the part of the TAC over 120 000 tonnes everything exceeding 108 000 tonnes goes to fishmeal. Letting $y1$ denote canning and $y2$ fishmeal, the following relationship appears:

$$y1(y) = \begin{cases} y & \text{if } y < 60 \\ 12 + 0.8y & \text{if } 60 < y < 120 \\ 108 & \text{if } y > 120 \end{cases} \quad (7)$$

$$y2(y) = \begin{cases} 0 & \text{if } y < 60 \\ -12 + 0.2y & \text{if } 60 < y < 120 \\ -108 + y & \text{if } y > 120 \end{cases} \quad (8)$$

Over 90% of the canned sardine is exported to South Africa, but there is a limit to how much the South African market can consume, and this depends on the South African harvest of sardine. Over time the limit varies between 60 000 and 120 000 tonnes, and for the time being it is believed to be close to 60 000 tonnes (personal communication with Les Clark at the Ministry of Fisheries and Marine Resources, Windhoek). This is reflected in the function above. The distribution between canning and meal described here is only relevant for the period from the 1990s and onwards.

Figures 2 and 3 show observed prices and quantities for canned sardine and fishmeal. The year 1996 has been removed as total harvest this year was very small. In addition 1994 was removed for canning and 1995 for meal. These observations were removed after visual inspection of the data before running the regressions. An explanation why 1995 was an outlier may be that about half of the fish was caught in Angolan waters, and, due to the distance, this caused low quality and a low ratio of canning to meal.

The prices have been converted to fixed 1998-prices. These observed prices are used to estimate functions of the form in (5). The data for fishmeal give better estimates than for canning. For fishmeal we have estimated both the minimum price and the maximum price as well as the parameter a. For canning the maximum price has been pre-determined to be 2.2 N$/kg (per-

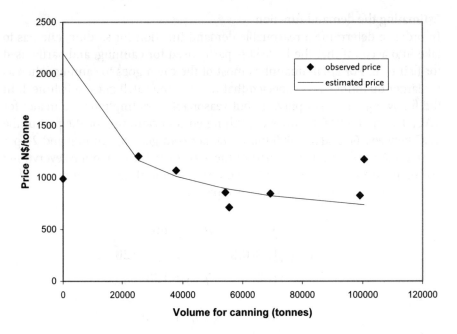

Figure 2. Observed and estimated prices for canning against volume from 1990 to 1997 (1994 and 1996 are removed as outliers).

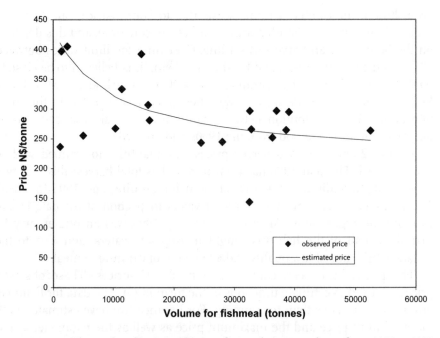

Figure 3. Observed and estimated prices for meal against volume from 1980 to 1998 (1995 and 1996 are removed as outliers).

sonal communication with Les Clark at the Ministry of Fisheries and Marine Resources, Windhoek), and the minimum price and parameter a have been estimated. The statistical properties of the estimations are given in Tables (2) and (3).

With the values from these tables we have defined the price of canned sardine as a function of the volume of canned sardine,

$$p1(y1) = \frac{16.3 \cdot 2.2 + y1 \cdot 0.5}{16.3 + y1},$$

and the price of fishmeal as a function of the volume going to fishmeal,

$$p2(y2) = \frac{9.8 \cdot 0.44 + 0.21 \cdot y2}{9.8 + y2}.$$

However, $y1$ and $y2$ are functions of the total allowable catch, y, given in (7) and (8), and therefore $p1$ and $p2$ are functions in y only, too. Hence the average price defined as total value divided by total volume is given by

$$p(y) = \frac{p1(y1)y1(y) + p2(y2)y2(y)}{y}. \qquad (9)$$

This function is a piecewise, non-linear function as depicted in Figure 4.

Calibrating the cost function

The cost function is calibrated on the assumption that there is a break-even point where gross revenue meets the cost function. We have assumed that this break-even point is at a harvest level in the region 1.1 to 1.4 million tonnes. This assumption is based on observations of the fishery in the mid-

Table 2. Statistical properties of the estimation of the demand for canned sardine.

Parameter	Value	t-value	
A	16.3	2.24	$R^2 = 0.74$
p_{min}	0.5	2.69	$F = 11.1$

Table 3. Statistical properties of the estimation of demand for fishmeal.

Parameter	Value	t-value	
A	9.83	0.78	$R^2 = 0.52$
p_{max}	0.435	6.7	$DW = 1.4$
p_{min}	0.212	3.52	$F = 7.45$

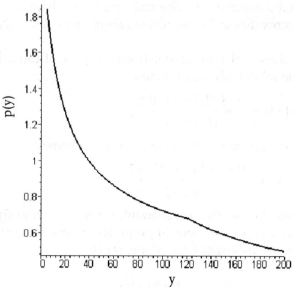

Figure 4. Illustration of p(y).

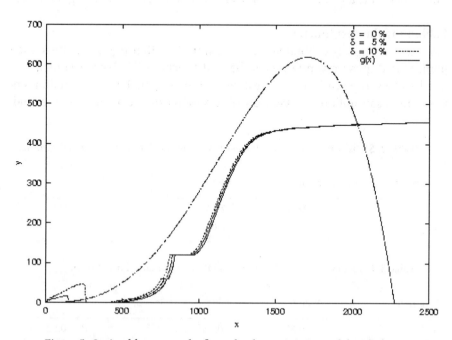

Figure 5. Optimal harvest paths from the deterministic model with discount rates of 0%, 5% and 10% together with the surplus production.

sixties when the system, for all practical purposes, can be treated as open access although some regulations (licenses, TAC) were in place. According to the theory, open access leads to exhaustion of the resource rent. With this assumption the value of the parameter k in (6) is 0.00034. The value of b is a pure guesstimate and is set to 0.05. In other words, the specification of the cost function is:

$$c(x,y) = 0.00034 \frac{y^2}{x^{0.05}}. \tag{10}$$

The weak part of the economic model is the cost analysis. Sensitivity analysis with respect to the cost parameter on this model was performed by Sandal and Steinshamn (1999). There it was found that a 50% increase in the cost parameter k implies 5% increase in the optimal steady state stock and 33% decrease in optimal steady state harvest whereas a 50% decrease in the cost parameters implies 15% decrease in the stock and 36% increase in harvest.

THE EMPIRICAL MODEL

The empirical model applied here is the one where the demand function is given by Equation (9) and illustrated in Figure 4. The cost function is given by (10), and the biological model is the surplus production model based on data from 1969 - 2000:

$$g(x) = 3.9E-7 \cdot x^3 - 1.6E-10 \cdot x^4. \tag{11}$$

Now stock and harvest are measured in 1000 tonnes instead of millions. The optimal harvest paths with these assumptions and different discount rates are illustrated in Figure 5. The case with zero discounting is of special interest as sardine is a renewable resource, and therefore we want to put as much emphasis on the future as on the present. We can see, however, that optimal harvest paths are not very sensitive to changes in the discount rate. This is always the case as long as the discount rate is small relative to the intrinsic growth rate of the stock. A detailed discussion of the deterministic model can be found in Sandal and Steinshamn (1999 and 2001b).

Note that for stock levels below 500 000 tonnes a harvest moratorium is recommended. It is interesting to note that this is very close to the level proposed by Fossen et al. (2001) although their recommendation is based solely on biological considerations. Even if the implications on employment of a harvest moratorium are taken into account, the moratorium is still optimal because the long-term effects by far outweigh the short-term effects. In other words, to increase employment temporarily by having higher quotas than the optimal ones will severely decrease the possibility of having a high employment in the future due to the adverse stock effects. As we shall see, this effect is reinforced when stochasticity is introduced.

Changing the discount rate

So far the study of optimal harvest paths has been performed with zero discounting. Optimal paths, however, are not very sensitive to changes in the discount rate as long as the discount rate is small relative to the intrinsic growth rate of the stock. This has been shown both theoretically and in practice (Sandal and Steinshamn, 1997a and c). It is further illustrated for Namibian sardine in Figure 5 where optimal harvest paths are drawn for a discount rate of 0%, 5% and 10% based on the same basic model as earlier.

With zero discounting it is optimal to implement a moratorium on harvest for stock levels below a certain limit. When the discount rate is increased to 5% and 10% we observe positive harvest for small stock levels followed by a small region of harvest moratorium. As the discount rate increases there will be competition between the discount rate and the intrinsic growth rate of the stock. This means that it may pay off to increase harvest in order to put the money in the bank instead of letting the fish grow in the ocean, for very small stock sizes. This is no problem as long as the optimal harvest is below the growth curve in a deterministic model because the stock will increase towards the optimal steady state anyway, although slower than with a smaller discount rate. If the stock level is so small that the harvest curve is above the growth curve, on the other hand, it is economically optimal to wipe out the rest of the resource according to the model. This is, of course, not an acceptable strategy in practical management. In practice it is probably not even possible as it is usually very difficult to find a statistically significant relationship between spawning stock biomass and recruitment for small pelagic fish. An ideal model should also include functions that represent the social and political costs of extinction. The validity of the biological model is limited close to the origin, and one should be careful, especially when high discount rates are used. For most values of the stock, optimal harvest is quite insensitive to reasonable changes in the discount rate. For small stock sizes, however, a change in the discount rate may imply the difference between the existence of a moratorium stock level or not. In other words, it is not the optimal harvest that is sensitive to the discount rate, but the moratorium stock level can be sensitive. Generally, higher discounting implies less conservative harvesting.

RESULTS FROM THE STOCHASTIC MODEL

So far it has been assumed that the biological growth is a deterministic process within the various periods we look at. It is, however, a well-known fact that the Benguela upwelling system is a highly uncertain and irregular system. This calls for inclusion of stochastic processes in the optimisation models. The interesting question is whether such processes have major impacts on the optimal paths and in what direction the paths are affected. In other words, what are the quantitative and qualitative implications of including stochasticity? This question will be answered both in the case of zero discounting and with positive discounting, but first a short description of the stochastic model will be given.

THE STOCHASTIC MODEL

In the stochastic case we want to maximize expected returns defined as

$$\max_{y \geq 0} E \int_0^\infty e^{-\delta t} \Pi(x, y) dt$$

where E is the expectation operator. The dynamics of the stock x is given by

$$dx = [g(x) - y]dt + \sigma(x)dw$$

where $\sigma(x)$ represents the stochastic term (or the volatility function) and dw is a standard Wiener process increment. The term $g(x) - y$ represents the deterministic part of the dynamic equation. The volatility function, $\sigma(x)$, can be any function of x in principle. Here the conventional approach of linear functions is used; that is, the uncertainty is increasing proportionally with the stock size.

Dynamic programming is used to solve this problem, and technical details on the numerical solution of such problems can be found in Kushner and Dupuis (2001). Here we only concentrate on numerical solutions of specific cases and not on analytical and theoretical aspects of the problem.

RESULTS FROM THE STOCHASTIC MODEL WITHOUT DISCOUNTING

The results from the stochastic optimization with three different stochastic processes are illustrated in Figure 6. The stochastic process is given as

$$\sigma(x) = \sigma_0 x,$$

and the three values of σ_0 that have been applied here are $\sigma_0 = 0.15$, $\sigma_0 = 0.3$ and $\sigma_0 = 0.45$. It is difficult to estimate the parameter σ_0 from data, but there

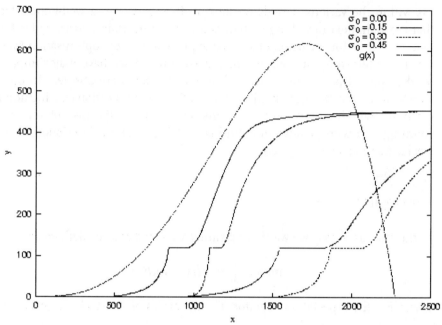

Figure 6. The deterministic optimal path and three stochastic optimal paths together with the surplus production when the discount rate is zero.

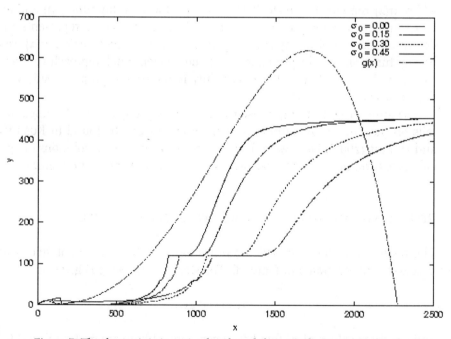

Figure 7. The deterministic optimal path and three stochastic optimal paths together with the surplus production when the discount rate is 5%.

are indications that the noise in this system is large. An indication that σ_0 is reasonable is that the standard deviation of g and $\sigma_0 \cdot x$ for a representative x-value are in the same order of magnitude. The standard deviation of g from the data is 213 whereas a typical x-value is half of the carrying capacity, namely 1200. Thus all three values of σ_0 suggested above are reasonable.

With more data more sophisticated functional forms of σ than the simple linear one could have been applied. The ideal situation would have been to have replicated time series such that g and σ could have been estimated simultaneously (see McDonald and Sandal, 1999). In the case when g and σ are estimated simultaneously, we get a different g from the one obtained when g is estimated in a purely deterministic context.

There are several interesting points to note from Figure 6. First we observe that the optimal control curves are constant at certain levels. This effect is due to the piecewise definition of the demand function. We also observe that the two lowest values of σ_0 imply a more careful policy than the deterministic model. They also introduce harvest moratorium on substantially higher stock levels. When the noise parameter is increased from 0.30 to 0.45, however, optimal harvest increases. This indicates that there exists a value of σ_0 between 0.3 and 0.45 that yields the most conservative policy. Finally we note that for very high stock levels, much higher than the carrying capacity, the three stochastic control curves and the deterministic policy all approach the same level. This level is the harvest level that maximizes net revenue as a function of x. In other words, the harvest defined by

$$\frac{\partial II(x,y)}{\partial y} = 0.$$

RESULTS FROM THE STOCHASTIC MODEL WITH DISCOUNTING

In the description of the empirical model it was seen that in the deterministic case, optimal paths are not very sensitive to changes in the discount rate. In the stochastic case, however, this is no longer necessarily so. The optimal paths for three different values of σ_0, when the discount rate is five per cent, are illustrated in Figure 7.

We observe positive harvest in the region close to zero, and the stock may be driven to extinction. When σ_0 is small we also note that the moratorium level is reduced considerably. For large values of σ_0 there is no moratorium level at all except zero. The main reason for this is high volatility combined with a biological growth function with depensation. It can be shown that the stochastic process implies self extinction almost surely when σ_0 is large. We may say that this is an "induced critical depensation" process. By induced critical depensation is meant a stochastic model *without* critical depensation that behaves like a deterministic model *with* critical depensation. Therefore it is economically optimal to mine out the rest of the resource in this case and invest the money as it will go extinct anyway. For moral and political reasons this is, of course, not an acceptable possibility in practice. For small σ_0 levels the self extinction probability is less than one, depending on the initial stock level. When the initial stock is small, the self extinction probability may be large even for σ_0 close to zero. Again this is due to the low growth rate in this region. The positive self extinction probabilities may also explain the substantial shifts in the moratorium level when stochasticity is introduced. It is now very important to prevent accidental low stock levels where the stock can wipe itself out even with zero harvest.

In addition, there is discounting and therefore there is a marginal trade-off between leaving the fish in the ocean and let it grow or taking it up, selling it and putting the money in the bank. At the margin these two alternatives should be equal. A well-known result from this trade-off is that it may pay off to harvest even the last fish and put the money in the bank if the intrinsic growth rate of the resource is much smaller than the discount rate. Remember that in this context the discount rate is interpreted as the alternative rate of return defined as the rate of return on the best alternative investment.

Usually it is argued that stochasticity should be taken into account when optimizing in order to avoid risk. In a strict dynamic optimisation framework the result may be, as we have seen, that it is optimal to wipe out the whole resource because there is a certain probability that the resource may go extinct anyway. This probability is of course zero in the deterministic version of the model (unless one is faced with critical depensation, see Clark (1990)). On the other hand it is widely agreed that extinction is not a realistic option both for political and for moral reasons.

What is there to be learned, then, from this exercise? The most important result is that introducing uncertainty implied more careful exploitation of the resource, but only for modest noise levels. With high levels we may sometimes get induced critical depensation if the surplus production function has depensation, and the optimal controls change accordingly. They may still be correct if we really believe that the stock faces self extinction. In addition the study confirms the well-known result that discounting is another factor that drives towards less conservative management.

SUMMARY

The aim of this study has been to find optimal harvest (total allowable catch strategies) for Namibian sardine when socio-economic as well biological conditions are taken into account. A bio-economic model has been developed for this purpose and the method used is dynamic optimization. In bio-economic models of the kind used here, only long-term, aggregated biological models can be used. The detailed biological models used currently appear to need some improvements because they are based on short time horizons, and therefore the parameter values in them are only relevant for a limited time. In bio-economic models we need average parameter values, valid in the long run. Therefore some effort was spent in the first part of the chapter in order to develop a suitable biological model. Although there are indications that a collapse took place in the mid seventies, data from 1969 to 1994 can be used to estimate a rather good relationship between stock size and growth.

The apparent collapse in the mid seventies may have been due to depensation in the growth function more than to a new biological regime. The presence of depensation, which is supported by the data, may explain why the stock has been trapped at low stock levels since then. We find that a moratorium on harvest is recommendable in order to bring the sardine stock out of the biological trap.

The economic submodel of the bio-economic model consists of two parts: revenue analysis and cost analysis. The aim of the revenue analysis has been to estimate a reasonable demand function for sardine. Demand is divided in two: demand for canned sardine and demand for fishmeal. The volume allocated to each use depends on the total volume. For each demand function there is a maximum and a minimum price that has been estimated together with other parameters. This yields a piecewise non-linear demand function.

The outcome of the bio-economic model is optimal harvest paths as a function of the stock size. In other words, it is an operational model that can be used to determine total allowable catches. Based on what we think is the most realistic version of the model, the main result is that in the present situation, that is for stock levels less than about 500 000 tonnes, a harvest moratorium should be instituted immediately. The optimal harvest path is then a piecewise curve due to the demand function. The optimal long-term steady state is a stock at around two million tonnes, and the economic optimal harvest is around 450 000 tonnes at this stock level. This may, however, be sensitive to changes in the cost function, but under no circumstances should the harvest be higher than the maximum sustainable yield, which is estimated to about 600 000 tonnes. However, before this model could be used for actual management of the stock, a much more thorough empirical analysis of the input functions would have to be made, especially with respect to the cost function.

Applying a stochastic version of the model, it was seen that optimal harvest paths were sensitive to the assumptions about stochasticity as well as the discount rate. With large uncertainty or discounting we find that it may be optimal to wipe out the stock. This is obviously not acceptable for social, political and moral reasons. In future work we will study how political harvest constraints affect the optimal policy.

The overall conclusion, then, is that in the present situation severe measures have to be taken in order to get the sardine stock out of the trap where it is at present. These measures may include a harvest moratorium for several years. If the right measures are taken, the sardine stock may give a valuable contribution to the Namibian economy in the future, otherwise it may disappear for a long time, possibly for ever. However, 2002 is the first year since 1981 with a zero TAC, indicating that drastic action is now beginning to take place.

REFERENCES

Arnason, R., Sandal, L.K., Steinshamn, S.I., Vestergaard, N., Agnarsson, S. and Jensen, F. (2000): Comparative Evaluation of the Cod and Herring Fisheries in Denmark, Iceland and Norway, TemaNord: 2000:526 (Nordic Council of Ministers, Copenhagen), 82 pp.

Butterworth, D.S. (1983): Assessment and management of pelagic stocks in the southern Benguela region. In: Proceedings of the Expert Consultation to Examine Changes in Abundance and Species Composition of the Neritic Fish Resources, San Jose, Costa Rica, April 1983. (G.D. Sharp and J. Csirke eds.): FAO Fisheries Report No. 291 (2): 329 - 405.

Clark, C.W. (1990): *Mathematical Bioeconomics: The Optimal Management of Renewable Resources*. John Wiley & Sons, New York.

Fossen, I., Boyer, D.C. and Plarre, H. (2001): Changes in some key biological parameters of the northern Benguela sardine stock. In: *A decade of Namibian Fisheries Science* (A.I.L Payne, S.C. Pillar and R.J.M. Crawford, R.J.M. eds.), *South African Journal of Marine Science* 23: 111 - 122.

Grafton, R.Q., Sandal, L.K., and Steinshamn, S.I. (2000): How to improve the management of renewable resources: the case of Canada's northern cod fishery. *American Journal of Agricultural Economics* 82: 570 - 580.

Kamien, M.I., and Schwartz, N.L., 1991, *Dynamic Optimization: The Calculus of Variations and Optimal Control in Economics and Management*, North-Holland, Amsterdam.

Kushner, H.J., and Dupuis, P., 2001, *Numerical Methods for Stochastic Control Problems in Continuous Time*, Springer-Verlag, New York.

McDonald, A.D. and Sandal, L.K. (1999): Estimating the parameters of stochastic differential equations using a criterion function based on the Kolmogorov-Smirnov statistic, *Journal of Statistical Computation and Simulation* 64: 235 - 250.

McDonald, A.D., Sandal, L.K., and Steinshamn, S.I. (2002): Implications of a nested stochastic/deterministic bio-economic model for a pelagic fishery. *Ecological Modelling* 149: 193 - 201.

Sandal, L.K. and Steinshamn, S.I. (1997a): A feedback model for the optimal management of renewable natural capital stocks. *Canadian Journal of Fisheries and Aquatic Sciences* 54: 2475-2482.

Sandal, L.K. and Steinshamn, S.I. (1997b): A Stochastic Feedback Model for Optimal Management of Renewable Resources. *Natural Resource Modeling* 10: 31-52.

Sandal, L.K. and Steinshamn, S.I. (1997c): Optimal Steady States and the Effects of Discounting. *Marine Resource Economics* 12: 95-105.

Sandal, L.K. and Steinshamn, S.I. (1999): Adaptive Management: The Case of the Namibian Pilchard Fishery. SNF Report no. 71/99 (Institute for Research in Economics and Business Administration, Bergen, Norway), 89 pp.

Sandal, L.K. and Steinshamn, S.I. (2001a): A simplified feedback approach to optimal resource management. *Natural Resource Modeling* 14: 419 - 432.

Sandal, L.K. and Steinshamn, S.I. (2001b): A Bioeconomic Model for Namibian Pilchard. *South-African Journal of Economics* 69: 299 - 318.

14 INSTITUTIONAL AND INDUSTRIAL PERSPECTIVES ON FISHERIES MANAGEMENT IN NAMIBIA

Bendigt Maria Olsen

Abstract

The Sea Fisheries Act (SFA) of 1992 as well as the new Marine Resources Act (MRA 2000) provide for consultations with the fishing industry in the Sea Fishery Advisory Council and more recently in the Marine Resources Advisory Council. Besides, the industry participates in working groups, fora for information and discussion together with Ministry officials and scientists, in particular about research and stock assessment. The mandate and composition of the two Advisory Councils as well as Terms of Reference (TOR) for the working groups are examined. Based on interviews in 1997–98 and again in April-May 2000 the major part of this Chapter presents the experiences and expectations of the industry regarding their role in these institutions. Theories within the analytic framework of 'new institutionalism' are explored and found to have explanatory value. The state as a rational actor has created appropriate institutions for implementation of its objectives, while the industry also, from a rational, wealth maximising point of view, concur with the deal. Greater participation in the management process is, however, much requested by the fishing industry, emerging as a strong interest group. Their demands are legitimate in the tradition of participatory democracy, which so far is only at the outset in the Namibian society, characterised by a strong central power.

INTRODUCTION

While the world society long awaited the end of apartheid rule in Namibia, the UN system, with FAO as the driving force, in the 1980s started exploring alternatives for planning a new fisheries management regime for independent Namibia. This task resulted in the report: *The fisheries of Namibia and options for its management and development in the first period after independence* (UNDP/FAO, 1989). Prior to Namibia's independence, the management and regulations of inshore fisheries were the responsibility of the South African administration in Windhoek, whereas the International Commission for the

Southeast Atlantic Fisheries (ICSEAF) tried to regulate the offshore fisheries. However, ICSEAF was without the necessary powers to enforce required regulations set according to scientific advice. Virtually uncontrolled fishing by European and South African fleets led to depletion of the main commercial stocks, sardine as from the beginning of the 1970s and hake in the 1980s (Hamukuaya, 1994). Built on proposals presented in the UNDP/FAO document, the Government presented the White Paper: *Towards Responsible Development of the Fisheries Sector* (RoN, 1991), which has been guiding the Namibian fisheries policy ever since. In 1992 these policies were translated into a legal framework, the *Sea Fisheries Act* (SFA) (RoN, 1992). At independence the new Government took full control of the fisheries sector in accordance with Chapter 11 *Principles of State Policy,* Article 95 (l) in the *The Constitution of the Republic of Namibia* (RoN, 1990), which *inter alia* states "utilisation of living natural resources on a sustainable basis for the benefit of all Namibians, both present and future". To a large extent fisheries can be seen as an exception to the general conservative policies in other economic sectors, e.g. land, mines and in the case of taxation (Leys and Saul, 1995; Rakner, 2001) where status quo was more or less let to remain in order to reassure and encourage private investment. However, also for the development of a viable fishery industry, private capital was certainly indispensable.

Soon after independence in 1990 the new Namibian government, realising the great economic potential of the fishery resources, declared a 200 mile Exclusive Economic Zone (EEZ). The introduction of an efficient fisheries management regime to rebuild resources was urgently needed. For this purpose the authorities did not hesitate in choosing the 'state model' (Hersoug and Holm, 2000) for future management of Namibian marine fisheries. In 1990 and still today this model is dominant among fishing nations in the western world. The 'market model' (Hersoug and Holm, 2000) was certainly by that time discredited while a 'community model' (Hersoug and Holm, 2000) was of little relevance since no Namibian traditional fishery community of any commercial significance existed[1]. For the new government it was of utmost importance to control the exploitation of the resources. A new

---

[1] The 'state model' refers to a centralized and bureaucratic form of fisheries resource management, especially appropriate when resources are over-exploited. Control of fishing effort has priority. The 'market model' or the ITQ (individual transferable quota) model, emphasizing efficiency, uses the tools of the market for quota allocations. This model is now applied in several countries as a means to solve problems of overcapacity, underdevelopment and inefficient allocation of resources. The 'community model', based on local structures and co-management, stresses equitable and fair access to resources, thereby enhancing the legitimacy of resource regulations (Hersoug and Holm, 2000).

management regime was therefore defined to enable the state to take out the resource rent as well as promoting reallocation of access rights to allow new entrants from Namibia. In 1991 the new Ministry of Fisheries and Marine Resources (MFMR) was established and designated to be responsible for protection, monitoring, control and surveillance of all marine living resource within the EEZ. The Ministry was further given the responsibility to set up an institution for fisheries resources research, the tasks of which would include advice on catch quotas and other measures connected with fishery regulations (RoN, 1991). Already in 1989, however, research into large-scale fisheries started when a research group was established in Swakopmund. From independence responsibility for planning and execution of routine surveys with Research Vessel *Dr. Fridtjof Nansen* was shared with the Institute of Marine Research, Bergen (Sætersdal *et al.*, 1999). In 1993 the Swakopmund group became the core of the new National Marine Information and Research Centre (NatMIRC).

The importance of good communication with the public and the industry was recognised, and accordingly the *Sea Fisheries Act* (SFA) of 1992 (RoN, 1992) stipulated an institution, the Sea Fishery Advisory Council (SFAC), for consultations *inter alia* with representatives from the fishing industry and their employees. This provision was considered unsatisfactory by the industry, which requested more involvement in management matters than provided for within the mandate of the Advisory Council. In particular for the development of the new orange roughy fishery, it was acknowledged that more direct co-operation between the Ministry's scientists and the industry was required. A Deep Water Fisheries Working Group (DWFWG) was established in 1995 (Boyer *et al.*, 2001), and later working groups for hake, monk, horse mackerel and rock lobster were also planned. On 1 August 2001 there entered into force a new act, the *Marine Resources Act* (MRA, 2000) (RoN, 2000), which does not recognise working groups as statutory bodies. Legally the Minister is not obliged to consult with the working groups on management issues of living resources, but according to a ministerial officer they are now functioning again and are good for building trust and a feeling of ownership of research[2]. In his *Annual statement on the status of the Namibian fisheries sector*, January 2002, Dr. Abraham Iyambo, Minister for Fisheries and Marine Resources, states *inter alia* that the Ministry of Fisheries "habitually consults extensively with the industry on certain matters before action is taken to adjust any aspect of policy" (Iyambo, 2002).

This chapter presents and discusses the views of the industry with regard to their experiences of, and expectations for, participation in the management process within the institutions of consultation. The bulk of informa-

---

[2] Communication with Dr. H. Hamukuaya, MFMR, March 2002.

tion is obtained by interviews with managers of the main fishing companies. In all, over 40 interviews with this group of respondents were carried out in the years 1997-1998 and in April-May 2000. Open-ended questions were applied in order to let the interviewee freely discuss the subject. For the purpose of understanding the Government's choice of management model and the Ministry of Fisheries and Marine Resources' interpretation of it, theories within the analytic framework of 'new institutionalism' are explored, such as rational choice theory and historical institutionalism. The request of the industry for more co-operation can on the other hand most likely be traced and explained within the paradigm of participatory democracy. Before turning to the theoretical framework a presentation is given of the Sea Fishery Advisory Council (SFAC) and its successor the Marine Resources Advisory Council (MRAC) as well as of the working groups and the leading associations of the fishing industry. The major part of this study relates to the opinions of the respondents regarding industry's participation in these institutions as well as their communication with the Ministry. A discussion of results and concluding remarks wind up the review.

INSTITUTIONS FOR CONSULTATION WITH THE INDUSTRY

Comparison between the Sea Fishery Advisory Council (SFAC) and the Marine Resources Advisory Council (MRAC)

The main objective of the new council remains the same as for SFAC, namely to "advise the Minister in relation to any matter on which the Minister is required to consult the advisory council in terms of this Act and any matter which the Minister refers to the advisory council for investigation and advice" (MRA, 2000, Part V Article 24). Both the old and the new Act stipulate consultation with the Advisory Council before the Minister determines a total allowable catch (TAC). The MRA in addition specifies that such decision shall be taken "on the basis of the best scientific evidence available", thereby reinforcing the principles of responsibility and sustainability (MFMR, 2000a). However, the clause in SFA that the Advisory Council in addition should advise the Minister in relation to "the management and development of the sea fisheries" is omitted in the new act, which could be interpreted as if the initiative lies only with the Minister to take up issues for discussion in MRAC. With regard to the constitution of the MRAC the total number of persons to serve on the council is reduced from 16 to 12. Besides the Permanent Secretary only one staff member from the Ministry of Fisheries and Marine Resources will serve on the council and there will no longer be any obligatory representatives from other ministries. Instead the Minister will appoint five persons with knowledge in matters relating to marine fisheries and expertise of relevance to issues, which have to be referred to the Advisory Council. And lastly, five persons are to be appointed, who according to the Minister "fairly represent the fishing industry or employees in the fishing industry". The main difference with regard to this point is that while the SFA states that "five shall be persons who, in the opinion of the Minister, have had experience in or shown capacity in any matter relating to any branch of the fishing industry" (MFMR, 2000a), the MRA, however, stipulates appointment of five persons who *represent* the fishing industry. The new MRAC, established in September 2001, consists of only one representative of the trade unions while the Minister chose four representatives nominated by the industry to sit on the council[3]. The Advisory Council may determine times and places for its meetings which should take place "at least once a year", in contrast to the old council which should meet "not less than twice a year". However, the chairperson is obliged to convene a meeting of the Advisory Council not only when the Minister requires so but according to the new Act also upon the request of at least four members. Moreover, for quorum only seven members present are necessary instead of ten in SFAC and a majority vote will continue to be the decision of the Advisory Council. An additional section in MRA specifies the discretionary power practised by the Council to permit the attendance of any person and participation in discussions with regard to specific interests and matters on the agenda. Such persons are not entitled to vote though. This paragraph has reference for instance to the Ministry's scientists who often are requested to attend the meeting of the Advisory Council for clarification of the TAC recommendations. In addition scientists engaged by the industry were earlier permitted to present their results for the Advisory Council. This new section defines the status of these non-members of MRAC (MRA, 29(7)) (MFMR, 2000a, pp. 31-32). In order to promote efficiency and transparency the new Act contains clauses of keeping records and establishing procedures for the Advisory Council meetings, as well as those of committee meetings. The Advisory Council may namely from time to time establish committees to perform certain of its functions (MFMR, 2000a). The Minister in his statement of January 2002 in fact reports about the sub-committee of the MRAC, which has begun an in-depth assessment of the economic repercussions of applying a more equitable division of freezer and wet hake quota (Iyambo, 2002).

Terms of Reference for the Working Groups

A Deep Water Fisheries Working Group (DWFWG) was established in 1995 by the Ministry of Fisheries and Marine Resources as a forum for guidance and advice to the authorities on efficient management of the deep-water

---

[3] Information received from Dr. H. Hamukuaya, MFMR, March 2002.

fisheries. The proposed Terms of Reference (ToR) (MFMR, 1998) intended to formalise the role of the DWFWG to ensure its full integration and complementary role to the management of the deep-water species of Namibia. According to the TOR the aims of the working group were to promote rational development of the Namibian deep-water fisheries, assuring that economic and social benefits are optimised and accrue to Namibia. The resources should be utilised to their full potential while safeguarding long-term sustainability of the stocks through proactive research and co-management strategies. Besides permanent members representing both the Ministry, its scientists and the industry, other persons might be co-opted by the DWFWG as consultants. Standing sub-committees would deal with management, research and compliance. The management responsibilities of the working group would include making recommendations on annual TACs based on the outputs from operational management procedures (OMPs) (MFMR, 1998, p. 3). Since recommendations on TACs are the prerogative of the Advisory Council, this transgressed the initial objectives of the working group. In 1999 the Minister therefore suspended the Deep Water Fisheries Working Group as well as the hake working group and others that were in embryo.

In 2000 the Ministry approved *Fisheries Working Groups: Terms of Reference* (MFMR, 2000b), which is a more generalised form of terms of reference designed to accommodate all working groups. While this finally formalised the working groups it should be noted that they are not part of the fisheries management regime and thus not enshrined in the MRA of 2000. In March 2002 there were working groups for orange roughy, hake, horse mackerel, monk and rock lobster. The overall aims as stated in the new Terms of Reference are identical to those referred to the Deep Water Fisheries Working Group of the nineties, namely to promote the rational development of Namibian fisheries and so forth. The objectives are to "develop an appropriate Operational Management Strategy for the optimal and sustainable utilisation of the Namibian Resource based on the results of biological stock assessment and social-economic research" (MFMR, 2000b, p. 1) The following specification under point 5 that the Working Group shall "analyse the socio-economic status of the relevant Fishery to facilitate a better understanding of management options" will satisfy the industry, which has asked for more attention towards socio-economic aspects in fisheries management. This point is well substantiated in the interviews. The main difference refers to point 2 of the Terms of Reference of 2000, that the Working Group shall "*deliberate* on the stock biomass in order to assist MFMR researchers in making recommendations on annual TAC based on outputs from the operational procedures" (MFMR, 2000b, p. 2) instead of *make* recommendations on annual TAC. As stated in the previous document, consensus shall be aimed at when the Working Group is conducting its business and "the number of representatives from the different Parties will be balanced to maintain a good, open and efficient working environment in the Working Group" (MFMR, 2000b, p. 1). Hence, no doubt the intentions of the Working Group are good.

MAIN ASSOCIATIONS OF NAMIBIA'S FISHING INDUSTRIES

The Namibian Hake Association was founded in 1992 and its objectives are to protect and further the interests of the Namibian hake fishing industry and to provide a forum for collaboration and discussion between members. The Hake Association will also, as and when required, negotiate on behalf of the entire hake fishing industry with all government departments. In addition its aim is to encourage rational fishing by its members following practices that are appropriately concerned about recognised fishery conservation concepts (Fishing Industry Handbook, 2001, p. 197). The Namibia Tuna and Hake Longlining Association was also founded in 1992 with the same purpose as that of the hake association. In particular it co-operates closely with the government to obtain a reasonable share of the quota on migratory and straddling fish stocks appearing in Namibian and international waters (Fishing Industry Handbook, 2001, p. 197). In 1994 holders of Namibian fishing rights for the catching, processing and marketing of products derived from midwater trawling formed the Midwater Trawling Association of Namibia. Its objectives, in addition to serving as a forum for communication, are to actively co-operate at national and international level to achieve optimum sustainable utilisation of the midwater fishing resource (Fishing Industry Handbook, 2001, p. 195). The Namibian Monk and Sole Association dates back to 1996 and is open for all holders of Namibian fishing rights for monk and sole. This association also serves as a forum for communication among members and with government. Optimum sustainable utilisation of the resource is also here the objective reached by participation in local and international projects (Fishing Industry Handbook, 2001, p. 197). The Walvis Bay Pelagic Fishing Companies serves as a forum for the pelagic companies to discuss matters of common concern. In addition to promoting the optimum sustainable utilisation of the sardine resource, the members actively participate in research surveys (Fishing Industry Handbook, 2001, p. 197). Besides these main associations there is the Ad Hoc Committee, which functions as an umbrella body of the fishing industry. While the Committee has no formal structure it is acknowledged by the Ministry and often consulted.

NAMIBIA'S FISHERIES MANAGEMENT REGIME: AN INSTITUTIONAL INTERPRETATION

The characteristics of the resource, the constitutionally decided redistribution of the benefit accrued from the resource and the Government's international obligations in relation to global management are convincing reasons for continued state control of the exploitation of the Namibian fishery resources (Manning, 1998). What can 'new institutionalism' tell us about the choice taken by the Namibian Government at the time of independence? One of the foremost proponents of 'new institutional economics', Douglass North, incorporates institutions in rational choice analysis and constructs a theory of institutions by combining a theory of human behaviour with a transaction cost theory of exchange (North, 1990a). According to North, "institutions consist of informal constraints and formal rules and of their enforcement characteristics. Together they provide the rules of the game of human interaction" (North, 1990b). While North sees institutional change as overwhelmingly incremental, he admits that also discontinuous institutional change indeed may occur in the form of revolution (North, 1990b, p. 397). In the case of Namibia it is easy to equate independence with a revolution where a new government with its base in a black majority 'overthrew' the old white elite supported by and part of an alien occupant. The new government took control of the fisheries resources by creating a fisheries management regime in line with Koeble's (1995, p. 240) formulation: "rational individuals design them (institutions) to help them achieve certain ends". Margaret Levi, also a rational choice institutionalist, concludes her chapter *A Logic of Institutional Change* (Levi, 1990) by recapitulating that the power of institutional decision-makers depends on their capability to barter benefits for compliance and their ability to monitor and coerce the noncompliant. Imperatively is also their capability to derive trust by proving that the bargain is good and persisting. The analysis of the interviews may answer the question whether the representatives of the fishing industry, also rational actors, confirm this bargain.

When studying fisheries management in Namibia, and especially the role of the Ministry of Fisheries and Marine Resources, it is evident how old structures and culture still strongly prevail within the decision-making process. To perceive the causes of this, one may turn to *historical institutionalism*. This version of 'new institutionalism' focuses on common-sense concepts of formal institutions such as legislatures and bureaucracies and the role of ideas in defining institutions (Peters, 1999). Ideas, such as the objective of Namibianisation, are also a major component in fisheries management. In sum, institutions, according to the historical institutionalists, are forming actors' goals and not only their strategies as in rational choice. Moreover institutions, by arbitrating actors' relations of co-operation and conflict, shape political situations and outcomes (Thelen and Steinmo, 1992).

The fishing industry's quest for more influence in the decision-making process can find its *raison d'être* within the conceptions of participatory democracy. This democracy model emphasises involvement of key stakeholders in contesting and debating planning as well as implementation of public policy. Formal consultative agencies such as deliberation councils might be established to channel the influence of organised interest groups. There is a strong instrumentalist cause for greater participation in macroeconomic policy manifested in enhanced legitimacy of the decision-making process, improved rate of return and sustainability of programmes (Robinson, 1998). Thus, in addition to the central institutions of liberal democracy - that is competitive parties, political representatives, periodic elections - *inter alia* interest-group competition in governmental affairs can further the principles of participatory democracy. One of its key features is accountability of party officials to membership (Held, 1997). The fishing industry's representatives in the new Advisory Council are not formally representing any association. However, since they are appointed by the Minister from a list of candidates provided by the Ad Hoc Committee, the industry clearly expects them to be accountable to their respective associations. Departing from the concept of participatory democracy, extensive literature and numerous studies on 'co-management' have been published, especially with regard to fisheries management, among others by Jentoft and McCay (1995) and Hutton (2002). This debate will, however, not be reviewed within the scope of this chapter.

VIEWS EXPRESSED IN THE INTERVIEWS WITH REPRESENTATIVES OF THE FISHING INDUSTRY IN NAMIBIA[4]

Experience with the old Sea Fishery Advisory Council (SFAC)

This theme is of great concern to the industry and gave rise to many comments with regard to the mandate, composition and credibility of the Council. Several respondents point to the fact, that the Council does not have enough power and that the proceedings are confidential. While the SFAC is not transparent and its proceedings are supposed to be confidential they affect the industry. Rumours from SFAC create unstableness in the industry a representative claims. SFAC should be an open forum for industry, scientists and management. Members ought to be Namibian citizens, and politi-

---

4   These views are recounted in the present tense, as expressed in the interviews, to make them authentic.

cians are accepted as long as they are fair-minded and people of standing. The same interviewees also accept that the discussion should be confidential, it is an opinion, an advice, and the Minister should listen to them. SFAC should be a responsible body and consensus reached within the meetings and the recommendations should be acceptable to the Minister. However, the credibility of SFAC is not there, since the advice is not taken. One representative admits that the members of the SFAC often get frustrated, trying to give constructive comments but never knowing if the Minister takes these and forwards them to the cabinet and accordingly the SFAC only functions as a sounding base. This summarises the opinions of several interviewees.

A couple of other respondents, also members of the SFAC, have a more positive attitude to their role in the Council and to their fellow members. One mentions that also politicians from opposition parties sit in the Council, "toeing the line of national interest" and making SFAC much more democratic. Another member stresses that although he is not representing any association but being appointed in his personal capacity, he has proposed that when a member has interest in a question being discussed he cannot vote, and this has been accepted as a principle. With regard to TACs, the SFAC members think it is important and therefore endeavour for joint decision, pulling scientists' recommendations and industry's recommendations together.

Concerning the influence of the unions, the answers are that they have very little input. In general some of them are not well prepared and not bargaining in good faith. Demand for more money is their priority, not the most important, which is housing, according to an experienced member of the SFAC. Over the last three years the industry's Ad Hoc Committee has dealt with special issues as fuel, tuna etc. and given recommendations to SFAC. However, there is not always quorum, causing cancellation of SFAC meetings. You can over-democratise by rules requiring too many people for a quorum. The Council is able to meet, though, when it is necessary to discuss responsible and constructive suggestions.

The request for a democratic representation is the foremost concern according to most interviewees. The main task of SFAC is to be in touch with the industry. The general perception is that SFAC is dominated by the larger companies and appointed by the Minister. To deal with this perception, there should be a quota system: each sector of the industry should be adequately represented, whereas at present there are e.g. only trawling companies and no longliners. Although having one representative in SFAC the rock lobster industry is not content but is going to suggest that there will be two from their industry. Thus, most of the interviewees are of the opinion that the industry should have the right to nominate candidates to SFAC, which ought to truly reflect all sectors of the industry, "we are not satisfied with the present set up; the fishing industry looks forward to restructuring of SFAC". The Minister will still decide, but there will be consultations, a democratic process with more representatives nominated by the industry's associations to avoid the Minister's appointment of members, and opening up for the various sectors to formulate their points of view. One member of SFAC recommends that the Minister still should have the right to appoint specialists in certain fields pertaining to fisheries and who are deemed to make a contribution. A proposal is that also other persons coming from broader economic spheres as e.g. Chambers of Commerce should be appointed. Those recommended should be contented and the public would be allowed to participate as observers, with the overall purpose of making SFAC a credible body.

However, there are a few exceptions to the firm view that the various fishery associations should nominate candidates to be appointed members of SFAC. One person argues that if the Minister looks for advice it should be unbiased. Therefore it is not easy to find somebody from an association who could be selected on merits. In SFAC all have vested interests; it would be unrealistic to expect a person to be unbiased and to represent the full interest of e.g. the Hake Association's 25-30 concessionaires. SFAC handles classified information and insight is not open. So far SFAC has not been an independent body; there are conflicting views and hence somebody above interest in the industry should be member. Furthermore, the Minister is not forced to take the advice. Another respondent stresses that associations cannot legitimately claim representation as SFAC is an advisory board to the Minister. Instead different co-operative bodies should be created together with scientists resulting in combined recommendation to SFAC.

Expectations of the new MRAC

At the time of renewed interviews in April-May 2000 the industry was informed that the draft of the new act, the Marine Resources Act, suggested only five members to represent the fishing industry in the new MRAC. Ahead of appointments, consultations would take place with industry as well as with trade unions. According to the drafting officer in the Ministry, the outcome could theoretically be a mix of: 3-2, 4-1 or five persons from the industry or five representing the employees[5]. Nevertheless the industry had been advised to nominate eight persons of whom the Minister would appoint four. These were the conditions and information given and which the industry was going to follow.

The number of persons finally representing the whole industry in the new Council is, however, considered far too small by most of the respondents, one characterising it as 'murder, suicide'. The industry had argued

---

[5] Information received from W. Scharm, MFMR, April 2000.

that all the eight nominees should get in. The representation might be given in accordance with the various fisheries' contribution to GDP as hake, horse mackerel, pelagic, monk/sole. This option is, however, not ideal as the small concession owners, longliners e.g. will not be represented. All small sectors of the industry are scared according to one respondent. And how will the interest of the newcomers and the divide between Walvis Bay and Lüderitz be taken care of? The new Council will not be representative enough of fishing industry per se. It is also argued that the orange roughy fishery, being a major troubled industry, might need a representative in MRAC. Spokesmen for this industry are doubtful, though, about chances to get their own delegate in the Council. They argue that the Minister will most likely choose representatives from big capital, the important industries. The nomination process with industry drafting a list of eight candidates out of which the Minister will choose four is not really democratic. The orange roughy industry has expressed its concerns several times and discussed at length with the Permanent Secretary (PS). The appointments according to the new act are not satisfactory, forcing them to rely on another industry to speak on their behalf. Neither has the monk industry so far been represented in the SFAC, and will there be anyone to represent the industry in the new Council, is the question. This spokesman as well is disillusioned about the nomination process, foreseeing that the Minister is going to choose on his own decision and that all is decided beforehand and nothing is going to change. Contrary to this statement, one respondent declares that at the Council people from the industry should be neutral and work in the interest of the whole industry. And the Minister has been quite selective in choosing members. The same attitude was noticed among some of the 'old' members of the SFAC.

Several managers of the hake and pelagic industries are on the other hand more positive to the new procedures. The Hake Association has been extended with a branch in Lüderitz and each branch has its own chairman and vice chairman. The Hake National Committee meets every second month and several respondents express their satisfaction that the company Pescanova, the biggest hake quota recipient, has joined the Association. Constitutional, democratic elections will take place when selecting the nominees of the industry. A formula will be made to let the major sectors of the industry vote for their representatives, and from the list of these candidates the Ad Hoc Committee will send eight names for the Minister to choose four. The Ad Hoc Committee of the fishing industry has no formal constitution but works as an umbrella body and is acknowledged by the Ministry. "We are not far from coming under one roof" and it is also anticipated that in the future only few persons representing the fishing industry in the MRAC will be necessary as more and more companies consolidate. The new system is welcome by the Hake Association and will be a step forward, having been pushed by the industry right from the beginning. A representative from the pelagic industry also expresses confidence in the new Council. So far this industry has been well represented in SFAC with as many as four persons. The nomination through the Ad Hoc Committee for the restructured Council will entail two candidates from each sector. Let the Minister do as he wants if it is only fair and as long as the pelagic industry is represented. The four large sectors will have one representative each assures our spokesman, and maybe the Minister will address the other smaller industry.

A general positive comment given by several respondents is the expectation that the appointed members will be representing the industry on an accountability basis, which is welcome. With regard to the status of the new MRAC it depends on the ability of somebody to discuss and talk for the industry. The question is if the Council really has teeth, i.e. whether it affects the decision of the Minister. The MRAC will probably carry more weight and will be a change for the better in comparison with the old SFAC, which was a 'non event' according to one critical voice.

Raison d'être of Working Groups[6]

What is needed is that both the Ministry and the industry take responsibility at Working Group levels, one respondent stresses, claiming that this is the view of the higher officials in the Ministry. For each fishery sector one should operate through a working group. The Deep Water Fisheries Working Group (DWFWG) is a test case. It is nice being in a small group, being accountable to each other. The industry now pulls together scientists of their own in the DWFWG and the Ministry's scientists have to justify the acoustic survey. A good co-ordinator of the WG is necessary and both sides should take 'ownership' of decisions made, which is possible with more frequent meetings of working groups. In order to take away as much of the uncertainty as possible, consensus should be reached in the WG. There is still disagreement regarding documents and recommendations, but at the end of the day one reaches consensus[7]. The top management in the Ministry wishes a united document to be presented to the SFAC. Bio/economic data would be needed and added to the agenda of the DWFWG and it should allow an element of flexibility, e.g. regarding over-catch: let's manage it ourselves. The structure is so rigid, companies go under with over-catch fees, and orange

---

[6] The following views were expressed in the years 1997-98 referring particularly to DWFWG, which was the pilot working group. While speaking of the management responsibilities of DWFWG the respondents seem to take for granted that the mandate also included recommendation on TACs. In fact this objective was part of the proposed first Terms of Reference (TOR) of 1998 as discussed above.

[7] Since 2000 consensus has not been reached (Boyer and Oelofsen, this volume).

roughy are often caught together with alfonsino. Now industry and Ministry are working together, starting to understand each other better - as a team to work up a database. Yet another representative from the deep water fisheries is expressing his content with the working group as a means to co-operate better, a way to go. The Ministry considers the DWFWG as a test for managing the resource in a responsible way. The DWFWG, however, is not formalised, but functions inside the Ministry with scientists and industry. The working group is too new, it is all an experiment and there are only three companies and two more as observers. With few companies it is easier to work than in comparison with e.g. the hake industry. Socio-economic questions are included in the mandate of the working group. The aim is to make recommendations to the Ministry with regard to three aspects: 1. research, 2. management i.e. TAC, and 3. compliance. In addition to suggesting policies and advice to the Ministry and to the SFAC, the idea behind working groups is simple according to this respondent: the Ministry is tired of the conflict between the industry and the scientists, therefore the establishment of working groups has come through with the address: "try to solve the issue between you". It is now up to everyone if it will work, if players are willing. A working group is a less formal body and we in the industry want the working group to work.

The industry has no say in directing government resources but contributes to 10 per cent of GDP and NatMIRC say that they do not have resources, a representative of the hake industry claims. This industry wants an input in a working group as DWFWG, which the industry has requested for a couple of years. Our spokesman totally supports co-operation in actual research. Another respondent says that the hake industry is pushing for a new meeting with the hake working group provided it is not toothless as the working group for orange roughy, a 'family shop', is the comment by a representative from this industry. The ultimate goal is to set targets on an annual basis, not provisional and to give industry the feeling of participation. In the past there was no relations, it was a one-side affair, the industry was just told; now the industry and the scientists sit together and the industry believes it makes a contribution to the utilisation and preservation of the resources on a sustainable basis.

Working groups are a wonderful exercise, another respondent asserts, but they should be given a better, proper and legitimate status. They are breaking new ground, hopefully eliminating mistrust. It would also be easier for the Minister to be faced with more realistic issues as socio-economic and national concerns. An almost euphoric statement comes from a nestor in the industry foreseeing the birth of a new era in Namibian scientific research and a lot of scope for mutual co-operation and peaceful fisheries coexistence in the management of Namibian fishing resources. Co-operation is the solution and Namibia could be an example for the rest of the world. A bold statement comes from another respondent questioning the necessity of SFAC if working groups are established and consensus on TAC arrived at. It is doubted that SFAC would 'shoot down' an agreement reached by scientists and industry. Working groups are useful to sort out day-to-day problems, but scientists of international standing should be selected to form a panel for final decision. This is necessary says the spokesman as fishing is not a marginal industry and its future important for the Namibian economy. Representatives of the monk, rock lobster industries and the longliners especially would welcome to have their own working groups involving a collective, joint effort not least in practical research.

At the renewed interviews in 2000 many of the same arguments were presented but certainly with more weight after another couple of years of experience. A representative of the horse mackerel industry points to the high TAC for several years as a reason why there has been no need for the industry to disagree and argue with the scientists. However, this is the best time to start a working group our spokesman stresses. There is a need for a forum to be called in when it is a crisis, insist representatives of the hake industry. In addition it is important that working groups function for responsible discussions with scientists of common interests and to sort out problems on a lower level. Working groups are also a place for socio-economic questions for which information is lacking. As regards the pelagic industry there has not been any working group, just a sardine-pelagic workshop in 1997. Nothing actually came out of it as the recommendations were not implemented one respondent declares. With no fish the industry is not pushing for a working group at this particular time but awaiting the initiative of the Ministry. Representatives of the orange roughy industry are all convinced of the benefit of having a joint forum in the form of a working group. They are confident that the DWFWG will be established again as a permanent body with periodic meetings, every two to three months. The terms of reference have been redrafted and the new working group will have two aims: recommend research for management and consider socio-economic aspects, which in the future will carry more weight when resources are short. The feeling of responsibility is also strong within the industry, which is willing to continue to assist Ministry's scientists by spending resources on research. They have a duty to lead the process, one respondent claims.

Communication with the Ministry
The relations and communications between the industry and Ministry of Fisheries are carried out through several channels and the experiences are both negative and positive. Many complain about lack of regular contacts with Ministry, which only occur on an ad hoc basis. The industry almost has

to force itself in and the communication with the scientists is not enough. The Minister came to Walvis Bay and he wants to co-operate but the officials in the Ministry are not carrying it out. An example given refers to the change of quota year, a decision taken by the Management Committee in Windhoek alone. The Ministry and the industry must work as a team, the respondent stresses. The biggest stumbling block over the years is lack of trust and confidence between the Ministry and the industry. The respondent admits that they have improved, moved ahead, but there is still a lot of ground to be covered. The same representative also complains that issuing of licences and other paperwork takes time with the Ministry. Communication should certainly be better.

Among those referring to positive contacts some say that officials at all levels in the Ministry are available and that there are no problems with communication. Depending on the question, the contacts go through the associations when it is general and direct to the Ministry or the Minister when it is company related. Some industries have a direct line to the Minister through their chairman, particularly when there is a need for more contacts in the establishment phase or the special profile of an industry is making it a showcase. There have been no secret doors, a respondent assures. When having problems one company does not go through their association but straight to the Minister of Finance or to politicians. Previously they had contacts on a regular basis with the Prime Minister and the President, but it is not a personal relationship. Another representative states that the company has a good communication with the Permanent Secretary and down at administrative levels, but declares that he does not like to talk specifically with the Minister but sees the President very often, admitting that "it is a question of style." There are certain managers who have frequent informal contacts with the Permanent Secretary, as much as twice a week according to one representative who assures that he himself sees the PS only twice a year. This same spokesman appreciates the present incumbent who permits an intellectual level of discussion. In contrast another respondent declares that after the shift of Permanent Secretary and the administrator dealing with the industry, relations with the Ministry have improved immensely. Earlier the industry was met with a dictatorial type of attitude, a typical South African attitude. The new staff and not least the Minister stand for an open door policy. One respondent would like to see a representative from the Ministry at the monthly meeting of the Hake Association where matters dealt with are not that confidential. A radical suggestion put forward is that more functions should be moved to Walvis Bay – also the Maritime Affairs, which goes hand in hand with the fisheries. As it is, the authorities sit in Windhoek and they do not know about seagoing problems. Our spokesman recommends: leave the Permanent Secretary and the Minister in Windhoek and move the directorates to Walvis Bay. Windhoek is isolated from the industry and if in Walvis Bay there would be daily contact. In a similar vein comes the suggestion of establishing regional advisory councils under chairmanship of the Prime Minister. Such councils could be of advantage for SFAC since the executives in Windhoek do not understand what is happening at the coast. However, in Namibia people feel uncomfortable with decentralisation, our proposer reckons.

DISCUSSION AND CONCLUDING REMARKS

North's (1990b) definition of institutions as consisting of informal constraints, formal rules and of their enforcement characteristics is indeed a valid description of Namibia's fishery management regime as an overarching institution. The two first characteristics have been exposed in this chapter while the monitoring, control and surveillance components of the regime are not the subject of this review. Levi (1990) stresses that revenue maximisation is the main goal of any rational government. Also in the case of Namibia the objective of utilisation of the fishery resource on a sustainable basis is the prerequisite for rent maximisation in this sector[8]. In accordance with Levi's (1990) prescription for successful decision-makers, the Ministry of Fisheries and Marine Resources barter benefits, *id est* fishing rights and individual non-transferable quotas for compliance by the industry. Further, the Ministry indeed has the capacity to monitor and coerce the noncompliant (Bergh and Davies, this volume). The government, acknowledging its dependence on private capital for the development of the sector, created institutions of consultation and co-operation with the industry. Thus, within the model of state control of the exploitation of the marine resources, these institutions the Sea Fishery Advisory Council/Marine Resources Advisory Council and the working groups are established. The purpose is to involve the industry and get advice in matters of concern to the fisheries. As stated by an officer in the Ministry the working groups are fora for building trust and a feeling of ownership of research. Or in Levi's words: a means to derive trust by proving that the bargain is good and persisting.

However, in the interviews the respondents express general dissatisfaction with the institutional arrangements for their participation in the decision-making process. This relates especially to their experiences with the Sea Fishery Advisory Council, and several managers question whether their advice is at all taken into account. Further, the lack of transparency causes ru-

---

[8] This goal is, however, weighed against the objectives of Namibianisation and job creation, constituting the social rent from the exploitation of the fisheries resources.

mours and the appointed members from the industry are accused of only looking after their own interests. Yet, a few companies feel well represented in the Council and it is evident that those interviewees who themselves have been appointed as members to the Advisory Council are much more positive. Such statements correspond with the notion of participatory democracy emphasising the instrumentalist cause for stakeholders' involvement in consultative agencies. The effect is improved legitimacy of the decision-making process and sustainability and support for programmes. The establishment of the new Marine Resources Advisory Council gives rise to expectations of better representation and influence of the various sectors of the industry, at least for the four main ones with highest contribution to the GDP. In the end the smaller sectors have to rely on them as spokesmen. The interviewees voice quite some pride when explaining the nomination procedure of representatives from the various sectors and then the final choice to be taken by the industry's Ad Hoc Committee. It is also stressed that in the future the representatives are expected to be accountable to their respective associations. Thus, the change in rules from the old SFAC to MRAC will in fact promote participatory democracy through the nomination process, and accountability, an important pillar of participatory democracy, will also be fostered. In addition, the statute requesting that records should be taken at the meetings of the Marine Resources Advisory Council will promote democracy and hopefully prevent rumours. It is pertinent to point out though, that the fishing industry only nominates their candidates and the Minister chooses them. The quest for more democracy is obvious, both within the respondents' own organisations and in relation to the state, here the Ministry of Fisheries and Marine Resources. This pursuit is first of all expressed in the request for participation in working groups. As one respondent explains, working groups are important for responsible discussions with scientists and for sorting out problems on a lower level (also pointed out by Boyer and Oelofsen, this volume). Such a statement is in line with the postulate of historical institutionalism, i.e. that institutions, by arbitrating actors' relations of cooperation and conflict, shape political situations and outcomes. A similar criterion could possibly also be applied to the SFAC and MRAC. Although the working groups are not statutory bodies according to the new Act, the revival of them seems to work to the satisfaction of the industry. It should be noted, however, that as in most of the world's main fishing nations with long-standing established consultative institutions, the Minister has the final say in fisheries management. In relation to the general weakness of the civil society and lack of strong parliamentary opposition in Namibia (Diener and Graefe, 2001) the fisheries industry, being an interest group, is emerging as a strong partner to the Government within the established institutions for consultation. This is in contrast to most other interest groups, where structures and connections linking them to the state are not formal and obvious (Diener and Graefe, 2001). It is foreseen that in a not too distant future, the various associations of the fishing industry will have come under one roof in Walvis Bay and will speak with one voice[9]. The Minister in his Annual Statement 2002 in fact called for a cohesive voice from the Hake Association in connection with a management question (Iyambo, 2002).

Nevertheless, the industry representatives complain about lack of communication with the Ministry. They regret not being more involved, consulted and informed by the Ministry and NatMIRC about new plans, changes in management, evaluations, etc. Their opinions are incompatible with the statement of the Minister, Dr. Iyambo, as referred to in this introduction. He says that the Ministry "habitually consults extensively with the industry on certain matters before action is taken to adjust any aspect of policy".

The respondents find that the structures are so rigid and often inefficient and it takes such a long time to get a response. Some even feel maltreated, slapped as a naughty boy. Again historical institutionalism by stressing ideas, in this case control, may explain the continued dictatorial approach that members of the fishing industry have perceived in their contacts with the Ministry. The bureaucracy of the Ministry of Fisheries, at independence consisting of the old guard of civil servants, who in accordance with the new Constitution retained their posts and privileges, has held on to many of the features characterising the colonial power (Melber, 2000). Centralisation and control still to a great extent prevail not only towards the industry, its clients, but also within the Ministry itself. As an example the post of the Permanent Secretary (PS) implies great discretionary powers, in contrast to that of comparable posts in for instance Scandinavian public administrations. The personal contacts that the representatives of the industry keep with officials in the Ministry, or even with the President, is a facet of the neo-patrimonial state. Another example is the habit noticed that the managers of the fishing companies themselves turn up, well dressed, at the Ministry to deliver their annual quota application, not to the Minister or the Permanent Secretary but to the officer in charge of registration. Historical institutionalists will certainly also agree with Gretchen Bauer (1999) in tracing this authoritarianism not only to the legacy of hundred years of colonial rule but as well to the bequest of the exiled liberation movement.

This chapter has presented industry respondents' opinions and perceptions of their experiences and expectations for participation in the institutions of consultation. As pointed out above, the interviews were made in 1997-98 and again in 2000, at a time when the first phase of the experiment

---

[9] Comment by L.Clark, MFMR, May 2000.

with working groups was discontinued. From the industry's perspective the ideal objective was obviously co-management of the resources, as indicated in the first proposed terms of reference (TOR) of the DWFWG. However, neither the working groups nor their mandate was formally approved by the Minister when the interviews took place. The cardinal mistake made by the DWFWG was clearly the recommendations for TACs, a prerogative of the Advisory Council at that time as well as under the new act (MRA 2000). Evidently a number of the opinions of the respondents are mirroring misconceptions and lack of understanding of the management system applied, but it has not been the task of this review to judge and point out what is correct or false in their statements. Boyer and Oelofsen discuss this problem, in particular with regard to the sardine industry, in their contribution to this book.

However, with the restoration of the working groups it is presumed that some of the frustrations among the interviewees have eased and greater transparency and co-operation will develop between the Ministry, its scientists and the industry. Although not explicitly referred to above, the representatives of the industry communicated their general acceptance and even support of the objectives of Namibia's fishery policy. As rational actors they accept the 'bargain', which no doubt is a profitable one and it is a fair conclusion that fisheries management in Namibia has largely worked according to the intentions. The institutions, SFA/MRA, SFAC/MRAC, the Ministry of Fisheries and Marine Resources and NatMIRC are indeed created to fulfil the objectives laid down in the Constitution, thus the state as a rational actor has designed the institutions to achieve its ends.

References

Bauer, G. (1999): Challenges to democratic consolidation in Namibia. In: *State, Conflict and Democracy in Africa*, (Joseph, Richard, ed.), Lynne Rienner Publishers, Boulder: pp.429-448.

Boyer, D.C., Kirchner, C.H., K. McAllister, M., Staby, A. and Staalesen, B.I. (2001): The Orange Roughy Fishery Of Namibia: Lessons To Be Learned About Managing A Developing Fishery. *A Decade Of Namibian Fisheries Science* (A.I.L. Payne, S.C. Pillar, and R.J.M. Crawford eds.), *South African Journal of Marine Science* 23: 205-221.

Diener, I. and Graefe, O. (2001): Introduction. In: *Contemporary Namibia. First Landmarks of a Post-Apartheid Society* (I. Diener and O. Graefe eds.), pp. 19–33 Gamsberg Macmillan Publishers, Windhoek.

Fishing Industry Handbook (2001): *South Africa, Namibia and Mocambique*. 29th edn. George Warman Publications, Cape Town.

Hamukuaya, H. (1994): Research to determine biomass of hake. In: *Namibia Brief*, No. 18 1994, Namibia Foundation pp. 73-74.

Held, D. (1997): *Models of Democracy*. 2nd

edn. Polity Press, Cambridge.
Hersoug, B. and Holm, P. (2000): Change without redistribution: an institutional perspective on South Africa's new fisheries policy. *Marine Policy* 24 (2000): 221-231.
Hutton, T. (2002): Industry-government co-management arrangements in the South African offshore demersal hake fishery. In: *Waves of Change: Coastal and Fisheries Co-management in South Africa* (M. Hauk and M. Sowman eds.), pp 199-225. University of Cape Town Press, Cape Town.
Iyambo, A. (2002): Annual statement on the status of Namibian fisheries sector. Ministry of Fisheries and Marine Resources. January 2002.
Jentoft, S. and McCay, B. (1995): User participation in fisheries management. Lessons drawn from international experiences. *Marine Policy* 19 (3): 227-246.
Koeble, T. (1995): The new institutionalism in political science and sociology. *Comparative Politics* 42 (1): 231-247.
Levi, M. (1990): A Logic of Institutional Change. In: *The Limits of Rationality* (K. Schweers Cook and M. Levi, eds.), pp. 402-418. The University of Chicago Press, Chicago.
Leys, C. and Saul, J.S. (1995): *Namibia's Liberation Struggle*. James Currey, London.
Manning, P.R. (1998): Managing Namibia's Marine Fisheries: Optimal Resource Use and National Objectives. A thesis submitted in partial fulfilment of the requirements for the degree of Doctor of Philosophy. London School of Economics and Political Science, 313 pages.
Melber, H. (2000): Public sector and fiscal policy. In: *Namibia A Decade of Independence* 1990-2000 (H. Melber, ed.), pp87-108. Namibian Economic Research Unit (NEPRU), Windhoek.

MFMR (1998): Deep Water Fisheries Working Group. Terms of Reference. NatMIRC. Updated: January 1998.
MFMR (2000a): Synopsis of Sea Fisheries Act, 1992 and the proposed Marine Resources Bill.
MFMR (2000b): Fisheries Working Groups: Terms of Reference.
North, D.C. (1990a): *Institutions, Institutional Change and Economic Performance*. University Press, Cambridge.
North, D.C. (1990b): Institutions and Their Consequences for Economic Performance. In: *The Limits of Rationality* (K. Schweers Cook and M. Levi, eds.), pp. 383-401. Chicago University Press, Chicago.
Peters, B.G. (1999): *Institutional Theory in Political Science. The 'New Institutionalism'*. Pinter, London.
Rakner, L. (2001): The politics of revenue mobilisation; explaining continuity in Namibian tax policies. *Forum for development studies* 28 (1): 125-146.
RoN (1990): The Constitution of the Republic of Namibia. Ministry of Information and Broadcasting, Windhoek.
Robinson, M. (1998): Democracy, participation, and public policy. The politics of institutional design. In: *The Democratic Developmental State Political and Institutional Design* (M. Robinson and G White, eds.), pp. 150-186. Oxford University Press, Oxford.
RoN (1991): Towards Responsible Development of the Fisheries Sector. Ministry of Fisheries and Marine Resources, Windhoek, December 1991.
RoN (1992): Promulgation of Sea Fisheries Act 1992 (Act 29 of 1992) of the National Assembly. *Government Gazette of the Republic of Namibia* No. 493, Government Notice No. 135, Windhoek, 1 October 1992.
RoN (2000): Promulgation of Marine Resources Act, 2000 (Act 27 of 2000)

of the Parliament, *Government Gazette of the Republic of Namibia* No. 2458, Government Notice No. 292, Windhoek, 27 December 2000.

Sætersdal, G., Bianchi, G., Strømme, T. and Venema, S.C. (1999): The DR. FRIDTJOF NANSEN Programme 1975-1993. Investigations of fishery resources in developing countries. History of the programme and review of results. FAO Fisheries Technical Paper No. 391. Rome, FAO. 1999. 434 pp.

Thelen, K. and Steinmo, S. (1992): Historical institutionalism in comparative politics. In: *Structuring Politics Historical Institutionalism in Comparative Analysis* (S. Steinmo, K. Thelen and F. Longstreth, eds.), pp. 1-32. Cambridge University Press.

UNDP/FAO (1989): The fisheries of Namibia and options for its management in the first period after independence. Mission Report August-September 1989. Bergen, September 1989.

15 AGAINST ALL ODDS: TAKING CONTROL OF THE NAMIBIAN FISHERIES

Per Erik Bergh and Sandy Davies[*]

Abstract

Monitoring, Control and Surveillance (MCS) is concerned with compliance of fishers to the rules and regulations that support the fishery. Over recent years Namibia has gained a solid reputation in MCS: a reputation that is supported by the findings of this chapter. The dramatic initial enforcement of the EEZ; the personnel and resource constraints that faced the Ministry; the national level policies; the positive and negative aspects of a complex legal framework; the influence of the geographical location and the impact of a strong reliance on donor support are identified as important factors in the shaping of the Namibian MCS system. The MCS system is evaluated relative to the stated objectives for three fisheries; the demersal, pelagic and midwater. Conclusions were drawn in relation to compliance levels, deterrence value, impact of the economic viability of fishery, legitimacy and norms and morals. The overall results indicate that the demersal fishery maintains a high level of compliance, the pelagic is variable but within acceptable levels, while the midwater fishery is unacceptably non-compliant. In relating these results back to the objectives it is concluded that the first objective, to restrict fishing activity to those entitled to do so, has been fully achieved; the second objective, to ensure that fishing activity is conducted within legal and administrative guidelines, has been partially achieved; and the third objective, of ensuring that revenues from landings are correctly calculated, has not been achieved. The cost of MCS is calculated and found to be at an acceptable level (around 42% of the income to Government), but concern is raised over the future cost implications with two new patrol vessels and possibly a new plane and helicopter on the horizon, at the same time as withdrawal of substantial donor support. Careful planning and streamlining will be required if the MCS organisation is to continue to develop a successful and sustainable operation; options for this are identified.

---

[*] The authors would like to thank the staff of the MFMR for assistance in compiling data and information. They would like to thank Mr S. Ambabi, Mr M. Block, Mr M. Koopman, Dr B. Oelofsen and Dr V. Wiium for their continued assistance during the preparation of the chapter.

INTRODUCTION

Namibia is a significant player in the international fishing industry, and this is reflected in the importance that is placed on ensuring healthy fisheries. Monitoring, Control and Surveillance (MCS) is concerned with compliance of fishers to the rules and regulations that support the fishery. The MCS organisation can therefore be seen as the official overseer of fishing operations and related activities. Over recent years Namibia has gained the reputation of having one of the few functioning and effective MCS systems in Africa.

This chapter aims to describe the events that have led to this achievement and the current structure and components of the MCS system. The system is analysed from three approaches in order to ascertain its underlying strengths and weaknesses and to view these in relation to its success and sustainability. The approaches used are: a critical analysis of the factors that have influenced the system; an evaluation of the stated objectives and the levels of compliance achieved; and an assessment of the cost of the system compared to the revenue and value of the fishery. Future challenges that the MCS organisation will face nationally, regionally and internationally are discussed and options for tackling these considered.

THE MCS SYSTEM

Historical perspective

Following independence, Namibia inherited the marine zones implemented by the South African administration; these included a fishing zone, but no 200 nautical mile exclusive economic zone (EEZ). An attempt had been made in 1983 to claim the EEZ, covering 560 000 km^2, by the United Nation Council for Namibia, but this claim was rejected by South West Africa Peoples Organisation (the exiled government) and the former Soviet Union. In 1990, the new Namibia ratified the United Nations Convention on the Law of the Sea and made full claim to the Namibian EEZ (Walvis Bay remained South African territory until 1994). Resources to enforce national jurisdiction in the zone were almost non-existent and the country had virtually no experienced personnel to carry out the required operations. At this time the offshore fishing was thriving, with many vessels illegally catching large amounts of demersal and midwater species: a situation that required immediate action by the Government.

When it became clear that diplomatic efforts had failed, Namibia turned to tough practical action. A helicopter was launched to locate and arrest the poachers. Twelve Spanish and one Congolese trawler were arrested and severe sanctions issued. These actions stated Namibia's will and ability: actions that generated remarkable international attention and set the scene for future MCS.

MCS structure and objectives

Having addressed the immediate concern of illegal fishing the next step was to develop an MCS organisation. The Ministry of Fisheries and Marine Resources (MFMR), established in 1991, was tasked with the management of living marine resources within Namibian waters. Two directorates covering resource management and operations were created: the directorate of operations was allocated the responsibility for MCS.

The principal objective of MCS is the regulation of fishery sector activities. For operational purposes this is divided into three objectives: the restriction of fishing activities to those entitled to do so; ensuring that fishing activities are conducted within legal and administrative guidelines; and ensuring that the revenue from landings is correctly calculated. In order to meet these objectives, an integrated system (with the senior management in the national capital of Windhoek and the operational bases in the fishing ports of Walvis Bay and Lüderitz and at the aerial surveillance base of Arandis) was formed. The operational stations are tasked with roles such as the deployment of fishery officers to air, sea or land operations; the deployment of fishery observers onto fishing vessels; the compiling of fishery statistics; the calculating of revenues; and the analysing and planning of operations.

Spatial dimensions

Spatially there are four possible dimensions to an MCS system: sea, air, land and remote sensing. Validating information across these dimensions is a strength. Namibia has a complex system with elements in each spatial dimension. The sea dimension, where the fishing operations take place, offers opportunities for monitoring fishing events and catches, ensuring that these are within legal guidelines and for face-to-face contact with the fishers. Random inspections at sea, the patrolling of boundary and closed areas, hot pursuit of poachers and the arrest of vessels all form elements of this dimension. These activities are facilitated through a fleet of two MFMR-operated patrol vessels, currently the P/V *Tobias Hainyeko* and P/V *Nathaniel Maxwilili* (for a comparison of their specifications see Table 1). A third patrol vessel with similar specifications to P/V "*Nathaniel Maxwilili*" will supplement and strengthen this fleet in the near future. The patrol activities are supported by a cadre of 200 fishery observers who are deployed onto the larger fishing vessels of the demersal, midwater, crustacean and large pelagic fleets to observe, record and report. Observers provide a unique opportunity to make *in situ* observations of compliance to regulations (such as dumping or discarding, fishing in closed areas, pollution, misreporting of catches, retention of prohibited catch, or the use of illegal gear) and to collect scientific information on catches. Fishing activity logbooks that captains complete on a daily basis also form part of the at-sea dimension.

Aerial surveillance is the principal method used to monitor, locate and track fishing vessels and thus provide a complete overview of activities in the EEZ. Aerial operations can photographically record violations such as fishing in closed areas and poaching, but arrests of vessels are impossible, unless helicopters are used. In the last two years operational difficulties have been experienced with the Cessna F406, Namibia's fixed wing aircraft, but prior to these difficulties an annual average of 500 hours was flown. The plane itself, although 14 years old, has been reliable and flexible, offering a cruising speed of 240 knots, a flight range of 1,600 n.m. and an endurance of more than 8 hours.

MCS activities on land vary considerably from fishery to fishery. All fishing vessels are required to undergo a port inspection in order to receive a clearance certificate at the start of each fishing season. At the end of a fishing trip, fishery inspectors monitor the unloading of fish as the catch is brought ashore or transhipped in the limits of the harbour (transhipping at sea is not permitted). This provides information on catches, which can be cross-checked against the logbook data and the data provided by the vessel operators, in order to facilitate the calculation of levies, quota control and scientific assessment. Coastal and inland patrol operations are carried out throughout the year in an attempt to control trade in seafood and to monitor and regulate the coastal recreational fishery.

Remote satellite tracking of vessels through vessel monitoring systems offers a modern and cost-effective way to provide a baseline of movements and activities of licensed vessels. A system is currently under implementation that will ensure that fishing fleets operate only in authorised areas, and

Table 1. A comparison of Namibian MCS patrol vessel specifications.

Specification	P/V *Tobias Hainyeko*	P/V *Nathaniel Maxwilili*
Gross tonnage (t)	652	1 400
Length over all (m)	50.0	57.6
Breadth, moulded (m)	10.5	12.5
Maximum draught (m)	3.7	4.2
Maximum speed (knot)	17	17
Range at 11 knot (n.m.)[*]	5 500	6 600
Other		Helicopter deck

[*]1 nautical mile (n.m.) = 1.852 km

will provide regular information on the location of all licensed fishing vessels. This overview is important to assist with the deployment of MCS platforms (such as patrol vessels, observers and plane) and to gain a better understanding of the fleet dynamics.

Coordination and system links

Coordinating such a complex mix of MCS platforms and activities is a challenging task for the MCS organisation, a task made no easier by the geographical spread of operational centres. Coordination and information links take the form of: quarterly and annual reports; briefing and debriefing sessions; mission reports; the compilation of statistical summaries, compliance registers, a vessel register; and a wide area computer network that links Windhoek, Walvis Bay and Lüderitz. It is envisaged that the vessel monitoring system, once implemented will provide the framework to electronically monitor the operational activities of the MCS system and to link these directly to the fleet activities. This will provide an electronic planning, management and analysis tool that is currently not available.

Human resources and training

Early in the development of the MCS organisation, training was identified as the key factor required to build up essential knowledge and experience. This approach has been implemented with the assistance of co-operating partners and in 2002 it has resulted in a total MCS workforce, including fishery observers, of 353 (Table 2). The training itself has taken many forms, but three programmes stand out as having been particularly important: the fishery inspector and observer course, a nine month course (that has now trained 64% of the inspectors and observers); the training of observers in catch and scientific monitoring through the commercial sampling programme; and the

Table 2. Human resources of the Namibian MCS organisation in October 2002.

Location	Profession	Number
Walvis Bay	Inspectors	68
	Patrol vessel crew	52
	Air crew	9
	Observers	122
	Others	15
Lüderitz	Inspectors	15
	Observers	65
	Others	5
Windhoek	Management	2
Total		353

training of ships officers and crew for the patrol vessels. The last of these is an example of an eleven year progressive training programme that, with the assistance of foreign (Norwegian and Danish) officers and crew, enabled patrol vessel operations to start in 1993 while training was under way and to then slowly instate with qualified Namibians.

ANALYSIS OF THE NAMIBIAN MCS SYSTEM

How to analyse MCS systems has fuelled much debate from MCS and fisheries managers alike. With even a brief glance at the literature, it is clear that assessing the benefits, effectiveness and cost efficiency of MCS systems is still an issue that has yet to be successfully addressed (Furlong, 1991; Bergh and Davies, 2001; Kelleher, 2002). It is therefore not surprising that there has been little effort made to analyse the Namibian MCS system. However, this type of assessment, even if imprecise, is required if the future of the system is to develop in line with the real needs and demands of the fishery and in support of the management system in place. In the following sections an attempt is made to analyse the system, giving consideration to factors that influence the system, achievement of the stated objectives and the costs of the system.

Key factors that have influenced the MCS system

Both internal and external factors have played a role in shaping the development of the system: an assessment of these factors is presented below.

Historical inheritance and national policy. - The MCS organisation is not only a link in the fisheries management chain, it is also part of a more complex picture related to national policy and transformation. Therefore, the objective of fisheries management has to contribute to these higher social and political priorities of Namibianisation, employment creation and capacity building. In the short term these policies left managers, who themselves were new to MCS, faced with a team of unskilled and untrained personnel. The size of the team was also large, with a hierarchal structure that put seven ranks between an inspector and the director, thus making the delegation of responsibility and associated decision-making a cumbersome task. In the short term, it was evident that these policies accrued limited benefits to fisheries management. However, in the longer term, it is evident that these policies will lead to the provision of a large trained and dedicated national fisheries workforce.

Legal framework. - A supporting legal framework for MCS systems is the first element required to ensure that management measures are enforced. Namibia has always benefited from a strong fisheries legal framework and this was recently enhanced further by the enactment of the Marine Resource Act of 2000. The negative aspect of such a relatively complex legal system is that at times it is not fully understood by MFMR staff or fishers. This can result in various interpretations of the rules and regulations, and ultimately differing penalties being awarded for similar violations. This factor is a common symptom of legislation that covers more than one fishery: a symptom best addressed through awareness campaigns and training, backed up by guidelines interpreting legal instruments into simple everyday language. Namibia also benefits from a fair and just judicial system that ensures transparency to any fisheries offences that proceed to court. However, a lack of understanding by legal personnel of the potential gains made by illegal fishing has meant that very low fines have been given for serious violations, resulting in high gains for the companies (Table 3).

Geographical and physical factors. - Namibia is an extraordinary country in terms of its long, harsh, desert coastline punctuated by only two ports. These physical characteristics have been ideal in terms of designing an MCS organisation: the two ports allow for monitoring of all landings, the coastline offers no havens for illegal transhipment or landings, and the harshness of the environment and the nature of the fish stocks have discouraged the development of artisanal fisheries. These factors combined with good infrastructure have all contributed to the success of the MCS system. A limitation to the system has been the physical distance between the different operational sections of MCS and other sections of the fisheries organisation, including management that is placed 400 km away from the coastline. Modern trends in fisheries management are to link components together in order to support checks and balances across the system. It is therefore important that the links between the inspectorates, management and research are strengthened at every opportunity.

MCS information and the fisheries information management system. - Internal information related to MCS operations is generated by all the MCS platforms and activities. The aim is that this operational information is compiled, analysed and distributed via the operation centre in Walvis Bay. However, recent reviews (EBCD and GOPA, 1996) noted that there is little compilation and analysis of this information and it is not readily available in a form useful for operational control or planning. The MFMR also utilises a fishery information management system that is an integrated database system developed for Namibia. However, it is reported that not all fishery data is incorporated into the system and this includes the MCS data (Blondal, 2000; Iversen and Gilja,

2001). Due to complications in the main fishery database the MCS organisation has been reluctant to develop its own database tool. Such a system would provide the opportunity to analyse information and to plan more cost effective and efficient operations, as well as contributing to the overall knowledge of the fisheries.

Donor support - Donor support has been instrumental in the development and financing of the MCS system. The most critical period was from independence until 1997, but the support continues to play an important role today. It is a known fact that donor support rarely comes with no ties attached: this point was stated by the Honourable Minister for Fisheries and-

Table 3. Convictions for fishery offences in Namibia since 1996 with level of sanction given.

Year	Name of vessel	Infringement	Fine (N$)
1996	F/V Juno Warrior	Fishing inside 200 m depth contour. Use of undersized codend.	50 000
1996	F/V Alsu	Failed to comply with the order of a fisheries control officer. Use of illegal attachment to the codend. Attached bottom chaffer wrongly.	30 000
1996	F/V Nivensikoe	Failed to comply with the order of a fisheries control officer by destroying the fishing gear.	3 000
1996	F/V Marshall Novickov	Fishing inside 200.m depth contour. Submission of false information to a fisheries control officer.	3 000
1996	F/V Alsu	Use of restricting construction on the net.	30 000
1996	F/V Ofelia	Fishing inside 200 m depth contour. Submission of false information to a fisheries control officer.	22 534
1997	F/V Marshall Yakobovski	Use of undersized codend. Use of undersized round straps.	10 000
1997	F/V Rosendo da Villa	Use of undersized codend.	50 480
1999	F/V Roselyn	Use of foreign fishing vessel without a permit or license.	Escaped
2000	F/V Juno Warrior	Fishing inside 200 m depth contour.	127 154
2001	F/V Weskus 7	Illegal use of vessel as a fishing vessel.	1 800
2001	F/V Weskus 5	Illegal use of vessel as a fishing vessel.	1 800
2001	F/V Weskus 4	Illegal use of vessel as a fishing vessel.	1 800
2001	F/V Sonia	Illegal use of vessel as a fishing vessel.	1 800
		Average fine for the period	**25 644**

Marine Resources, Dr Abraham Iyambo in a press statement in September 2002: "Aid should be given with an open heart. It should take into account the history of Africans. Those providing aid, even to beggars should listen to the recipients' needs and aspirations. Pre-conditions should not be imposed by the powerful nations on the economically weak nations. Aid should not be given with strings dangling. Aid should be aid". Although the donor-funded activities within MCS have been sensitive to the cultural, political and social needs, it is evident that the donating country's preconceived ideas and expectations have at times played a driving force. For example, one could question the appropriateness of acquiring two new patrol vessels at the same time as implementing a vessel monitoring system, or whether the development of the fisheries information management system has improved the accessibility of information or if the objectives of the Southern African Development Communities MCS programme really addressed the regional concerns of Namibia. These questions and others are often politically sensitive and not easy to answer.

Evaluation of the success of the Namibian MCS system

An evaluation of the success of the MCS organisation must be made against the operational objectives as defined in the MFMR strategic plan (MFMR, 1999).

Restrict fishing activity to those entitled to do so - This objective deals with poaching by non-licensed vessels within the EEZ. Since the initial dramatic arrest of 13 vessels in the early 1990s (described in the section: the MCS system, historical perspective), only one unlicensed large pelagic vessel has been caught fishing within the EEZ (Table 3). A significant benefit was gained from the international exposure that followed these arrests, and the reputation that Namibia takes poaching very seriously has helped to deter potential poachers. The plane and patrol vessels are the means to regularly patrol the edge of the EEZ in the areas where transboundary fisheries occur. In order to ensure that levels of poaching do not increase, a deterrence presence is required and the reputation of the country's capacity and capability must remain high.

Ensure that fishing activity is conducted within legal and administrative guidelines. - This objective aims to ensure that licensed fishers comply with the regulations. The issue of compliance is vital: non-compliance is the principal reason for the failure of fishery management regimes. Therefore assessing the level of compliance is extremely important both as an indicator of the effectiveness of the regulatory element of the management regime and ultimately as an indicator of the success of the overall system.

To estimate compliance an assessment of the total violations is required. This, for various reasons, is a very complex task (Furlong, 1991). Difficulties are faced in areas such as defining the type of offence (e.g. from a minor mistake in reporting to a deliberate misuse of gear), in assessing the level of sampling, or in determining if all violations are detected during one inspection. As a result of this, all assessment methods show weaknesses (Kelleher, 2002). Two principal methods exist: one is to compare the number of detected infringements in relation to the percentage of the population being sampled (vessels, fishers, gears, etc.), the second is to use surveys. This chapter presents results of the first formal assessment of compliance within the Namibian fishery. Both methods for assessment of violations have been used, with data being sourced from MFMR records of violations detected by patrol vessels and observers on the demersal, midwater and small pelagic fleets and results from a written questionnaire survey, specifically designed for this assessment. Eighty-four completed survey responses were obtained in the survey: 78% of these were from MFMR staff and 22% from the companies in the fishing industry.

A general look at the average number of violations per inspection from the patrol vessels for the three fleets provides a range of figures with the maximum average number of detected violations per inspection being 4.67 in the midwater fleet in 1993 to a minimum figure of 0.08 in 2001 for the demersal fleet (Figure 1). The midwater fleet has continuously been caught with at least one violation per inspection over the last 10 years and frequently over two. These high levels of violations are also reflected in the observer

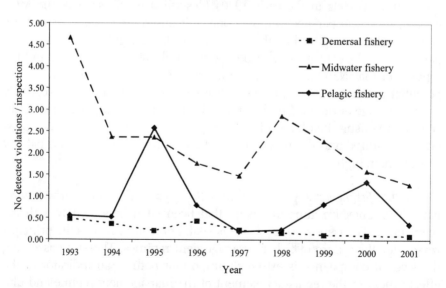

Figure 1. The average number of violations detected per inspection from the patrol vessels for 3 fisheries in Namibia.

data (Table 4), where between 31 and 74 violations per 100 observer days have been reported for this fleet between 1999 and 2001. The most common violations recorded are for the use of short roundstraps, fishing within the 200 m restricted zone, inaccurate or missing diagrams depicting vessel storage areas, or illegal stowing of gear. The pelagic fleet has generally had less than one violation per inspection in all years, except for 1995 and 2000 when peaks in the violations occurred; no observer data are available to validate these figures. The demersal fleet (the largest and most financially important of the Namibian fleets) produced a profile with much lower levels of violations than the other two fleets: from 1993 to 2001 a range from 0.48 to 0.08 average violations per inspection were detected. The 2001 figure of 0.08 reflects a generally compliant fleet. Observer data confirm this trend, with only 3 violations detected per 100 observer days in 2001. It has already been stated that accessing compliance is not a precise science: therefore the consideration of trends over time is very important. For the midwater and demersal fisheries the trend over the last 10 years is of an overall reduction in violations, while for the pelagic fishery no clear trend emerges.

The survey addressed the issue of perceived compliance by fleet and management measure. The estimate of compliance from the survey (where no significant difference was detected between industry and MFMR results) supports the earlier conclusion that the demersal fleet is perceived to be the most compliant fleet with an estimate of non-compliance at 26%, the midwater fleet is perceived to be the most non-compliant with an estimation of non-compliance at 42%, and the small pelagic fleet was perceived to be 32% compliant (Figure 2). It is of interest to note that the survey results, in comparison with the patrol vessel and observer data, appear to have overestimated compliance on the non-compliant fleets and underestimated compliance on the more compliant fleets. In relation to management measures, the industry perceived compliance to be higher than the MFMR staff (Figure 3). This possibly reflects concern by the industry in presenting themselves as a non-complaint body. However, the results show that the perception of compliance to management measures is higher than that found from analysis of actual inspections and observer observations.

Table 4. Number of violations detected by observers by fishery per 100 observer days in Namibia.

Fishery	1999	2000	2001
Demersal	8	5	3
Midwater	74	48	31
Other fishery	27	12	6

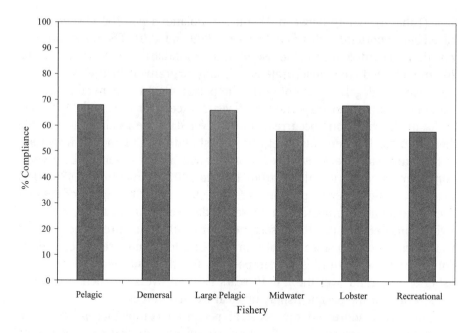

Figure 2. Compliance by fishery in Namibia, as perceived by MFMR and industry survey respondents

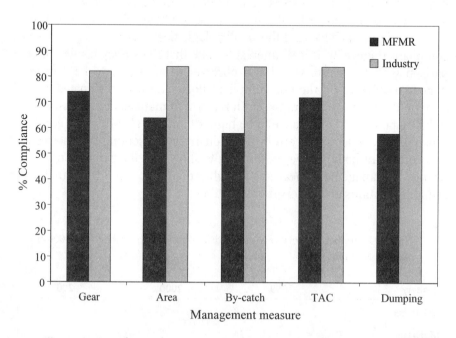

Figure 3. Complience by management measure in Namibia, as perceived by MFMR and industry survey respondents.

The survey yielded two further results that contribute to the understanding of overall compliance levels. The first related to the effectiveness of different MCS elements and platforms for the purpose of enforcement. The results indicated that all platforms were perceived as above average in their effectiveness. Aerial surveillance and the issuing of sanctions were perceived to be the least effective elements for enforcement; this is a correct reflection of the practical situation. Finally, in considering where violations take place, the survey indicated that 88% occur while fishing, 7% during transhipments and 5% during landings. This result, if correct, supports the use of the data from patrol vessels and observers as an indicator of the compliance levels in the fishery.

The theoretical basis for compliance. - Having evaluated the general level and trend of compliance in the main fisheries, it is now necessary to identify which factors are playing an influential role in causing non-compliant behaviour. Becker (1968) produced the first theoretical framework for a deterrence model (based on explanations of criminal activities). This laid the basis for the development of simple models combining deterrence and bio-economics (e.g. Sutinen and Anderson, 1985). Sutinen et al. (1990) made the first real attempt at structurally analysing fishers' compliance behaviour; this was later followed and built on by others (e.g. Kuperan and Sutinen, 1998; Hønneland, 1999; Raakjær Nielsen and Vedsmand, 1999). In summary, fishers can be classified as either chronic or moderate violators. Chronic violators, who generally constitute 2-5% of the population, will always violate regulations if there is a gain to be made, while moderate violators will only bypass regulations if the economic gain is high enough (relative to the economic situation of the fishery and the potential sanctions) and the chance of detection low enough. For moderate violators there are two secondary influences that may affect their decision to violate or not: these are the legitimacy of the regulation (and fishery management organisation) and the norms of behaviour, including both the general behaviour of the fishers and the moral code of the individual fisher. In recent literature the trend has been to place an increasing importance on the aspects of legitimacy and norms as major factors affecting the level of compliance in fisheries (Raakjær Nielsen, 2000).

The influence of economic factors on compliance - Individual fishers are more likely to violate regulations if the financial situation relating to fishing is bad. Calculating financial indicators over time is a complex task that at best can only be indicative of the general economic situation of the fishery. Economic indicators were evaluated for the three fisheries by considering the landed value per vessel and the total landed value. For the midwater fishery a correlation was found for both measures (Figure 4) when a substantial increase in landed value of the horse mackerel (over the last 4 years) did correlate with a decrease in the number of violations recorded. The pelagic fishery showed an erratic pattern for both measures with no clear trends emerging. The general trend of the demersal fishery supported this economic theory, although an anomaly occurred in 1996 (Figure 5). Therefore it can be concluded that the pelagic fishery that has gone through severe financial difficulties in the last years did not provide any evidence of a link between the economic return of the fishery and the level of violations. However, the midwater and demersal fisheries both yielded evidence in support of this

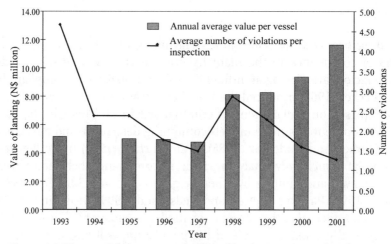

Figure 4. Midwater fishery – annual value of landings per vessel and average detected violations per inspection in Namibia.

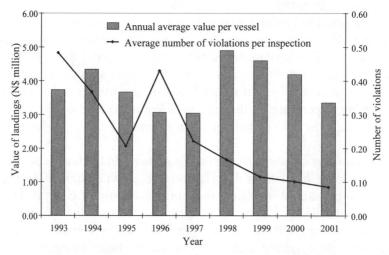

Figure 5. Demersal fishery – annual value of landings per vessel and average detected violations per inspection in Namibia.

theory that with an increase in economic return from the fishery a decrease in violations can be anticipated.

The influence of deterrence and sanctions on compliance - If the chances of detection are low then a fisher is more likely to violate regulations. The survey results indicate that the MCS platforms are perceived to be as important for deterrence as enforcement. Patrol vessels were perceived as the most effective deterrent while the plane was the least effective. The patrol vessels have maintained a steady and high level of patrolling over the last 10 years and thus been a visible presence on the fishing grounds while the plane did not operate in 2001 and only flew for half the normal hours in 2000; this has dramatically reduced the impact of the plane on MCS operations (Figure 6).

For the midwater fishery the patrol vessel inspections have remained at about 2 per vessel per year for the last 4 years (Figure 7). This has coincided with a decrease in the number of violations. It is likely that the presence of the patrol vessel on the fishing grounds and the fact that each vessel expects to be inspected about twice per year has contributed to the reduction of violations. The pelagic fishery yielded an interesting correlation between the two years where high average levels of violations were detected (2.57 and 1.33 per inspection) and the two years when each vessel had less than 20% chance of being inspected (Figure 8). The low level of inspections may have influenced the non-compliant behaviour or it may be that the inspections targeted non-compliant vessels making the sampling non-random; no data was available to verify this. Inspections on the demersal fishery have steadily reduced in

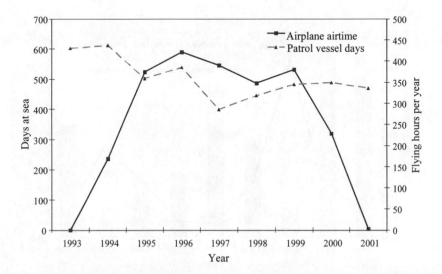

Figure 6. Time spent patrolling for the Namibian fisheries patrol vessels and plane, 1993 to 2001.

the last 4 years (from 2.2 to 1.3 per vessel per year) as have the number of violations over the same period. Therefore it cannot be concluded that an increase in the number of inspections has reduced the number of violations in the fleet. However, it may be that by maintaining the expected number of inspections per vessel above one, the deterrence value is maximised. This observation merits further investigation as it would assist in the setting of operational plans that are linked to a predefined compliance level.

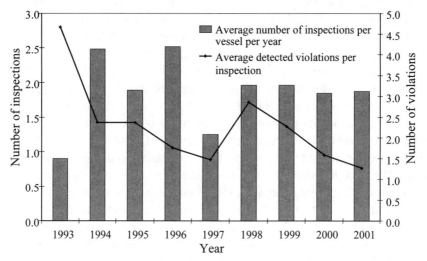

Figure 7. Midwater fishery average number of inspections per vessel per year and the average number of detected violations per inspection in Namibia.

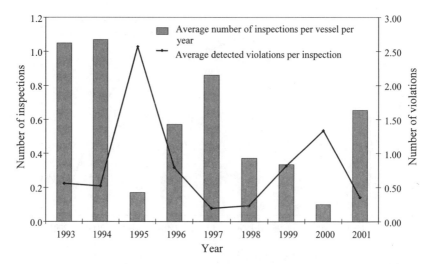

Figure 8. Pelagic fishery average number of inspections per vessel per year and the average number of violations detected per inspection in Namibia.

Once an offence has been detected it is generally considered that immediate action is required to reinforce the deterrent aspect of the punishment and that the punishment should be high enough to deter other potential violators. Survey results showed that the sanctions issued were not considered effective as a means of deterrence, due to very low fines and the lack of immediate action. Very few fines of over N$300 have been issued (Table 3) as only the courts can issue higher fines. Even when higher fines were issued they were also relatively low in comparison to the potential gain from the crime; this reduces any deterrence value.

The influence of legitimacy on compliance. - The consideration of legitimacy equates to how strongly the fishers believe in the regulatory system and the regulations imposed on them. Various studies have suggested that the stronger this belief, the less likely the fisher is to violate regulations (e.g. Copes, 1986; Tyler, 1990). Issues that build legitimacy include: the level of fisher participation in management, especially in the choice of management measures; meaningful regulations that are compatible with the fishing method; equitable distribution of the resources; impartial targeting of enforcement activity; faith in the assessment that underpins management measures; confidence in fair judicial systems; and a good understanding of the regulations in relation to sustainable management of the resource.

The MFMR has recently been encouraging participation of the industry in the management of the fishery. Efforts include annual meetings with the industry, involvement of the industry in scientific working groups and in the Marine Resources Advisory Council. The issue of confidence in management measures is routinely addressed through the Marine Resources Advisory Council and participation in developing new laws and regulations. All of these factors contribute positively towards creating a sense of legitimacy in the regulatory system. This was confirmed by the survey: an average score of 71% (good to very good) was awarded to the level of co-operation between the government and the industry in the areas of fisheries science (72%), legislation (74%), management (70%) and enforcement (70%). The level of perceived professionalism of the MFMR was acceptably high, but as with all aspects of the regulatory system there is still room for improvement: 79% of those questioned stated that increased participation from the industry in the MFMR would increase compliance.

The influence of morals and norms on compliance. - The influence of norms and what the fisher considers as moral behaviour are important considerations when analysing non-compliant behaviour. Norms can be considered as the characteristic actions, attitudes and expectations among fishers concerning the behaviour and attitude of their peers: in other words social pressure

(Giddens, 1984). Moral on the other hand is based on personal evaluation: what is personally considered right or wrong.

Namibia has a multi-culture, multi-language population and the fishing industry demonstrates this mix along with an additional foreign element. This makes analysis of norms and morals quite complex. The survey asked the question 'are Namibians more complaint than foreigners?'; the answers were 41% yes, 11% no, 39% the same and 9% don't know. This range of answers indicates no real trend in the manner that different stakeholders view the norms of Namibians and foreigners. It is common to assume that foreigners are less compliant because they do not identify with the long-term gains of good management practice. However, with the long-term rights allocations of some joint venture companies in Namibia the picture of cultural norms is not clear. One exception is that of the midwater fleet, that has been identified as the most non-compliant fleet: this fleet has the highest level of foreigners at 80% and is also a fishery that commonly uses flag of convenience.

Ensure that revenue from landings is correctly calculated. - The collection of landings data leading to the calculation of revenue is done manually at the factories by fisheries inspectors before it is entered into the database. The process starts in the moment the catch is landed and ends with a reconciliation of landings information with the vessel operators. Various reviews of the manual and electronic systems have taken place (Blondal, 2000; Iversen and Gilja, 2001). Weaknesses have been identified that make the final calculation of revenue very unreliable. These include: cumbersome work routines when data are collected and registered; an inaccurate reconciliation process leading to almost a 100% confidence in industry figures; large backlogs in data entry; software problems and inadequate training in the use of the database. It was concluded that the present system could be more accurate and timely if work routines were improved and by the implementation of a more user-friendly and reliable software programme.

Lost revenue to the Government due to the weaknesses in the present process was estimated in 1999 to be N$700 000 in lost bank interest (Blondal, 2000). The loss in terms of inaccurate reconciliation and thus underreported catches is impossible to estimate, but may be considerable. This uncertainty around the catch figures will also have an impact on the scientific calculation of the TAC and may imply that the TACs are over caught as not all caught fish are recorded in the system.

Cost of MCS operations

From a practical point of view, neither complete compliance nor perfect enforcement are realistic objectives for a fishery (described in the section:

analysis of the Namibian MCS system, evaluation of the success of the Namibian MCS system). The costs that would be incurred in perfect enforcement (and thus complete compliance) would be far greater than the economic revenue resulting from it (Sutinen and Andersen, 1985). Based on these principles it is generally agreed that the overall aim in relation to the financing of fisheries management is to strike a balance between the costs incurred and the benefits to be obtained from the fishery. Namibia has followed this principle and over the last six years the revenue from the combined fisheries has remained higher than the operating costs incurred by the government (Figure 9).

When making comparisons it should be kept in mind that profit or benefits can be defined in many ways: analytically the direct revenue generated to government is considered, but in economic terms, profit includes indirect benefits to the country such as employment opportunities, food security, foreign exchange earning and business synergy. If the wider range of social-economic benefits is considered, then the revenue generated would be considerably higher than that shown in Figure 9. It is also possible to include capital costs and donor contributions in the cost calculation. The capital costs, for example the financing of a new patrol vessel or a new inspectorate building, are costs falling under the national development budget and are therefore not included as MFMR operating costs. Donor assistance to the fishery sector has been considerable in the last years (Figure 10), but it is also not considered as operational costs. The inclusion of the full capital and donor contributions, or even inclusion of the full costs of servicing these

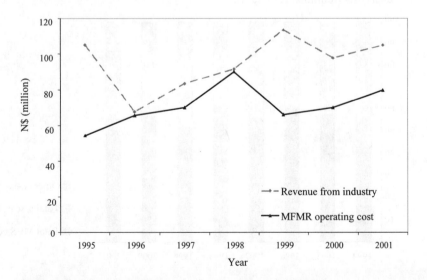

Figure 9. The cost of MFMR operations in relation to revenue (Namibia, 1995 - 2001).

inputs, in the operating cost of the MFMR would give a misleading picture. This is because, firstly, many of the capital and donor supported investments have been one-off events or programmes that are required when developing a new fishery administration and they can be treated as sunk costs, and secondly, some of the donor contributions are not considered fungible in Namibia and thus do not have opportunity costs. Therefore, for the purpose of analysing the MFMR and MCS costs, the comparison presented in Figure 9

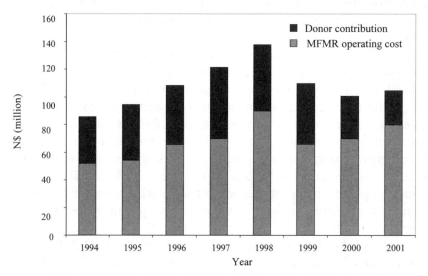

Figure 10. The financial contribution by donors in relation to the MFMR operational costs (Namibia, 1994-2001).

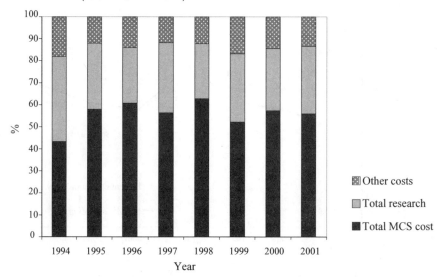

Figure 11. The division of financial resources within the MFMR (Namibia, 1994 – 2001).

is considered to reflect a realistic approach. In the last two years the MFMR has utilised 72% and 76% of the revenue from the industry for fisheries management and 41% and 42% of the industry revenue has been used for the MCS organisation. These figures are verging on the high side in comparison with fishery management organisations globally, but what is important is that these costs are covered by the revenue.

Enforcement is frequently the most costly element of fishery management, accounting for one quarter to one half of all expenditure (Sutinen and Kuperan, 1994). In Namibia the MCS organisation has utilised an average of 56% of the MFMR operating costs between 1994 and 2001 (Figure 11). This figure is higher than that found by Sutinen and Kuperan: a possible explanation for this is the requirement for inherently more expensive enforcement due to the predominance of output controls (e.g. quotas, no discards) as management measures and to the complexity of the multiple monitoring and enforcement platforms (e.g. landings inspections, observers, plane and patrol vessels) used to manage each fishery. Options to streamline operations will become an important consideration for the MFMR if they do not want to see costs escalating above revenue.

In order to plan the optimal MCS operations within a specific financial framework, for example with targets set as a percentage of revenue (Figure 9) or as a percentage of total MFMR expenditure (Figure 11), the MCS manager must collect the costs of the different components of MCS and evaluate these in relation to the compliance levels aimed for and those achieved within each fishery (Kelleher, 2002). In extracting and compiling the MCS operational costs (1999-2001) the average annual cost was nearly N$40 million. This can be divided by MCS component (Figure 12): operation of the patrol vessels in order to achieve at-sea inspections accounts for 32%, the monitoring of landings and transhipments accounts for 29%, the placing of observers onto fishing vessels accounts for 23%, while the remaining costs are shared between recreational fishery patrols and air patrols (at 8% each).

This figure allows a comparison of the cost of enforcement by MCS component across the entire Namibian EEZ, which is useful as an indicative figure for planning overall operations. However, in order to compare compliance and enforcement with costs, the average cost per event within each component is required. Using the same time period an average cost per event has been calculated (Table 5). It should be noted that determining these figures was fraught with complications as the recording of costs and events is different within each element of MCS. However, the figures provide useful values for comparative purposes.

The earlier section covering the analysis of the Namibian MCS System, evaluation of the success of the Namibian MCS system attempted to determine compliance levels for three fisheries; demersal, midwater and pelagic.

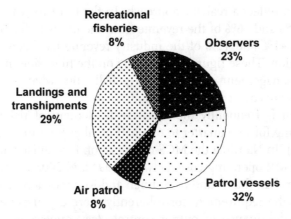

Figure 12. The division of MCS operational costs per MCS component in Namibia (based on an average 1999-2001).

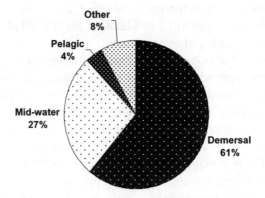

Figure 13. Percentage of MCS operational costs by fishery (Namibia, 1999-2001).

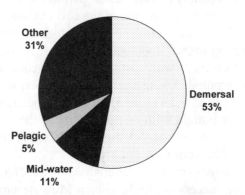

Figure 14. Percentage of landed value by fishery (Namibia, 1999-2001).

Using the figures from Table 5 the average cost of enforcement by component within each fishery has been calculated (Table 6) and the total by fishery as a percentage of the total MCS operational costs is indicated in Figure 13. It is clear that the allocation of the financial resources has a direct impact on the compliance levels observed in each fishery (as described in the earlier section entitled analysis of the Namibian MCS system, evaluation of the success of the Namibian MCS system) with the demersal fishery being the most costly to enforce, but showing the highest compliance level. In planning future activities it is also interesting to compare the cost allocation with the landed value by fishery to gain an indication if the cost is appropriately divided (Fig. 14). Other social-economic factors, such as employment and Namibianisation, should be considered. For example, the low financial commitments in terms of MCS compared with the higher landed value for the midwater fleet is a reflection of the low social economic importance of this fishery.

Table 5. MCS events per year in Namibia, and average operational cost per event (1999-2001).

MCS Event	Average cost per event (N$ 1999-2001)
Observer at-sea day	240
At-sea inspection of fishing vessel (from patrol vessel)	36 000
Aerial observation of fishing vessel (from airplane)[*]	2 000
Monitoring of a landing or transhipment	2 400
Coastal patrol mission (for recreational fisheries)	3 200

[*] The calculations for aerial observations do not include 2001 data as no flights took place during the year.

Table 6. Average total costs (1999-2001) of MCS by components by three fisheries in Namibia.

Component	Demersal (N$)	Midwater (N$)	Pelagic (N$)
Observers	6,453,557	1,515,221	262,460
Patrol vessel	8,292,000	1,728,000	372,000
Aerial observations	1,320,000	240,000	270,000
Landings or transhipments	5,280,000	960,000	1,080,000
Total cost by fishery	21,345,557	4,443,221	1,984,460

FUTURE CHALLENGES

The present MCS system is generally performing well and it is clear that Namibia deserves the international reputation the country has achieved. It is, however, important not to become complacent with this situation, but rather to aim to improve MCS efficiency and effectiveness while remaining flexible enough to meet future challenges.

Setting realistic compliance levels to guide MCS development and operational planning

Perfect MCS with 100% compliance does not exist. It is therefore important to explore what level of compliance is required and realistic for each fishery. These can then be compared with the actual levels of compliance being achieved. The earlier section evaluating the success of the Namibian MCS system presented estimates of compliance levels across three fisheries; this information was then linked to factors that influence compliance (economic situation, deterrence and sanctions, legitimacy, and norms and morals) in order to determine which forces may be driving the non-compliant behaviour within each fishery. This information, combined with knowledge of the main external forces on the system, an assessment of the costs incurred for the MCS operations, an assessment of the overall cost of MCS in relation to income, and a consideration of the scientific risk associated to the sustainability of the stock, provide the basis for setting practically and physically realistic targets for compliance levels.

The first challenge is to estimate present levels of compliance in all fisheries; secondly, to analyse these estimates in relation to the factors discussed in the previous section entitled analysis of the Namibian MCS System; thirdly to set realistic target compliance levels for each fishery; and finally to use these target figures as the shaping influence on the day to day distribution of resources, the operational plans and on future developments of the MCS system.

Improving the efficiency and effectiveness of MCS operational platforms

Improving what is already in place is a constant challenge to managers. Patrol vessels are extremely costly platforms (Table 5, Figure 12); maximising their performance is therefore essential. The use of night inspections and multiple boarding teams are feasible options that may require more than one inspector on each vessel but this is a minor cost compared with the current cost per inspection of N$36 000. With good information on the location of fishing vessels, strategic patrolling and a quick response is possible. This information will soon be available via the vessel monitoring system, which will provide an opportunity for improving the deployment of the patrol vessels and the targeting of specific fishing fleets or vessels.

The patrol plane consumes 8% of the annual MCS costs (Figure 12). Aerial surveillance is useful for effective surveillance of the EEZ and border violations. The plane can cover up to 45% of the EEZ during a 5 hour patrol in comparison with the patrol vessel that can only patrol 1-2% of the EEZ in one day. In 2001 the plane was grounded due to operational and human resource problems; this had been noted by the industry as they scored the plane the lowest for deterrence ability (survey results). The operational issues could be addressed through the privatising of the plane and the crew. Efficiency of the aerial surveillance would increase if night patrols existed; this requires the use of night vision equipment in integration with the vessel monitoring system.

Fishery observers represent a valuable and important MCS component. Over the last three years 67% coverage of the fleet has been achieved; although this falls short of the MFMR's objective of 100% it is an acceptable coverage rate. Due to this presence on fishing vessels, at the actual time of fishing, observers have the potential to become one of the most effective and cost-efficient tools in MCS. However, in the survey they were perceived as less effective for compliance and deterrence than the patrol vessels or landings control. The newly established Fisheries Observer Agency therefore has an opportunity to improve this perception and the performance of the fisheries observers.

In the previous assessment of the MCS system, the least successful element was identified as the calculating of revenue from the industry and the quota control. The result of this weakness was identified as not only causing an undetermined loss in revenue but also offering a substantial threat to stock sustainability by the possible over catching of quotas. Both the cumbersome working practices for landings control and data management and the database software system were identified as contributing factors to the system failure: a total redesign of working practices and information management is required, including consideration of an MCS information system.

In the section analysing the key factors that have influenced the MCS system the low level of sanctions given to serious offences was seen as reducing the value of the penalty system as a punishment measure. It is vital that crime doesn't pay and that the penalty is greater than the potential economic gain from the crime. Correcting this imbalance may boost the deterrent effect of penalties enough to allow a reduction in other more costly areas of MCS operations.

Voluntary compliance is the cheapest and most effective MCS solution available. The earlier section analysing the key factors that have influenced the MCS system demonstrated the links between legitimacy, norms, morals and voluntary compliance. Building on the legitimacy of the system and increasing the participation of the industry in the management cycle may provide an opportunity for increasing the compliance of fishers at little cost.

Facing future financial implications

Changes in the fiscal framework of the MFMR and the MCS organisation are inevitable. The earlier analysis shows that the management costs in 2001 were comfortably within the bounds of the revenue generated from the fishery (Figure 9) and that the MCS costs comprised almost 60% of the management figure (Figure 11). Future changes may be driven by fluctuations in fish stocks, changes in market demands, global political or social events or changes in the priorities of the Namibian government, to name just a few. Whatever is the driving force, the result may bring higher landings, and thus a greater demand on the present resources, or lower landings and thus a reduction in revenue and consequently in the funds available for MCS operations. Donor support to the MFMR is on the decline and will be dramatically reduced over the next years. As Namibianisation of the industry progresses, a reduction in the revenue due to tax incentives is expected. Research findings are forecasting a reduction in some fish stocks in the medium term. These are all signals that call for caution in the near future to maintain a well-proportioned and balanced fisheries management organisation.

The MCS organisation must be sure to contribute to this conservative approach: the previous section relating to future challenges has focused on options to facilitate this. However, the MCS organisation will face large capital repayment costs and increased running costs for MCS platforms in the coming years. A new patrol vessel has just been acquired and a second vessel is soon to arrive. There are plans to acquire a new surveillance plane and potentially a helicopter. This hardware represents an estimated capital investment of more than N$300 million with predicted running costs of N$20 - 30 million per year. Optimal management of these new resources is vital if they are going to be cost effective investments. This management will rely heavily on the analysis and assessment of the performance of the MCS organisation in relation to compliance levels.

International and regional challenges

Namibia has taken great pride in participating in international and regional fora and organisations to enhance fisheries management locally and globally. This is exemplified through: the new Southern African Development Community Protocol on Fisheries; the active participation in the establishment of the South East Atlantic Fisheries Organisation (SEAFO); membership to the International Convention on Conservation of Atlantic Tunas (ICCAT), and the Convention on Conservation of Antarctic Marine Living Resources (CCAMLR); and the efforts made to implement the principles of international legal instruments such as the FAO Code of Conduct for Responsible Fisheries.

The holistic approach of ecosystems management is becoming an increasingly more popular tool used for ocean and coastal management. The World Summit on Sustainable Development held in Johannesburg in 2002 united to reaffirm this trend and to place significant emphasis on the need for this type of approach. Efforts to date have focused on the research challenges related to defining transboundary issues and resources and to developing legislation to support regional management. However, soon a serious test will be given to the MCS organisation when they will be required to implement associated management measures both internally and in neighbouring waters. This will create a demand for qualified personnel such as fishery observers and inspectors to work outside of their home waters, hardware for regional patrolling such as vessels and planes, and the implementation of port state and flag state responsibilities. Systems to professionally deal with these challenges are not yet in place. However, a solid foundation exists that will facilitate the preparation of an implementation plan to allow Namibia to fully participate in these regional and international entities.

CONCLUSION

Following a dramatic initial enforcement of the EEZ in the early 1990s the newly formed MFMR was then faced with developing an MCS organisation with limited resources and an unskilled and inexperienced workforce. Over twelve years, Namibia has developed a complex, multi-dimensional and modern MCS organisation. The system has been shaped by the national-level policies of Namibianisation and transformation; it has been supported by a strong legal framework – although at times the complexity of the law has confounded compliance; the isolated coast of Namibia has facilitated MCS efforts; while the distance between operational units has been an obstacle for communication; the lack of an information system for MCS has reduced the capacity to plan and cross-check information; while finally, the strong reliance on donor support has on the one hand assisted the development, but on the other hand shaped progress in ways that may not always be objective.

Evaluation of the success of the system in relation to the three strategic objectives concluded that the first objective, to restrict fishing activity to those entitled to do so has been fully achieved, while the second objective, to ensure that fishing activity is conducted within legal and administrative guidelines has been partially achieved, and the third objective, to ensure that the revenue from landings are correctly calculated has not being achieved.

Evidence supports that compliance has generally improved over the last decade, although levels vary considerably across fisheries. Data from patrol vessels, observers and a questionnaire survey were used to assess compliance levels: results from the survey indicated that 88% of violations occur while fishing thus supporting the use of patrol vessel and observer data as indicators. It was observed that regular inspections by the patrol vessels reduced the number of violations. Analysis of the demersal fishery yielded very low violation rates, which were supported by survey results on perceived compliance levels. This fishery gave a strong correlation between economic return from the fishery and the level of violations, supporting the theory that financial viability of the fishery affects the behaviour of fishers. The midwater fishery on the other hand, a fishery of less social and economic importance to Namibia, is faced with unacceptably high levels of non-compliant behaviour. This fishery also provided evidence that an increase in economic return coincided with a decrease in violations. In analysing the norms and morals of the fishers it was evident that this, a predominantly foreign fishery (often utilising flag of convenience), was also the least compliant. The pelagic fishery that has gone through severe financial difficulties in the last years has remained with a steady level of recorded violations, with no evidence of a link between the economic return of the fishery and the violation level. Results indicated that progress in improving the compliance level across all fisheries is hampered by the low deterrence value of the fines imposed and the delay between crime and punishment.

The third objective, ensuring that revenue from landings is correctly calculated, is not successfully implemented. Evidence indicated that the calculation of revenue is very unreliable and that in 1999, N$700 000 was lost in bank interest, while the loss due to inaccurate reconciliation and underreported catches was impossible to estimate, but may have been considerable.

The cost of MCS in the last two years was 41% and 42% of the industry revenue: this is considered an acceptable level, as was the distribution of cost across MCS components. However, serious concern was raised over the future cost; with two new patrol vessels on the horizon, possibly a new plane and helicopter and a withdrawal of donor support. Careful planning and streamlining will be required if the MCS organisation is to continue to develop a successful and sustainable operation. Improvements are therefore required and various options have been identified to meet this challenge: the setting of compliance targets to streamline logistical operations and planning; the improvement of the performance of MCS platforms to increase cost effectiveness; a more analytical approach to balancing enforcement and voluntary compliance in order to unlock potential increases in compliance;

the shortening of the decision-making process to promote more immediate reactions to serious violations; an increase in fines to ensure that crime doesn't pay; the creation of an MCS information system to facilitate cross verification and improved planning; and finally and most importantly to redesign the working practices and information systems used to calculate landings in order to ensure that catch limits are not exceeded and that revenue is correctly calculated.

It is evident that, against all odds, in a period of twelve years, Namibia has taken control of its EEZ. Poachers have been removed and licensed fishing is managed through a combination of catch limits and technical control measures. A multi-platform, multi-dimensional MCS system has been developed to enforce fishery regulations: a system that, while meeting demands for effective enforcement, has also met the demands of a new social and political order. It can be said that Namibia not only deserve the international reputation it has gained for MCS, but that it is also ready to operate, sustain and develop the MCS system without external assistance. This is a quite considerable achievement.

REFERENCES

Becker, G.S. (1968): Crime and Punishment: an Economic Approach. *Journal of political economy* 76(2): 169-217.

Bergh P.E. and Davies, S.L. (2001): Monitoring, control and surveillance. In: *A Fishery Manager's Guidebook FAO Fisheries Technical Paper, No. 624*. Rome, FAO. (K.L. Cochrane, ed.), 175-204.

Blondal, J. (2000): Report on Namibia's Fisheries Information System. Ministry of Fisheries and Marine Resources, Windhoek. 28 pages.

Copes, P. (1986): Critical review of the individual quota as a device in fisheries management. *Land Economics* 62(3): 278-291.

EBCD and GOPA (1996): Feasibility study for SADC monitoring control and surveillance of fishing activities. Project No. 7, ACP RPR 484: Windhoek. 41 pages.

Furlong, W.J. (1991): The deterrent effect of regulatory enforcement in the fishery. *Land Economics* 61: 116-129.

Giddens, A. (1986): *The constitution of Society: outline of the theory of structuration*, Cambridge: Polity Press, Berkeley: University of California. 440 pages.

Hønneland, G. (1999): A model of compliance in fisheries: compliance behaviour in the fishery. Theoretical foundations and practical application. *Ocean and Coastal Management* 42: 699-716.

Iversen, F and Gilja, A. (2001): Internal Report on landings routines and management of landings data. MFMR, Windhoek. 37 pages.

Kelleher, K. (2002): *The costs of monitoring, control and surveillance of fisheries in developing countries*. FAO, Rome, 47 pages.

Kuperan, K., and Sutinen, J.G. (1998): Blue water crime: legitimacy, deterrence and compliance in fisheries.

Law and Society Review 32(2): 309-338.

MFMR (1999): *Planning in action 1999 – 2000 (Our strategic plan).* MFMR, Windhoek. 30 pages.

Raakjaer Nielsen, J. and Mathiesen, C. (2003): Important factors influencing rule compliance in fisheries – lessons from Danish fisheries. *Marine policy [Amsterdam]* Vol. 27. No. 5. September 2003, 409-416.

Raakjaer Nielsen, J. and Vedsman, T. (1999): User participation and institutional change in fisheries management: a viable alternative to the failure of "top-down" driven control? *Ocean and Coastal Management* 42(1): 19-37.

Sutinen, J. and Andersen, P. (1985): The economics of fisheries law enforcement. *Land Economics* 61(4), 387-397.

Sutinen, J. and Kuperan, K. (1994): A socioeconomic theory of regulatory compliance in economics and trade. In: *Proceedings of the VIIth conference of the international institute of fisheries economics and trade volume* 1 (D. Liao, ed.), 18-21 July: 189-203, Taipei, Taiwan. National Taiwan Ocean University Press.

Sutinen, J. Rieser, A. and Gauvin, J.R. (1990): Measuring and explaining non-compliance in federally managed fisheries. *Ocean Development and International Law* 21: 335-372.

Tyler, T.R. (1990): *Why People Obey the Law.* Yale University Press, New Haven and London, 273 pages.

Wiium, V. and Uulenga, A. (2003): Fishery management costs and rent extraction: the case of Namibia. In: *The Costs of Marine Fisheries Management* (W.E. Schrank, R. Hannesson and R. Arnason, eds.), 173-186, Aldershot: Ashgate Publishers.

16 MARINE FISHERIES MANAGEMENT IN NAMIBIA: HAS IT WORKED?

Paul Nichols

Abstract

The history of Namibia's fisheries is characterised by massive and uncontrolled fishing, primarily by European and Eastern bloc fleets, followed by near collapse of many stocks. This period was followed by a dramatic recovery of the resources following Independence in 1990, and the implementation of a resource management system, that incorporated a highly effective, cost-efficient system of monitoring, control and surveillance. Namibia's successful post-independence track record bears testimony to what a young developing nation can achieve if sufficient resources and political will are provided in support of fisheries management.

INTRODUCTION

Largely as a result of up-welling of the nutrient-rich Benguela Current, Namibia's waters are highly productive. Prior to Namibian Independence in 1990, uncontrolled fishing on a massive scale - perpetrated mainly by Spanish and Soviet vessels, and to a lesser extent vessels from Portugal, South Africa, Romania, Poland, Bulgaria, and Cuba - greatly reduced the abundance of all the major fish stocks.

During the 1960s South African factory ships undertook fish processing at sea outside the then 22-kilometre jurisdiction of Namibia's fisheries administration. Over-exploitation caused sardine catches to plummet; when the vessels turned to anchovy, that stock also plummeted.

From 1964 foreign interest in Namibia's offshore fishing grounds grew rapidly, with the advent of long-distance freezer trawlers (Bonfil *et al.*, 1998; Sumaila and Vasconcellos, 2000). For example, in 1964 a mere 47,600 tonnes of hake were caught, but by 1972 hake catches were reported to be 820 000 tonnes although the true catch figure was probably much higher.

The International Commission for Southeast Atlantic Fisheries (ICSEAF), established in 1969 with the intent of good management, was in reality used by many of its 17 member states to legitimise plundering of fish stocks in the southeast Atlantic, and particularly in Namibian waters. Namibia declined to become a member of ICSEAF at Independence, and the organization is now in the process of being formally disbanded.

BUILDING A MANAGEMENT REGIME

Finding itself at Independence with a heritage of systematically depleted fish stocks, the newly elected Government moved quickly to establish a fisheries administration - the Ministry of Fisheries and Marine Resources. The policy framework for the marine fisheries sector is set out in the White Paper of December 1991, titled "Towards Responsible Development of the Fisheries Sector". The White Paper sets the goal of fisheries management and development as being:

> "To utilize the country's fisheries resources on a sustainable basis and to develop industries based on them in a way that ensures their lasting contribution to the country's economy and overall development objectives."

This goal is pursued through four main strategies: (a) rebuilding stocks; (b) building a national fishing and processing industry; (c) Namibianisation, to ensure that the benefits of rebuilding stocks and building a fishing industry in Namibia accrue substantially to Namibians through increasing ownership of companies and vessels, new job creation and replacement of foreign labour by Namibian labour (Armstrong *et al.*, this volume); and (d) empowerment, to ensure an equitable balance of participation and increasing employment for Namibians, especially the previously disadvantaged.

Once a policy environment had been set forth, an appropriate legislative framework was put in place. One of the first acts of Parliament was the Territorial Sea and Exclusive Economic Zone of Namibia Act of 1990, underlining the importance attached to fisheries. In 1992 Parliament passed the Sea Fisheries Act. Namibia subsequently signed up to a number of international fisheries conventions, agreements and arrangements. These new international obligations prompted a revision of the 1992 Act, which was replaced in 2001 by the Marine Resources Act (2000). Key elements of the management system defined in the policy and legislation are outlined below, and are summarised in the Annex.

Management Measures

Fishing rights

Fishing rights, or rights of exploitation, are the central element of the fisheries management regime. The Marine Resources Act 2000 states "No person shall ... harvest any marine resource for commercial purposes, except under a right..." The main purpose of fishing rights is to limit entry to the fisheries sector in order to protect the fisheries resources and maintain sustainable operations. In 2002 there were 152 right holders in the various Namibian fisheries. Fishing rights are granted for a period of 7, 10, 15 or 20 years depending on various factors, in particular the level of investment and the level of Namibian ownership of the enterprise. Fishing rights are not freely transferable in Namibia, so as not to undermine the Government's goals of Namibianisation and empowerment within the sector. The total number of existing rights in 2002 was 163. The table below shows the number and duration of existing harvesting rights for each species.

Fishing rights were first introduced in 1994, for periods of 4, 7 and 10 years. In June 2001, duration of rights was changed from 4, 7 and 10 years to, respectively, 7, 10 and 15 years. A number of 4-year rights (awarded be-

Table 1: Number and duration of existing harvesting rights as at December 2002.

Fishery	Duration of rights						
	Four-year	Five-year	Seven-year	Ten-year	Fifteen-year	Twenty-year	Total
Hake	4	0	10	5	19	0	38
Monk	2	0	2	0	5	0	9
Horse mackerel	0	0	0	11	1	0	12
Large pelagic	4	0	1	3	11	0	19
Red crab	0	0	1	2	0	0	3
Rock lobster	5	0	0	1	15	0	21
Linefish	2	0	1	2	7	0	12
Orange roughy	0	0	3	2	0	0	5
Sardine	0	0	5	17	0	0	22
Mullets	0	17	0	0	0	0	17
Seals	0	0	3	1	0	0	4
Guano	0	1	0	0	0	0	1
Total	17	18	26	44	58	0	163

fore 2001) are still current and will eventually be phased out once they expire. Five-year rights apply only to guano and mullet fisheries and will similarly be allowed to expire.

Fishing licences

All vessels are required to obtain a licence in order to fish commercially within Namibia's 200-mile exclusive economic zone (EEZ). All vessels that fly the Namibian flag are required to have a specific licence to harvest any marine resources in waters outside the Namibian EEZ.

The number of licensed vessels operating in Namibian waters from 1998 to 2002 is indicated in the table below. A total of 335 vessels were licensed for commercial fishing in 2002.

Table 2: Number of licensed vessels by fishery, 1998-2002.

Fishery	1998	1999	2000	2001	2002
Small pelagic	35	33	30	26	25
Demersal trawlers	85	97	111	128	114
Longliners	6	20	24	38	10
Midwater	25	26	26	24	20
Deepwater	5	6	5	3	6
Large pelagic	47	54	56	68	71
Linefish	25	27	26	22	26
Crab	3	3	2	2	2
Rock lobster	29	27	29	29	38
Monk					23
Total	260	293	309	340	335
Percentage Namibian	84%	80%	80%	68%	71%

Total allowable catches

Total allowable catches (TACs) are set for seven species: sardine, hake, horse mackerel, red crab and rock lobster, orange roughy and monk.

TACs are established annually on the basis of the best scientific evidence available of the size and structure of stocks as determined by the fisheries scientists employed by the Ministry. The purpose with the TACs is to ensure sustainable fishing operations; that the level of fishing effort does not undermine the status of each stock.

Table 3: Total allowable catches, 1990-2002 in tonnes.

	Sardine	Hake	Horse mackerel		Red crab	Rock lobster	Alfonsino	Orange roughy	Monk
1990	40 000	60 000	150 000		n.a.	n.a.	n.a.	n.a.	n.a.
1991	60 000	60 000	465 000		6 000	1 200	n.a.	n.a.	n.a.
1992	80 000	90 000	450 000		6 000	100	n.a.	n.a.	n.a.
1993	115 000	120 000	450 000		4 900	300	n.a.	n.a.	n.a.
1994	125 000	150 000	500 000		4 900	130	n.a.	n.a.	n.a.
1995	40 000	150 000	400 000	(50 000)	3 000	230	n.a.	n.a.	n.a.
1996	20 000	170 000	400 000	(90 000)	2 500	250	n.a.	n.a.	n.a.
1997	25 000	120 000	350 000	(100 000)	2 000	260	10 000	12 000	n.a.
1998	65 000	165 000	375 000	(75 000)	2 000	300	0	12 000	n.a.
1999	45 000	275 000*	375 000	(50 000)	2 000	350	n.a.	6 000	n.a.
2000	25 000	194 000	410 000	(50 000)	2 000	350	n.a.	2 400	n.a.
2001	10 000	200 000	410 000	(50 000)	2 100	400	n.a.	1 875	13 000
2002	0	195 000	350 000	(40 000)	2 200	400	n.a.	2 400	12 000

Notes: n.a. means 'not applicable'. Figures in brackets indicate the portion of the TAC (column immediately to the left) of juvenile horse mackerel caught for fishmeal. *There was a change-over for the hake fishing year from a calendar year to the period May-April. As a consequence an interim TAC of 65 000 was given for the period January to April 1999, followed by a TAC of 210 000 for the new fishing year May 1999- April 2000.

Individual (non-transferable) quotas

Once a TAC has been set for a fishing season, it is distributed among the right holders in each fishery in the form of quotas. The main purpose with the quota allocation is to promote economic efficiency – to give companies sufficient knowledge about expected catch levels for the year for proper planning of their fishing activities. Quotas are not permanently transferable for the same reasons that rights are not transferable. Production of marine resources for the period 1998 to 2002 is given below.

The sardine stock remained low during 2002, and as a result a zero TAC was declared for the sardine fishery in 2002. The catch that appears in the table was taken as by-catch in the horse mackerel and anchovy-directed purse seine fisheries, despite strict controls that were in place to minimise sardine by-catch. Despite the low spawning stock biomass recorded in March 2002, recruitment from the 2001/2002 spawning season was excellent and the October 2002 survey estimated that the stock had increased to more than 360 000 tonnes, allowing a 20 000 tonne TAC to be issued in 2003.

Table 4: Harvest of the main commercial species, 1998-2002 (tonnes, except seals).

Species	1998	1999	2000	2001	2002
Sardine	68 562	44 653	25 388	10 763	4 160
Hake	150 695	164 250	171 397	173 277	154 588
Horse mackerel	312 422	320 394	344 314	315 245	359 183
Monk	16 429	14 802	14 358	12 390	15 174
Kingklip	2 211	3 706	3 922	6 607	7 210
Tuna	1 442	1 155	2 401	3 198	2 837
Crab	2 283	2 074	2 700	2 343	2 471
Rock lobster	350	304	365	365	361
Other fish species*	51 271	26 500	22 987	30 810	77 407
Total fish harvest	605 654	577 838	588 404	554 998	623 391
Seals (numbers)	29 475	25 161	41 753	44 223	40 000
Seaweed	8 973	6 600	829	800	500

* Other fish species include orange roughy, alfonsino, anchovy, sharks, sole, and line-fish species.

Fees

Fees form an important part of Namibian fisheries management. Their role is twofold: firstly, to earn revenue for the government, and secondly to create incentives that work towards the goals of the management system, both conservation and Namibianisation.

The most important are quota fees, which are payable on allocated quota. By-catch fees are applied in order to deter right holders from targeting species other than those for which they have been issued a quota. This is a feature of the Namibian management system that is not seen in many other countries. Such fees provide an incentive to avoid catching non-target species. By-catch fees are carefully balanced to discourage the capture of non-target species, but are also not so punitive as to encourage dumping. A certain percentage of by-catch is not levied, since a reasonable amount of by-catch cannot be avoided. A Marine Resources Fund levy is imposed per tonne of landed catch to finance fisheries research and training initiatives. Finally, licence fees are charged for all fishing vessel licences issued to vessels that fish within Namibia's waters.

Subsidies

The Namibian fishing industry is not subsidised. Namibia is strongly opposed to the subsidy policies pursued by other nations due to a belief that subsidies cause over-capitalisation, distort trade unfairly and ultimately lead to over-fishing and the encouragement of illegal, unreported and unregulated (IUU) fishing practices (Millazo, 1998; Munro and Sumaila, 2002). Namibia instead prefers a system of taxation, applied especially through the quota fees, and this was one of the main attractions for implementing a rights-based system. On the one hand, the application of a rights-based system has led to healthier stocks, improved compliance and an efficient industry that supports proper fisheries management and earns healthy profits. On the other hand, limiting access to the resource and fishing mortality for each participant has provided a basis for extracting some of the profits.

Giving effect to international fisheries agreements
For any fisheries or international agreements entered into by Namibia, the Minister is empowered to make regulations necessary to give effect to such agreements. Texts of all conservation and management measures adopted under any international agreement to which Namibia is a party are published in the national Gazette, and thus such measures are then deemed to be a regulation as prescribed under the Act.

Monitoring, control and surveillance
On the day in 1990 that Namibia's 200-mile EEZ was declared, more than 100 foreign vessels were fishing illegally in Namibian waters. When other small coastal states had found it impossible to effectively control such operations in their EEZs, they faced little real alternative to sanctioning continuation of the foreign operations through licensing arrangements that did not leave them in real control.

Namibia, however, decided to put in place measures to reap the gains from sustainable utilization of its fisheries. During 1990 and 1991, 11 Spanish trawlers and one Congolese trawler were arrested for illegal fishing and successfully prosecuted; most of them were forfeited to Namibia by the Namibian courts. These actions sent a clear message to the international fishing community that Namibia was serious about establishing sovereignty over its new EEZ. There were a few further incidents of poaching after this, but effective monitoring, control and surveillance (MCS) and enforcement deterred poachers and improved compliance by licensed vessels.

Namibia's MCS system has evolved over the years into what is today widely regarded by the international community as a very effective system (Bergh and Davies, this volume). A crucial element has been the financial, human and material support from the Namibian Government. The costs to government and industry of MCS and other management activities have been kept commensurate with the value of the sector. From 1994 to 1997, the full cost to the Namibian Government of fisheries management, including fisheries research and MCS, was about 6 per cent of landed value; that fell to 4.9 per cent in 1998 and 3.6 per cent in 1999, due to the increasing value of landed catch. This cost is appropriate to the economic value of the fisheries sector and reasonable when compared with the cost of other comprehensive and effective fisheries management systems elsewhere in the world (Sutinen and Kuperan, 1994).

An integrated programme of inspection and patrols at sea, on land and in the air ensures continuing compliance with Namibia's fisheries laws. The major features of Namibia's MCS programme are described below.

On-board observer programme - Emplacement of fisheries observers on board larger vessels serves to ensure both compliance and the collection of scientific data. Coverage rates range from 70% to 100%, depending on the fishery in question. The establishment of the new Fisheries Observer Agency under the Marine Resources Act (2000) should improve current capacities in this regard.

Sea, air and shore patrols - Systematic sea patrols aim to ensure compliance with fishing conditions by licensed vessels through regular at-sea inspections. Air patrols detect and deter unlicensed fishing vessels and monitor the movement and operations of the licensed fleet. Shore patrols ensure compliance by both recreational and commercial fishers with conservation measures for inshore resources.

Monitoring of landings - Complete monitoring of all landings at the two commercial fishing ports, Walvis Bay and Lüderitz, by onshore inspectors ensures compliance with quota limits and fee payments. Transhipping fish at sea between catching vessels and carrier vessels is prohibited – all fish must be landed at a Namibian port. This helps to ensure comprehensive monitoring of catches. The absence of an artisanal fisheries sector also helps to simplify monitoring of landings.

Vessel reporting - All vessels are required to supply EEZ exit and entry reports, as well as daily catch and effort reports via radio and in the form of vessel log-sheets.

Vessel monitoring system - Namibia is well advanced in implementing a national satellite-based vessel monitoring system (VMS). Once fully operational, the system will benefit fisheries management in real-time monitoring of vessel movement and activities. The system that has been chosen is already in use in the United Kingdom, Germany, United States, Morocco, and, closer to home, South Africa and Mozambique. Namibia is fully supportive of collaborating in the development of a cost-effective, regional VMS.

Socio-economic importance of the sector

Namibia's policy to encourage on-shore processing has seen the number of whitefish processing plants increase from zero in 1991 to around 20 in 2003. Direct employment in the sector has expanded to around 13 500 people. Of the 5 575 employed onboard fishing vessels, 68% are Namibians. The 8 000 shore-based workers are nearly all Namibians. In addition, Namibians now enjoy real economic prosperity through participation in the marine resources sector. Of the 163 rights in 2003, all except one are majority controlled by Namibians. At Independence, Namibians controlled only 17% of the hake quota – today Namibian control is around 96%. In horse mackerel the story is similar, rising from less than 14% to around 92%. In some fisheries, such as the small pelagic fishery and rock lobster, all quotas are in Namibian hands. The proportion of Namibian vessels increased from 50% in 1991 to 71% in 2002. At the same time the sector continues to attract foreign capital, skills and market access necessary for further development.

Although the contribution of income from marine resources to GDP has fluctuated over the years, mainly due to the unpredictable nature of the resource, it has shown an overall increase from N$288 million (4%) in 1991 to N$2 016 million (6.6%) in 2002. The value of fisheries production has also increased substantially since 1991, mainly due to an increase in the prices obtained in the export markets as well as value addition. Landed value has

Table 5: Indicative investments and socio-economic contributions made by right holders since Independence.*

Sub-sector	Investments (N$)	Socio-economic contributions (N$)	Total (N$)
Demersal	1 203 153 010	16 472 599	1 219 625 608
Monk	296 165 000	2 066 241	304 631 241
Midwater	141 700 000	6 264 000	142 164 000
Small pelagic	262 480 000	6 769 000	269 249 000
Large pelagic	146 000 000	1 196 000	147 196 000
Linefish	12 023 000	65 000	12 088 000
Crab	14 400 000	N/a	N/a
Rock lobster	6 395 772	828 862	7 224 634
Total	2 082 316 782	33 661 702	2 115 978 484

* The figures in this table indicate the minimum level of investments and social contributions.

increased four times from N$520 million in 1991 to N$2 596 million in 2002. Final value has increased more than four times from N$644 million in 1991 to N$3 395 million in 2002. Since an estimated 97% of total fish production is exported, the value of exports closely follows the same trend as final value and has also increased substantially from N$631 million in 1991 to N$3 311 million in 2002 (Lange, this volume).

The marine fisheries sector has consistently been the second largest sector in the Namibian economy behind mining in terms of export earnings. A major export for our fisheries and marine resource production is the EU. According to the EU Market Survey (2002) for Fishery Products, the EU imported 99 410 tonnes of fish and fish products worth ca €250 million from Namibia, making Namibia the fourth-largest developing country supplier of fish to the EU in 2000 after Argentina, China and Thailand. Namibia was the largest developing country supplier of hake at 82 251 tonnes, worth ca €180 million.

A worthy achievement of the sector that goes largely unnoticed is the generous and continuing contributions that our fishing companies have made to social development schemes throughout the country. On a continual basis, our fishing companies provide money and other forms of assistance for the construction of schools, clinics and other much-needed civic facilities. The contribution of our fishing industry to these worthy causes over the past 11 years runs in excess of N$33 million. The newcomer companies deserve special recognition. Despite being 'new' to fishing, they have managed to contribute in excess of about N$11 million to these worthy causes.

REGIONAL AND INTERNATIONAL COOPERATION

Regional cooperation in fisheries management is enhanced through a number of mechanisms. The Southern African Development Community (SADC) is implementing two regional programs of particular relevance: the Regional Fisheries Information System Program, which aims to capture and disseminate timely, relevant, accessible, useable and cost-effective information to improve the management of marine fisheries resources in the SADC region; and the Regional Fisheries MCS Program, which aims to improve national capacity for efficient, cost-effective and sustainable MCS and to enhance regional cooperation on MCS and fisheries management.

A recent initiative is the SADC Protocol on Fisheries, which aims to promote responsible and sustainable use of the living aquatic resources and aquatic ecosystems within the SADC region.

A convention to establish the South-East Atlantic Fisheries Organization (SEAFO) was signed by nine states in Namibia on 20 April 2001, the first such convention to be signed following the establishment of the 1995 UN Fish Stocks Agreement. SEAFO establishes a management regime for conservation and sustainable utilisation of fish, molluscs, crustaceans and other sedentary species in the high-seas portion of what is essentially FAO Statistical Area 47. It excludes those sedentary species that are subject to the fishery jurisdiction of coastal states, and tuna and tuna-like species that fall under the jurisdiction of the International Commission for the Conservation of Atlantic Tunas (ICCAT). Namibia joined ICCAT in 1999 and abides by its comprehensive management tools to curb IUU fishing targeting tunas.

As a member of the Commission for the Conservation of Antarctic Marine Living Resources (CCAMLR), Namibia complies fully with the CCAMLR catch documentation scheme to reduce IUU fishing in Antarctic waters.

PROSPECTS AND CHALLENGES

The recovery of stocks from their over-fished pre-Independence state has been variable. Hake and horse mackerel continue to recover. The sardine stock biomass, however, has been adversely affected by environmental factors since 1994 and is currently at a low level. Other stocks are responding positively to conservation measures and are either stable or growing slowly. However, the marine environment will continue to strongly impact fish stocks in a largely unpredictable manner.

The long-term outlook for the trading environment for Namibian fishing companies is generally positive. The healthy state of the demersal stocks (hake, orange roughy and monkfish) is fortuitous at a time when stocks in the waters of many of Namibia's competitors in the global whitefish market are clearly suffering from over-fishing. As a result of restricted supply, market prices are buoyant for high-quality products. Increased consumer awareness of the benefit of eating fish is also contributing to increasing world demand. Continued growth of demersal stocks and increased market demand suggest a very positive outlook for the sector. The position is less clear for the two other major fisheries: the pelagic (sardine) and midwater trawl (horse mackerel) fisheries. The midwater trawl fishery enjoys sustained catches but is facing market constraints and is consequently looking for new marketing opportunities. The future of the pelagic sector depends entirely on recovery of the sardine stock. Namibia's marine fisheries sector was badly affected by the strength of the Namibian dollar during 2003, as were all export-orientated industries, resulting in reduced profitability.

Namibia's marine fisheries will, of course, continue to develop and evolve. Potential exists for expansion within several fields of the marine fishing industry, e.g. value-added fish products and new fishing opportunities in foreign waters. There will continue to be room for new investment by both new foreign and new domestic investors, but competition is fierce from those already in the industry, including from the many new companies that have entered the sector since Independence. Investors who are interested in Namibia's fishing or processing industries need to be well prepared and highly capable if they hope to secure successful trade and investment opportunities. With the recent enactment of the Aquaculture Act, 2001, a number of right holders are investing in aquaculture (fish farming). Aquaculture in Namibia has considerable potential, especially for the commercial production of high-value marine species, in accordance with current policy.

For its part, the Ministry of Fisheries and Marine Resources will continue to apply a responsible approach to fisheries management, by which it is hoped that stocks will continue to grow and support increased sustainable yields. There are a number of challenges to be faced, however. Reduced catches in many other important fisheries of the world, combined with growing demand for high-quality fish products, is expected to increase the risk of IUU fishing. Consequently the MCS role of the Ministry will become even more demanding than it is today.

Quality control in food industries will become increasingly stringent. As consumer awareness regarding fish quality increases, it will be essential for Namibian fish products to continue to meet the highest international standards. Plentiful harvests of fish are worthless if no consumers are willing to buy. Namibia has a reputation of a clean and environmentally unspoilt country and that image must be maintained at all costs.

Perhaps the most daunting challenge is human resource development. Sustained development depends on an enlightened and well-educated work force. The Government, together with other stakeholders, will continue its good work in educating the people of Namibia and preparing them to work in all dimensions of the marine sector and aquaculture. Related to human resource development is the prevention and combating of the spread of HIV/AIDS. This is one of the greatest tasks facing southern Africa and a concerted effort is needed, not only at a national level, but also at the regional and international levels.

Conclusion

Namibia's policy and legal framework for the marine fisheries sector has allowed the application of management strategies that are appropriate to Namibia's specific circumstances. The result has been the development of a business environment that has facilitated the growth of a healthy fishing and processing industry that pays a fair price for the privilege of utilising Namibia's marine resources.

The Ministry of Fisheries and Marine Resources will continue to ensure that the environment for fisheries-related business and aquaculture is conducive to continued healthy rewards for those willing to invest. Namibia is keen to capitalise on the gains made since Independence to the greater benefit of Namibia and its citizens. Care will be required to ensure that the level of fishing is commensurate with the size of the stocks. If the current system of responsible management is maintained, Namibia's marine resources may be expected to yield sustainable benefits in perpetuity.

REFERENCES

Armstrong, C.W., Sumaila, U.R., Erastus, A. and Msiska, O. (2004): Benefits and costs of the Nambianization policy. In: *Namibia's Fisheries: Ecological, Economic and Social Aspects* (U.R. Sumaila, D. Boyer, M. Skogen, and S.I. Steinshamn eds.), pp. 203-214. Eburon, Delft.

Bergh, P.E. and Davies, S. (2004): Against all odds: Taking control of the Namibian Fisheries. In: *Namibia's Fisheries: Ecological, Economic and Social Aspects* (U.R. Sumaila, D. Boyer, M. Skogen and S.I. Steinshamn, eds.), pp. 289-318. Eburon, Delft.

Bonfil R., Sumaila, U.R., Munro, G., Valtysson, H., Wright, M., Pitcher, T., Preikshot, D., Haggan, N. and Pauly, D. (1998): Impacts of distant water fleets: an ecological, economic and social assessment. In: *The Footprints of Distant Water Fleet on World Fisheries*. Endangered Seas Campaign, WWF International, Godalming, Surrey, 122 pages.

EU Market Survey 2002: Fishery Products (volume II). Centre for the Promotion of Imports from Developing Countries (URL: *www.cbi.nl*)

Lange, G.M. (2004): Economic value of fish stocks and the national wealth of Namibia. In: *Namibia's Fisheries: Ecological, Economic and Social Aspects* (U.R. Sumaila, D. Boyer, M. Skog and S.I. Steinshamn, eds.), pp. 187-202. Eburon, Delft.

Milazzo, M.J. (1998): Subsidies in World Fisheries: A Re-examination. World Bank Technical Paper, No. 406, Fisheries Series, Washington.

Munro, G. and Sumaila, U.R. (2002): The impact of subsidies upon fisheries management and sustainability: the case of the North Atlantic. *Fish and Fisheries* 3: 233-290.

Sumaila, U.R. and Vasconcello, M. (2000): Simulation of ecological and economic impacts of distant water fleets on Namibian fisheries. *Ecological Economics* 32: 457-464.

Sutinen, J. and Kuperan, K. (1994): A socioeconomic theory of regulatory compliance in economics and trade. In: *Proceedings of the VIIth conference of the international institute of fisheries economics and trade volume 1* (ed. D. Liao), 18-21 July,: pp 189-203, Taipei, Taiwan. National Taiwan Ocean University Press.

Annex : Summary of major fisheries management measures

Sustainability Measures (for biological and economic sustainability):
- Limited Entry in all fisheries: no fish may be taken for commercial purposes without a right.
- Limits on catches: TACs/Quotas covering 90% of landings.
- Quotas are issued as Individual Quotas to right holders.
- Limits on fishing effort: in some fisheries, limits on the numbers and size of vessels.
- Closed seasons: to protect spawning fish.
- Closed areas: to protect juveniles; no trawling in depths less than 200 metres; lobster sanctuaries.
- Minimum sizes and bag limits of fish that may be kept in recreational (non-commercial) fisheries: also to protect juveniles.
- Gear restrictions: including mesh size limits to protect juveniles, no beam trawling; no driftnets; no formation trawling; selectivity grids are currently being introduced into the hake fishery.
- Levies on by-catches: to control targeting of by-catch species.
- No discarding is permitted: all edible, marketable fish caught must be landed.

Control Measures:
- No transhipment at sea: all fish landed in Walvis Bay or Lüderitz – facilitates monitoring of catches and calculation of levies payable by all right holders.
- 100% weighing of fish landed or transshipped.
- 70-100% observer coverage, depending on fleet. Administration of observer programme by autonomous Fisheries Observer Agency.
- EEZ entry and exit reporting required by all vessels.
- Catch and effort reporting by all vessels (logbooks, daily radio position and catch report).
- Satellite-based Vessel Monitoring System being implemented for major fisheries.
- Mandatory for any vessel wishing to fish outside Namibian EEZ.

Industry Development Measures:
- Rebate system for landings by Namibian vessels, Namibian crew, onshore processing (to enhance).
- Compulsory levels of onshore processing.
- Preference in allocation of rights and quotas for ventures beneficially owned by Namibians.
- Affirmative action applied to allocation of rights and quotas in support of Namibianisation and empowerment.

17 CO-MANAGEMENT: NAMIBIA'S EXPERIENCE WITH TWO LARGE-SCALE INDUSTRIAL FISHERIES – SARDINE AND ORANGE ROUGHY

David Boyer and Burger Oelofsen*

Abstract

Namibia has implemented a style of fisheries management that in recent years has often been recommended for large-scale industrial fish stocks. Some sectors of the fishing industry, through representative groups, participate in research, assessment and the management of their fish stocks. The mechanisms that have affected such participation are described for two fisheries: sardine and orange roughy. The biomass of both these stocks have declined to low levels in recent years, which has required some difficult decisions to be made. The role of co-management in the fisheries research and management process and in reducing conflicts is examined. It is concluded that incorporating co-management into the management process can be advantageous, but such a development needs to be implemented with caution.

INTRODUCTION

Co-management
Co-management encompasses various arrangements that formally recognise

---

* Our colleagues within the Ministry (both in research and management) and the stakeholders of the sardine and orange roughy industries are thanked for often unwittingly providing many of the ideas presented here. Jean-Dominique le Garrec, a company manager from the Namibian orange roughy fishery and member of the Deep Water Fishery Working Group, and Hugo Viljoen, a company manager in the sardine industry and Chairman of the Walvis Bay Fishing Companies, were asked to review and make comment on an earlier version of the manuscript in an attempt to ensure that a balanced account of the developments in co-management in these two fisheries is recorded here. Their comments were extremely useful and they are wholeheartedly thanked. Two anonymous reviewers also provided many useful suggestions and are likewise thanked.

the sharing of fisheries management responsibility and accountability (Jentoft, 1989). It implies that interested parties (usually users) are brought into the management process (including the scientific assessment of the stocks) in some formal way, such that their knowledge and experience can be incorporated into the decision-making process and so that they gain some ownership in the process and hence are more likely to abide by the outcome. However, ultimate responsibility for decision-making regarding rules and management measures often remains with the management authority, especially when dealing with large-scale industrial fisheries. By definition, co-management implies a degree of decentralisation.

Co-management covers a wide range of institutional arrangements used in fisheries management (Figure 1). It is not a management system in itself, but rather establishes an institutional framework in which management can occur (Cunningham and Maguire, 2002). The FAO Technical Guidelines for Responsible Fisheries (Anon, 1997) terms this *"management in partnership"* and states that *"Determining and implementing the actions necessary to enable the management authorities, the fishers and other interested groups, to work towards the identified objectives ...should be done in consultation with all interest groups"*. Indeed a recurring theme throughout this seminal document is that

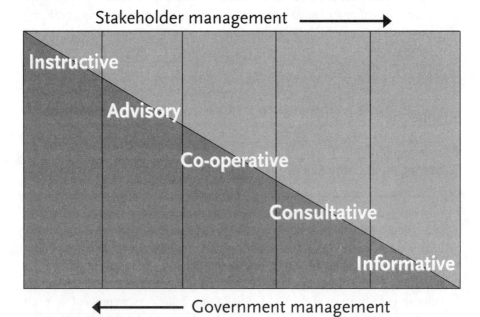

Figure 1. An illustration of the range of institutional arrangements of co-management; from "instructive" whereby government decides on all aspects of management and merely instructs stakeholders, through to "informative" where stakeholders take the decisions and inform government (adapted from Sen and Raakjaer Nielsen, 1996).

the efficient implementation of management measures is often highly dependent on the support gained from the interested parties. The United Nations Convention on the Law of the Sea provides the framework for co-management (Anon, 1983) while the FAO Code of Conduct for Responsible Fisheries (Anon, 1995; 1997) provides the moral obligation to implement such a process. It notes that management authorities should facilitate effective consultative and decision-making processes that enable all legitimate stakeholders to participate in setting objectives, management measures and all other matters that may affect the stakeholder (Articles 6.13, 6.16, 10.1.2, 11.3, 22.3.2).

Co-management has been proposed as one of the key tools for reversing the unsustainable practices that have dominated the world's fisheries during the past decades (FAO, 2002). One of the factors of unsustainability in fisheries is low participation by stakeholders in the management process (Garcia and Staples, 2000), while a recommendation of a recent FAO workshop on *Factors of Unsustainability and Overexploitation in Fisheries* for the reduction of over-fishing practices was to *"ensure participation of interested parties at all stages of the management process"* (FAO, 2002). A report to US Congress (Anon, 1999) states *"ecosystem approaches to management rely on the participation, understanding and support of multiple constituencies. Policies that are developed and implemented with the full participation of all stakeholders, including the interests of future generations, are more likely to be fair and equitable, and to be perceived as such"*. Wilson et al. (1994) believe that the well-documented demise of the Canadian east coast groundfish may have been prevented if the fisheries management system had been based on community government, while Stephenson (1997) argues that timeous management through dialogue with all vested interests would improve the management of the Atlantic herring stocks.

Involving stakeholders in the assessment and management process has been proposed for a wide range of countries and fisheries, e.g. Galicia, Spain (Freire and García-Allut, 2000); Mauritius (Hollup, 2000) and South Africa (Harris et al., 2002). Indeed, some major fishing nations have now incorporated co-management into the management process, e.g. USA, Canada, Norway and Japan (Pomeroy and Berkes, 1998), while others have not so far (Mikalsen and Jentoft, 2001). Namibia's neighbours, South Africa, have recently embraced co-management as part of their new fisheries management policy (Hauck and Sowman, 2001). However this is more to ensure equitable distribution of South Africa's marine resources, than an attempt to improve the sustainability of the resources. In fact, it must be noted that many of the proposed and existing examples of co-management are primarily for small-scale fisheries and not large-scale industrial fisheries as in Namibia's case (Pomeroy and Berkes, 1998; also e.g. Hauck and Sowman, 2001; Pomeroy et al., 2001). Such forms of co-management are often termed "community-based management".

The few published accounts of co-management available support the contention that bringing stakeholders into the management process yields substantial benefits. Yet reservations are frequently expressed (see Jentoft et al., 1998) as some attempts at co-management have failed (Hanna, 1996), while others note that co-management only works if legal and institutional instruments are implemented to facilitate such an arrangement (OECD, 1997; Pomeroy and Berkes, 1998). However, these case studies are rarely documented, creating a biased record of the success of co-management (pers. comm., J.Rice, Science Advisory Secretariat, DFO, Ottawa, Canada).

This chapter briefly documents the management systems of the orange roughy *Hoplostethus atlanticus* and sardine *Sardinops sagax* fisheries in Namibia, the former having implemented a form of co-management and the latter to a large degree not. We examine why these different approaches have developed and demonstrate that while co-management certainly has a place in large-scale industrial fisheries, it needs to be implemented with caution. In particular we examine the role of co-management in conflict situations (especially during periods of declining catches) and discuss whether co-management is the panacea it is often proclaimed to be or if it is simply another useful management tool.

Namibian fisheries - background
Namibia has a large fishing industry that makes an important contribution to the national economy, both in terms of income and employment. Fishing is the third-largest sector of the Namibian economy, behind agriculture and mining, and the second-fastest-growing industry in the Namibian economy (behind tourism), a growth achieved mainly through product value enhancement. The sector generates more than 10% of GDP, and had an export value of N$2 900 million for 2000 (N$1 ≈ US$10.00 in mid 2002), which makes the fishing sector the second-largest export earner behind mining. Employment in the sector was estimated at about 15 000 persons in 2000, with almost half working at sea.

The Namibian fisheries policy is to utilise the living marine resources on a sustainable basis for the benefit of the nation, and to manage these fisheries based on scientific information and principles. Ultimate responsibility for control measures rests with the state. All activities of the major fisheries are catch controlled, by total allowable catches (*TACs*) issued as individual non-transferable quotas, in conjunction with effort controls, primarily through limited vessel rights. These rights are issued for seven, ten, fourteen or twenty years dependent on a number of criteria. The longer rights are issued to companies who, *inter alia*, are majority owned by Namibians, employ Namibians, have a proven track record in the industry and have demonstrated a long-term commitment by investing in the fishing sector. See Oelofsen (1999) and Boyer and Hampton (2001) for more details.

Most of the primary research on fisheries resources is conducted by state-run research institutes, primarily the National Marine Information and Research Centre (NatMIRC) within the Directorate of Resource Management of the Ministry of Fisheries and Marine Resources. This research is largely funded by levies on commercial catches and recently has been supported by the use of commercial vessels to assist with resource surveys on hake, horse mackerel, orange roughy and sardine.

Scientific recommendations for the harvesting of all major resources are presented to the Namibian Marine Resources Advisory Council (MRAC) (Figure 2), which makes recommendations to the Minister of Fisheries and Marine Resources after considering socio-economic factors and the industry's perception of the state of the resource. MRAC is a broad-sectored group mandated to provide advice to the Minister on fisheries and fisheries-related matters. This Council consists of representatives of the major fishing industries (although they are appointed for their expertise and experience in the industry and not to represent their own interests), unions, the state conservation Ministry, financial institutions and the local university. The Minister,

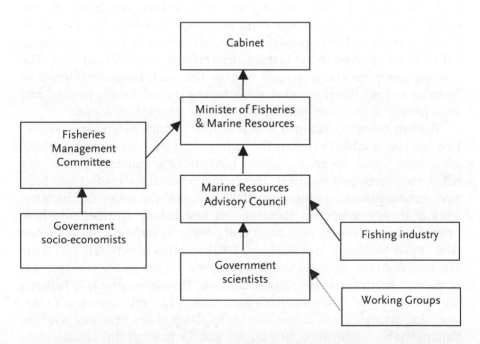

Figure 2. Diagram showing the flow of advice from research to decision-makers. Note that Working Groups do not exist for all fisheries (see text).

after consultation with the Ministerial Fisheries Management Committee and other senior managers within the Ministry (and often the scientists responsible for making recommendations), submits management recommendations to Cabinet for final endorsement.

Prior to Namibian independence in 1990, stakeholders were rarely involved in the management process and to a large extent this continued in the early 1990s. In a number of fisheries the industry was consulted informally, both in setting management objectives and for resource assessments, but this tended to be on an *ad hoc* basis. However, in the middle of the decade, declining abundance indices resulted in reduced *TAC*s in a number of fisheries (Boyer and Hampton, 2001) and brought the state and various sectors of the industry into conflict. The major bone of contention was over the scientific recommendations that tended towards caution, while the industry claimed that the stocks were still healthy and that the scientists were being over-pessimistic. In 1998 the then Minister of Fisheries and Marine Resources, Hifikepunye Pohamba, requested that, where possible, the divergent views of the scientists and industry should be reconciled, such that the decision-makers were presented with an agreed view on the state of each stock. This led to the formal approval for the establishment of working groups for each fishery, in which scientists and industry were represented to jointly conduct research and deliberate on the state and productivity of their respective stocks. To date, working groups have been established for the orange roughy, hake, monk, horse mackerel and rock lobster fisheries, while the other major fisheries (sardine, tuna, crab, recreational and subsistence fisheries) are involved in the management process in less formal ways. The working group for orange roughy was the first such forum established in Namibia and has therefore acted as the testing ground for this process, and subsequently as the model which other working groups have copied.

The two fisheries examined in this chapter, orange roughy and sardine, have both been subject to widely fluctuating catch levels during the past decade (Table 1) and the productivity of both stocks is currently considerably below their long-term potential. Hence the potential for conflict has been ripe. Various papers have investigated the biological and ecological characteristics of these resources in Namibia, and how fishing activities may have impacted on productivity (e.g. Boyer *et al.*, 2001a; Boyer *et al.*, 2001b; Fossen *et al.*, 2001; McAllister and Kirchner, 2001) and this is not dealt with here. The level of co-management that has been introduced into the management of the two species has been rather different. Orange roughy falls between "co-operative" and "consultative" (see Figure 1), with considerable co-operation through a formal working group (Deep Water Fisheries Working Group) established between the Ministry and the industry. In contrast, there is no formal arrangement for the sardine fishing industry to be included in the research or management process, and it is therefore more a form of "instructive co-management" *sensu* Sen and Raakjaer Nielsen (1996).

Table 1. Catch and TAC levels for the Namibian orange roughy and sardine fisheries since Independence.

Year	Orange roughy TAC*	Catch*	Sardine TAC	Catch
1990			63 000	89 000
1991			60 000	68 000
1992			80 000	82 000
1993			115 000	115 734
1994		3 315	125 000	116 483
1995		7 284	45 000	** 92 473
1996		13 136	20 000	2 372
1997	12 000	16 675	35 000	32 011
1998	12 000	6 845	65 000	64 000
1999	9 000	2 076	45 000	42 829
2000	1 875	1 209	25 000	26 496
2001	1 875	955	10 000	10 711
2002	2 400	603	0	2 614
2003	2 600		20 000	

* After 1996 the orange roughy fishing year extends from May to April, hence for example 2000 = May 2000 to April 2001. Also note that the TACs were for the quota management areas (QMAs), while the catches include fish taken outside of these areas. The amount caught outside of the QMAs is relatively small in relation to the total catch.
** Note that more than half of the sardine catch in 1995 was taken by Namibian vessels operating in Angolan waters and was therefore outside of the control of the Namibian authorities (see Boyer *et al.*, 2001a).

The following section describes and compares the development of these two different management systems and the Discussion then examines whether co-management has reduced, or helped to alleviate, the conflict situations that have arisen during the past decade.

CO-MANAGEMENT IN THE ORANGE ROUGHY AND SARDINE FISHERIES

The nature of the fisheries
The sardine fishery of Namibia developed into a large-scale industrial fishery more than 50 years ago. Indeed the northern Benguela stock of sardine was

historically one of the major clupeoid stocks of the world, supporting an average annual catch of over 700 000 tonnes throughout the 1960s (Boyer and Hampton, 2001). Since then the stock has declined, and annual catches decreased to around 50 000 tonnes between 1978 and 1989 and were only slightly more in the 1990s. In contrast, the orange roughy fishery developed relatively recently. An experimental licence was granted in 1993 for the exploration of the Namibian shelf-break region for unexploited deepwater resources. Exploration for orange roughy started in 1994 and it was only in late 1995 that several aggregations were discovered, and a viable fishery developed. The orange roughy fishery is a single species fishery operated by three companies with usually three medium-sized stern trawlers. One small processing factory was opened, but due to the low volume (but high product value) of orange roughy, this industry employs considerably fewer people than the other major fisheries sectors.

In comparison, the sardine fishery is considerably more complex. Firstly it is a multi-fishery fleet that in addition to sardine also harvests juvenile horse mackerel *Trachurus capensis*, anchovy *Engraulis capensis* and round herring *Etrumeus whiteheadi*. The fleet is currently reduced to about 12 purse seiners, which are unable to supply sufficient fish for three labour-intensive canneries, as well as for fish reduction plants. Hence, in recent years factory managers have pooled their resources and only one cannery plant has opened each year. Prior to 1996, however, more than 40 purse seiners were operating from five factories employing several thousand people. Thus in comparison with the orange roughy fishery the purse seiner fleet supports numerous jobs for land-based factory workers.

Management strategies
Another difference between the orange roughy and sardine fisheries is that the objectives of the orange roughy fishery have been fairly clear and well understood by all stakeholders, while those of the sardine fishery during the past decade have been somewhat contradictory. While a recovery of the sardine stock is obviously of importance, there have been conflicting aims towards rebuilding the stock. Such a rebuilding strategy implies reducing catches, or even implementing a moratorium, but at the same time the continued operation of the sardine fishery, even at very low levels, is seen as critical to safeguard employment and the prosperity of the harbour town, Walvis Bay. Scientists have recommended a spawner biomass limit reference point of 500 000 tonnes (and a target reference point of 1 000 000 tonnes), but this level is based on a rather poorly defined spawner-stock biomass - recruitment relationship and hence while the concept of such a reference point has been accepted in principle, it has not been applied in practice.

The aim of the orange roughy fishery has been simply to develop a profitable industry based on harvesting at a sustainable level. A management strategy was implemented whereby a 14-year fishing-down phase was to be followed by sustainable fishing once the stock approached the maximum sustainable yield biomass. As is evident from the catches, this strategy was not successful as the abundance of orange roughy on the fishing grounds declined considerably more rapidly than intended (see Boyer et al., 2001b for a discussion of the reasons).

The Deep Water Fisheries Working Group

An unofficial "orange roughy" working group consisting primarily of researchers and senior industry representatives was established in 1995. This forum was established early in the development of the fishery as it was realized that the capacity to sample, monitor and assess the resource was not available within the Directorate of Resource Management.

The primary task of this working group was initially to support and assist the Ministry's researchers in gathering detailed catch data, including length-frequency data, and to provide the means to bring in stock assessment advice, through consultants. Through the working group, government-employed fisheries observers were placed on all orange roughy boats to monitor catches, while annual biomass surveys were conducted from 1997 onwards. The working group also brought in fisheries scientists from other parts of the world (primarily those with expertise in orange roughy fisheries from New Zealand and Australia, but also from South Africa) to provide advice on management and assessment.

The working group functioned informally until late 1997. Until then, only a single company had a licence for orange roughy fishing, but in 1997 four additional licences were issued, although two of the new licensees were only given permission to conduct exploratory fishing (an option which to this day they have still not exercised). The other three were allocated shares in the TAC, although unequal, on the already established fishing grounds, and were given incentives to conduct further exploratory work. This effectively changed the orange roughy fishery from an experimental fishery to a fully commercialised fishery, while still maintaining a strong emphasis on exploration.

With the newly expanded participation in the fishery it was realised that an informal arrangement governing the cooperation between the Ministry and the industry could lead to unnecessary complications and misunderstandings, and therefore the working group was requested to formalise its role in the management process. The working group set itself the role of becoming *"a formal forum for providing guidance and advice on the efficient management of the deep water fisheries of Namibia to the appropriate manage-*

ment authorities". Furthermore the Deep Water Fisheries Working Group (as it was now called) aimed "*to ensure the long-term sustainable utilisation of the stocks exploited through proactive research and co-management strategies*" and "*to promote the rational development of the Namibian deep water fisheries, thereby ensuring that economic and social benefits of the fisheries are optimal and accrue to Namibia*". This greatly expanded the role of the working group, which had previously been primarily concerned with research, and it was not until early 2000 that the management authorities finally recognised the Deep Water Fisheries Working Group (DWFWG) in its formal role.

Regardless of this, the Working Group functioned as if it were formally recognised during the intervening period. From its early days in 1995 until present (2003), the Working Group has consisted of senior managers of the various fishing companies involved in the deepwater fishery, senior researchers from the Ministry, plus those researchers directly involved in deepwater research (Figure 3). While nominally part of the group, members of the Ministry's monitoring, control and surveillance Directorate and the Policy, Planning and Economics Directorate rarely participated in meetings, mainly because their role was not clearly defined and most of the deliberations were concerned with stock assessment and related matters.

The Working Group was funded by a jointly managed research fund whereby all concessionaires contributed in proportion to their quota. Major research costs, such as assessment surveys, were shared between industry and the state with both parties contributing vessels. Importantly, consultants were generally co-funded on an equal basis by the industry and the Ministry to promote impartiality.

This form of working group has since been implemented in a number of fisheries in Namibia, but to date not in the sardine fishery.

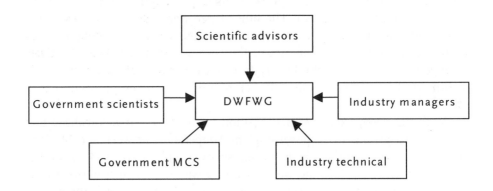

Figure 3. Diagram showing the flow of information within the Deep Water Fisheries Working Group.

Involvement of the industry in research

Both industries have over the past decade or so been involved to some extent in the research conducted on their species. The major difference has been that the research conducted on the sardine stock has been designed and conducted by the government researchers, although often with logistical support from the industry. In contrast, DWFWG guided the setting of priorities for the orange roughy research. This was largely through consultants appointed by the Working Group, and frequently much of the research was conducted by these same consultants.

Throughout the 1990s the sardine scientists made use of fishing vessels to scout for schools of fish during acoustic biomass surveys. In 1997, in an effort to standardise the survey design, this role was formalised whereby between two and four vessels accompanied the research vessel on each survey to find school groups and confirm that fish had not been missed close inshore or outside the area surveyed (Boyer et al., 2001a). In addition, experienced commercial skippers accompanied the research vessel to observe and critique the work of the scientists, although their lack of scientific training (and the inherent complexities of acoustic assessments) often resulted in a poor understanding of the methodology and may even have resulted in reduced confidence (and acceptance) in the results in some cases. Furthermore, the accompanying commercial vessels have often proved difficult to control so that searching effort varies to a certain extent between surveys. The industry, on the other hand, has expressed some frustration when they have proposed changes to the survey strategy and these have been rejected by the scientists. In particular, more flexibility in the survey timing has been viewed as important to the industry to ensure that the stock is surveyed when they believe it to be optimally available.

Since the early days of the surveys, in 1990, scientists have also volunteered to accompany commercial fishing boats to witness any large amounts of sardine that may not have been detected by a survey. More recently, the Ministry has also offered to make the government research vessel available at short notice to assess any such concentrations of sardine that the fleet may find. This was a further move to demonstrate the validity of the survey results, and to counter claims from the industry that the research surveys frequently missed large amounts of fish. While scientists have been called out several times to view concentrations of sardine the amount of fish has been considerably less than that found during surveys, and as a result the research vessel has never been required to make a scientific assessment of such concentrations.

The company that developed the orange roughy industry was managed by a person with a marine biology background and hence this industry was considerably more proactive in research than the sardine industry. In the early years (1997 and 1998) the orange roughy industry participated in surveys in a rather similar way to the sardine industry, providing commercial trawlers to assist. Initially orange roughy surveys used commercial vessels to conduct target identification trawls for an acoustic research vessel, but since 1999 the commercial vessel also collected acoustic data (Boyer and Hampton, 2001). An important difference was that through the Working Group, outside consultants were appointed to work alongside local researchers to assist and guide this research. Also stock assessment consultants developed and performed most of the stock assessment modelling conducted on this species, while any outside advice obtained for stock assessment of sardine has been entirely funded through the Ministry.

Involvement of stakeholders in the management process

While senior management of the processing and fishing companies of both industries are either formally or informally involved in the management process to some extent, many interest groups are either not consulted or are only included at a late stage. The unions have been represented at some consultative sardine management meetings, but other concerned parties, such as other sectors of the fishing industry, financial institutions and conservation bodies, have not been consulted.

The sardine fishery is characterised by fragmented stakeholders, partially driven by cultural heterogeneity; a consequence of Namibia's political history and the current attempts to redress the racial inequities of the past (Iyambo, this volume). There are several large companies that dominate the industry, but in addition, there are a number of private vessel owners who occasionally act as a unified group, thus increasing their power. These various players are however represented by a unifying group, a fishing association (currently referred to as the Walvis Bay Pelagic Fishing Companies), consisting generally of the fishing company managers and representatives of the small boat–owners. This in effect has become the *de facto* "sardine working group" with whom the Ministry has negotiated. The industry is a major employer and therefore has attracted considerable union attention. As in the orange roughy fishery, however, the unions are given little opportunity to be part of the formal management process. In addition, the fishers themselves, although important stakeholders, are totally reliant on the large companies for their jobs, and throughout much of the 1990s and early 2000s these jobs have not been in abundance. Thus the fishers tend to be deferential to the boat owners and are constrained in their input to the management process.

In contrast to the sardine fishery, the orange roughy fishery consisted of a single company until 1997, and even in more recent years only three companies have participated in the fishery. This has meant that communication within the orange roughy industry has been somewhat simpler and as a result the orange roughy industry tends to appear as a unified group with similar views and standpoints compared with the rather more diverse views presented by the sardine industry. This greatly facilitated the formation of the DWFWG.

Additionally, the orange roughy fishery was dominated by managers recruited from overseas who had experience in fisheries management and, crucially, fisheries research. In contrast, the sardine industry was in many cases managed by long-serving personnel who had an intimate and detailed understanding of the technical aspects of the fishery, but often lacked specific experience in fisheries management and research. This has hampered the industry's understanding of at least the research methods being used. For example, many participants in the sardine fishery had difficulty understanding the concept of "sampling" a population to determine abundance, and to this day standard acoustic survey methodology is met by some with scepticism because scientists only "see" a fraction of the population and do not "count" every fish. In recent years, some senior managers with an engineering background have been appointed, thus increasing the level of understanding of scientific concepts and management procedures.

Management procedures
The management procedures for both sardine and orange roughy (and indeed most of the commercially harvested marine fish and crustacean species in Namibia) are similar (see Namibian fisheries - Background). However, there are considerable differences in the involvement of the industry in the deliberations on catch levels and other related management measures. In recent years *ad hoc* meetings have been held between the Ministry and the sardine industry prior to management decisions being implemented, which may have far-reaching consequences to the stakeholders. The meetings have taken the form of periodical consultative meetings, where usually some proposed management action is explained to the industry by senior managers of the Ministry, but with limited opportunity for input (Olsen, this volume). Otherwise the industry (and other stakeholders) have no direct input to the management process for sardine.

In contrast, the formal working group that has been established between the Ministry and the orange roughy industry enabled the industry to participate in the management process (see next section). While this working group has been mandated to provide guidance in terms of management strategies, control and surveillance issues and research, the overriding preoccupation has been in determining stock size, productivity and hence future catch levels. The Working Group was explicitly restricted from making catch recommendations to senior management as this is seen as the function of the Ministry's Directorate of Resource Management. However, the assessments conducted under the auspices of the Working Group clearly defined optimal catch levels and therefore paved the way for *de facto TAC* recommendations by the Directorate of Resource Management.

Co-management and the orange roughy fishery. - With the development of the orange roughy fishery, and the fact that there was a lack of long term research knowledge to support the fishery due to its newness, co-management was an attractive means of delegating some of these responsibilities to the users. Initially the entire orange roughy fishery consisted of a single company, and therefore establishing formal channels of communication was relatively straightforward. Even after four new companies were given rights in 1997, the industry formed a cohesive unit, which meant that the functioning of the working group was relatively straightforward. In addition, as this was a newly developed fishery, there were clear management goals, while the participants in the working group tended to have a good understanding of management processes and scientific concepts.

The DWFWG therefore offered a structured way for the industry to be involved in the management process, especially at the assessment level.

Orange roughy research, especially in the early days, was dominated by consultants. This ensured that relevant and timeous research was conducted to make an assessment of the abundance and productivity of the stock, although this was not sufficient to prevent the subsequent declines in catches. It also enabled local, less inexperienced scientists the opportunity to learn. However, this also led to the feeling amongst these scientists that they were being used as data collectors, resulting in some resentment towards the more highly paid consultants.

Elaborate attempts were made in the assessments to estimate the abundance of the stocks and to quantify the risks associated with various harvesting levels (McAllister and Kirchner, 2001). One of the greatest difficulties for the Working Group was to formulate optimal catch levels in the face of conflicting interpretations of available data and assessment results, and their implications for resource status (McAllister and Kirchner, 2001). The spawning behaviour of orange roughy was, and still remains, poorly understood, and the dynamics of aggregation formation and dispersion are similarly unknown. Therefore, alternative hypotheses covering a wide range of biological possibilities were necessary in the formulation of the qualitative conceptual models of stock dynamics and behaviour, leading to several possible equally plausible interpretations of the results.

The annual debate on the state of the orange roughy stock and appropriate harvesting levels was generally protracted and difficult, as there were no protocols for setting harvesting levels. This forced an annual debate that was extremely costly and time-consuming for all involved. Some management procedures elsewhere (e.g. in South Africa, De Oliviera et al., 1998; Johnston and Butterworth, 2000) involve agreeing on pre-defined management decision rules for translating fishery or research data automatically into TACs or some other regulatory measure. Such pre-defined rules negate annual discussions and could be considered for the Namibia orange roughy fishery. However it has to be noted that such a management procedure can only be applied with confidence if the state of the stock and its productivity are well known; hardly the case in this example.

The orange roughy fishery catches averaged more than 10 000 tonnes per annum between 1995 and 1998, but these declined to less than 1 000 tonnes in 2001. This large decline meant that some difficult decisions had to be taken. In theory the Working Group system allowed for this, although in practice it was not without conflict. The industry, through the DWFWG, had considerable comments to make about the interpretation of the data, and in particular any trends in the data. Industry tended to focus on explaining away downward trends or negative perceptions, while high estimates or increases were not questioned. For example, a declining catch per unit of effort ($CPUE$) or low catch was frequently "accounted for" by the industry as due to operational problems, while increases in $CPUE$ were accepted as a true reflection of the state of the stock. In contrast, scientists were concerned that high estimates or positive trends were in some way biased (upwards) while accepting more negative perceptions of the state of the stock (the industry claims that the scientists were too "green" or conservative). As all parties had equal weight in decisions affecting the outcome of the assessments (in particular, weighting of input data), this tended to downplay any negative signs and probably resulted in more optimistic recommendations than would otherwise have been the case. Finally, in 2000 when consensus could not be reached, the government scientists submitted recommendations to MRAC based on the precautionary view that the stock was considerably lower than assumed by the DWFWG.

Co-management and the sardine fishery. - In contrast to the orange roughy fishery, the sardine fishery has a long history of activity in Namibian waters and hence the management structures that the Namibian government inherited in the early 1990s were relatively well entrenched. The industry was more fragmented than the orange roughy fishery, the understanding of scientific concepts was not as good (with notable exceptions), while management goals have never been clearly defined and were in some cases contradictory. The industry had, and to a large degree still has, no structured involvement in the management process, a process that has been largely instructive in character. Research projects are largely designed independently of industry and are frequently treated with scepticism by industry, especially when reduced TACs are recommended.

During the early years after independence, and until about 1993, the sardine stock was perceived as increasing in size and therefore TACs increased similarly. During this period the working relationship between the industry and fisheries authorities was excellent, and the research methodology and results were generally accepted.

As with orange roughy, there was a severe decline in stock abundance during the mid-1990s that required some very difficult decisions. Sardine catches increased between 1992 and 1995, averaging 100 000 tonnes, but from 1992 to 1996 the stock again declined and the lowest annual catch in the history of the fishery was taken in 1996. Recommendations based solely on biological grounds have called for a drastic reduction in catches since 1994 and indeed after 1996 proposed that a long-term closure of the industry would be wise to allow the sardine stock to recover. The labour-intensive nature of the fishery, however, meant that this would have resulted in substantial job losses, with serious social implications. Although there was a small increase in sardine abundance during the last three years of the decade, the stock remained at a worryingly low level and in 2002 a zero TAC was set for only the second time in the history of the fishery (the previous time was in 1980).

The industry increasingly questioned the science as the size of the TAC declined. With decades of local knowledge, they believed that the sardine had either moved away or were simply not available to the survey technique. The scientists rejected these claims out of hand, with little attempt either to investigate them or even to debate such issues with the industry. Outside consultants were brought in by the industry to question the survey methodology of government scientists, while considerable lobbying of decision-makers, both from the companies and labour unions, occurred. These appeals to the authority to support the beleaguered industry included attempts to discredit the science (and occasionally the scientists themselves). The uncertainties inherent in the surveying and assessment of any fish stock were highlighted and used to undermine the recommendations emanating from this research.

Eventually however, to the dismay of all, including the scientists, the warnings of the government scientists were vindicated as, despite a TAC of 20 000 tonnes, catches in 1996 consisted of a paltry 2 000 tonnes of by-catch sardine, almost 10-times lower than the previous lowest catch ever recorded. Good recruitment in 1996 saw the catches rise slightly in the subsequent seasons (Table 1), but in 2002, as noted above, the stock had declined to such a state that even the industry accepted that a complete closure was necessary.

The decline in the sardine abundance created tensions between the controlling authority and the fisheries, especially at the research level where the natural conservatism inherent in research (and the recommendations emanating from this research) has been viewed by some in the industry as a threat to the continued existence of some of the companies involved in this industry. This conservatism is now more formally acknowledged as part of the precautionary approach (Anon, 1995), which forms a central element of the Namibian fisheries management regime, even though it is yet to be fully implemented.

The industry has necessarily had to be concerned with its own day-to-day survival. As catches have fallen, the financial institutions supporting many of the fishing companies have put further pressure on them to at least meet their financial commitments. Thus, despite apparently incontestable evidence that the resource was in dire straits, the industry was reluctant to accept this, as it would have compromised their own survival.

Indeed, a difference in the basic concepts underlying successful utilisation of fish resources between the fishing industry and scientists has been highlighted during this period and is still to be resolved. The industry contended that overly conservative *TACs* may have pushed companies to bankruptcy and then there would be no fishery once the fish stocks recover. The scientists, whose mandate was restricted to making recommendations based on biological considerations alone, were not permitted to consider socio-economic issues, even if their recommendations, if implemented, would have meant severe hardship for the industry and its employees. The scientists also considered their recommendations as rational, objective and, considering the low level of stock biomass, non-negotiable, although they recognised the uncertainties in their estimates; an apparent contradiction. Obviously a balance between these two opposing views needs to be reached whereby risk up to a permitted level is considered acceptable to ensure the survival of the industry, but beyond that level the survival of the stock takes precedence. Attempts to introduce such concepts, in the form of biological limit and target thresholds, have so far met with little success, partly due to the uncertainties in the precise levels that these should be set at, but also because they would have required the fishery to be closed and thus their implementation was resisted.

It was during this period of conflict that the proposal for closer cooperation between the industry and scientists was developed, largely at the insistence of government managers.

Incorporating fisheries personnel and vessels into research surveys was intended as a mechanism to enhance the acceptance of the research, and to some degree was successful. However, as the people directly involved were skippers of commercial vessels who had spent many years at sea, but often had a poor understanding of the specific concepts used in acoustic surveys, this did little to increase the confidence of the industry in the research results. Offering to accompany fishers to view any significant aggregations of sardine also deflected some criticism from the researchers, but once again this strategy was probably not utilised as much as it could have been and therefore remained only partially effective in resolving the different perceptions of the state of the sardine stock.

Despite the unprecedented low catch in 1996, frequent deputations were made to decision-makers and the unions mounted public demonstrations, until there was no further question to the state of the stock and a zero *TAC* was implemented in 2002. Scientists were accused of lacking experience at sea and were therefore said to be unable to find sardine, while the capabilities of the research vessel to detect or catch sardine was questioned. Otherwise the industry simply contradicted the researchers, publicly stating immediately after a survey that there were considerable amounts of fish which scientists had been unable to find.

Throughout this period there was no mechanism to resolve these conflicting impressions of the state of the sardine stock, and scientists and managers from the fishing industry were expected to resolve their differences unaided. This resulted in entrenched positions being formed, and stereotyping of the two sides; industry were perceived by many as irresponsible and solely interested in their own short-term gain, while scientists were seen by the industry as conservationists merely concerned with preserving fish and having little regard for the welfare of fishers and the fishing communities.

In contrast to orange roughy, there was little formal debate over the state of the sardine resource. However, a management procedure is currently being considered for setting sardine catch limits. Such a procedure may benefit the orange roughy, although in fairness it has to be noted that the amount of data available to enable the state of the sardine stock and its productivity to be determined with some confidence is considerably greater than that available for orange roughy.

An area of potential conflict was the control of the catch of juvenile sardine. *TAC* restrictions on sardine have been the main control measure, but the capture of other purse-seined species, juvenile horse mackerel and anchovy, is also limited if the by-catch of juvenile sardine is considered too high (nominally above 5% by mass). This is one aspect that industry has largely taken responsibility for by controlling the by-catch of juvenile sardine through closure of areas if necessary. The process is monitored by the state to ensure that it is effectively implemented. The limitation of juvenile catches is a particularly sensitive issue because after a number of years of poor recruitment (Boyer *et al.*, 2001a), the rebuilding of the stock was reliant on the survival of juvenile fish. In addition, harvesting juveniles results in a reduced yield (due their small size) and, as they are rendered into low-value fishmeal, a considerable loss of potential income. Since the early 1990s, efforts by the industry to closely monitor the species composition of the catches and to voluntarily ban fishing in any areas where young sardine occur have been encouraged by the Ministry. As a result the industry successfully reduced the catch of juvenile sardine to acceptable limits and has shown that the sardine industry is capable to acting in concert and managing potentially complicated situations. Extending the management brief of the industry into other potentially more difficult issues, such as allowing them to control catch levels within the TAC, or even setting TAC levels, could be an interesting and worthwhile experiment, although clear and strict monitoring would need to be implemented to ensure that the situation was not abused.

Discussion

Co-management in Namibian fisheries
Fishers are often fishers, at least in part, because the form of lifestyle offers a level of independence rarely found elsewhere. Such people take poorly to top-down controls traditionally found in fisheries management (Jentoft et al., 1998), the type of control system that to a large extent is predominant in Namibia. Co-management seems to offer fishers (or at least the companies controlling the activities of the fishers) the chance to express that independence in a rational and responsible manner, thereby contributing to the successful management of the natural resources that they are reliant on.

The various fishing industries in Namibia have some say over their own activities. In most fisheries they are granted individual catch quotas rather than global quotas and can therefore decide when and where they will make their catches, at least within the season and areas limitation. Such a quota system also reduces the level of competition (at the catch level) with other fishers.

In recent years the Namibian authorities have attempted to incorporate some sectors of the fishing industry into the management process through formal arrangements, partly so that they can contribute to the process, but also to ensure that the industry has a part-ownership of any decisions that may be taken. The Namibian authorities have limited the concept of co-management to a rather narrow range of stakeholders, i.e. the managers of the fishing companies and, occasionally, employees, rather than the more accepted concept of including all who have an interest in the fishery, however indirect (Mikalsen and Jentoft, 2001). Fishers, factory workers, financial institutions, unions, other (competing) fisheries, conservation bodies etc. are either not formally consulted, or are only included very late in the process. On the other hand, the "stakeholders" from the state have been largely limited to scientists, with little input from mangers, policy-makers, or the monitoring, control and surveillance personnel.

The level of co-management that has been introduced into the management process of most of the important commercial fisheries in Namibia is rather limited. This is generally at the instructive level, although there are some aspects where the industry has taken control of a specific set of management decisions that are more normally controlled by the state, such as limiting the sardine by-catch and juvenile catches.

The main mechanism for enabling the industry to participate formally in the management process has been through the establishment of working groups, such as the DWFWG that have been formally mandated to assist with and participate in research and to develop management strategies consistent with the overall goals of each fishery.

Even more ambitious forms of co-management, such as allocating population stewardship rights as proposed by Gavaris (1996), have not been considered, despite their apparent attractiveness. The concept of population stewardship rights recognises that uncaught fish (including unborn fish) have a value. These uncaught fish are referred to as "natural capital" (Rees, 1991). This type of system allows the carry over of fish from one year to the next and allows for an increase in the quota with time due to recruitment and growth (with some reduction due to natural mortality). Thus one way of increasing a concessionaire's quota is to leave unharvested some of the allocated quota, thus increasing the contribution to future generations (= putting fish in the bank and then reaping the interest at some later date). Such a system requires quite detailed knowledge of recruitment, growth and mortality rates, and despite good advances in fisheries science in recent years (Iyambo, 2001) such detailed knowledge is still largely unavailable.

Co-management and institutional change

Co-management by definition implies that a number of parties are involved in the management and utilisation of fish resources. The industry has the opportunity to participate in the management of the resource, but such an opportunity will be more effective if the industry is conversant with fisheries management and research principles. Similarly government scientists also need to have a firm grasp of the fundamentals of managing a large company and in particular the specific problems of utilising a natural resource that is highly variable and unpredictable. Several of the differences highlighted between the Namibian sardine and orange roughy fisheries have been a result of the different levels of understanding of research and fisheries management concepts and the cohesiveness of the industries. The stakeholders of the sardine fishery, and to a lesser extent the orange roughy industry, will need to become more familiar with research and management principles if they are to participate effectively. As such, relevant awareness needs to be offered in both fisheries science and management to enable industry participants to interact effectively in this process. In addition, the entire industry (management, fishers, factory workers, etc.) needs to be able to act in concert. Indeed, unless an industry can demonstrate a maturity and competence to interact rationally and effectively, trust will soon be lost and their credibility may be reduced.

In addition, co-management implies that all stakeholders have a considerably higher level of responsibility than under a more traditional top-down management system (see also Pomeroy and Berkes, 1998). Short-term fishing strategies that are based on economic profit and are inconsistent with long-term sustainable fishing practices would not be acceptable. In effect this would require stakeholders to focus on the longer-term good of the fishery rather than their own more immediate shorter-term goals.

The Working Group concept resulted in a delegation of conflict resolution to the level of government researchers and fisheries managers. Government research staff are typically not trained or experienced in dealing with such issues, and state research institutions generally have insufficient capacity to handle this role (e.g. Brown and Pomeroy, 1999). If a co-management strategy is to be effective, capacity at the local level needs to be strengthened (Pomeroy and Berkes, 1998). By the same token that industry and other stakeholders need to become familiar with scientific concepts, scientists need to be fully supportive when a co-management strategy is implemented and need to develop an understanding of industrial management.

The Deep Water Fisheries Working Group allowed conflicts to be internalised within the Group, thus preventing more damaging public disputes to develop. Indeed many of the conflicts merely encouraged the proponents to improve their arguments by, for example, acquiring more data or more rigorous testing of their models. As such the Group has acted as a form of peer review, a process that has been sorely missing in sardine research. In contrast, attempts to introduce co-management into the sardine industry were made during a period of conflict, without any mechanisms for conflict resolution being implemented.

Ideally under a co-management style of management, conflicts are successfully resolved at the local level, or issues are solved before they develop into a conflict situation. However, as occurred in both the sardine and orange roughy fisheries during times of declining catches, conflicts that are not timeously resolved can escalate and by the time they reach higher levels within the government are considerably more difficult to resolve.

Concluding Remarks

Sardine would classify as "instructive" in the classification of institutional arrangements for co-management by Sen and Raakjaer Nielsen (1996) as this fishery is largely managed with a top-down approach, with very little input from stakeholders. In contrast, the orange roughy fishery is more "consultative", although all final decision-making is still retained by the state. While there may be room for greater involvement of both sectors of the industry in the management process, all stakeholders, including the government, would need to be aware of the implications of such a development.

Co-management tends to divert conflicts resolution to lower levels within the management organisation; in the Namibian situation this tends to be to the research level. Thus in order for co-management to be successfully implemented, the structure at these lower levels needs the additional resources, and also support from higher (management) levels, to be able to undertake this new role. As concluded by numerous other studies, the Namibian experience suggests that involving stakeholders in the research and management tends to lead to better acceptance (a feeling of ownership) of the outcome. However in some cases, for many stakeholders, this may require considerable effort as participants must have a clear understanding of biological concepts, research methods and management strategies.

In summary, Namibia's experience with co-management suggests that when times are good, the co-management process works well, but then so do most other forms of management. However, when times are poor, conflicts are likely to arise regardless of the management system used. Co-management certainly does not prevent these conflicts, although it may well serve to reduce them by providing a structured forum for discussion and resolution.

References:

Anon (1983): The Law of the Sea. United Nations Convention on the Law of the Sea. United Nations, New York. 242 pages.

Anon (1995): FAO Code of Conduct for Responsible Fisheries. Rome, FAO. 41 pages.

Anon (1997): FAO Technical Guidelines for Responsible Fisheries. No. 4. Rome, FAO. 82 pages.

Anon (1999): Ecosystem-based fishery management. A report to Congress by the Ecosystem Principles Advisory Panel. U.S. Department of Commerce, NOAA, National Marine Fisheries Service. 25 pages.

Boyer, D.C. and I. Hampton (2001): An overview of the living marine resources of Namibia. In: *A Decade of Namibian Fisheries Science*. Payne, A.I.L., Pillar, S.C. and R.J.M. Crawford (Eds.). *South African Journal of Marine Science*, 23: 5-36.

Boyer, D.C., Boyer, H.J., Fossen, I. and

A. Kreiner (2001a): Changes in abundance of the northern Benguela sardine stock during the decade 1990-2000, with comments on the relative importance of fishing and the environment. In: *A Decade of Namibian Fisheries Science*. Payne, A.I.L., Pillar, S.C. and R.J.M. Crawford (Eds.). *South African Journal of Marine Science*, 23: 67-84.

Boyer, D.C., Kirchner, C., McAllister, M., Staby, A. and B. Staalesen (2001b): The orange roughy fishery of Namibia: lessons to be learned about managing a developing fishery. In: *A Decade of Namibian Fisheries Science*. Payne, A.I.L., Pillar, S.C. and R.J.M. Crawford (eds.). *South African Journal of Marine Science*, 23: 205-222.

Brown, D.N. and R.S. Pomeroy (1999): Co-management of Caribbean Community (CARICOM) fisheries. *Marine Policy*, 23: 549 – 570.

Cunningham, S. and J-J. Maguire (2002): Factors of Unsustainability in large-scale commercial marine fisheries. Discussion paper prepared for FAO Workshop of Unsustainability and Overexploitation in Fisheries, Bangkok, Thailand.

De Oliviera, J.A.A., Butterworth, D.S., Roel, B.A., Cochrane, K.L. and J.P. Brown (1998): The application of a management procedure to regulate the directed and by-catch fishery of the South African sardine. In: *Benguela Dynamics: Impact of variability on shelf-sea environments and their living resources*. Moloney, C.L., Pillar, S.C., Payne, A.I.L., and F. A. Shillington. (Eds.). *South African Journal of Marine Science*, 19: 449-469.

FAO (2002): International Workshop on Factors Contributing to Unsustainability and Overexploitation in Fisheries. FAO Fisheries Report No. 672. Rome Italy.

Fossen, I., Boyer, D.C. and H. Plarre (2001): Changes in some key biological parameters of the northern Benguela sardine stock. In: *A Decade of Namibian Fisheries Science*. Payne, A.I.L., Pillar, S.C. and R.J.M. Crawford (Eds.). *South African Journal of Marine Science*, 23: 111-122.

Freire, J. and A. García-Allut (2000): Socioeconomic and biological causes of management failures in European artisinal fisheries: the case of Galicia (NW Spain). *Marine Policy*, 24: 375 – 384.

Garcia, S.M. and D. Staples (2000): Sustainability reference systems and indicators for responsible marine capture fisheries: a review of concepts and elements for a set of guidelines. *Marine Freshwater Res.* 51: 385-426.

Gavaris, S. (1996): Population stewardship rights: decentralised management through explicit accounting of the value of uncaught fish. *Canadian Journal of Fisheries and Aquatic Science* 53: 1683-1691.

Hanna. S.S. (1996): User participation and fishery management performance within the Pactific Fishery Management Council. *Ocean and Coastal management*. 28: 23-44.

Harris, J.M, Branch, G.M., Clark, B.M., Cockroft, A.C., Coetzee, C., Dye, A.H., Hauck, M., Johnson, A., Kati-Kati, L., Maseko, Z., Salo, K., Sauer, W.H.H., Siqwana-Ndulo, N. and M. Sowman (2002): Recommendations for the management of subsistence fishers in South Africa. *South African Journal of Marine Science*, 23: 111-122.

Hauck, M. and M. Sowman (2001): Coastal and fisheries co-management in South Africa: an overview and analysis. *Marine Policy*, 25: 173 – 185.

Hollup, O. (2000) Structural and sociocultural constraints for user-group participation in fisheries management in Mauritius. *Marine Policy*, 24: 407 – 421.

Iyambo, A. (2001): A Decade of Namibian Fisheries Science. In: *A Decade of Namibian Fisheries Science*. Payne, A.I.L., Pillar, S.C. and R.J.M. Crawford (Eds.). *South African Journal of Marine Science*, 23: 1 - 4.

Iyambo, A. (2004): Foreword to *Namibia's fisheries: Ecological, economic and social aspects*, Sumaila, U.R., Boyer, D., Skogen, M. and Steinshamn, S.I. (eds.), pp. XI-XIII. Eburon, Delft.

Jentoft, S. (1989): Fisheries co-management: delegating government responsibility to fishermen's organizations. *Marine Policy*, 13: 137 – 154.

Jentoft, S., McCay, B.J. and D.C. Wilson (1998): Social theory and fisheries co-management. *Marine Policy*, 22: 423-436.

Johnston, S.J. and D.S. Butterworth (2000): Current operational management procedure for South African West Coast rock lobster. Unpublished document submitted to the BENEFIT Stock Assessment Workshop, Cape Town, November 2000. BEN/NOV00/WCRL/5a: 41 pages.

McAllister, M.K. and C.H. Kirchner (2001): Development of Bayesian stock assessment methods for Namibian orange roughy *Hoplostethus atlanticus*. In: *A Decade of Namibian Fisheries Science*. Payne, A.I.L., Pillar, S.C. and R.J.M. Crawford (Eds.). *South African Journal of Marine Science*, 23: 241-264.

Mikalsen, K.H. and S. Jentoft (2001): From user-groups to stakeholders? The public interest in fisheries management. *Marine Policy*, 25: 281 – 292.

OECD 1997 – Towards sustainable fisheries: Issue papers. OCDE/GD (97) 54. 415 pages.

Oelofsen, B.W. (1999): Fisheries management: the Namibian approach. *ICES Journal of Marine Science*, 56(6): 999-1004.

Olsen, M.O. (2004): Institutional and industry perspectives on fisheries management in Namibia. In: *Namibia's fisheries: Ecological, economic and social aspects*, Sumaila, U.R., Boyer, D., Skogen, M. and Steinshamn, S.I. (eds.), pp. 267-288. Eburon, Delft.

Pomeroy, R.S., and F. Berkes (1998): Two to tango: the role of government in fisheries co-management. *Marine Policy*, 21: 465 – 480.

Pomeroy, R.S., Katon, B.M. and I. Harkes (2001): Conditions affecting the success of fisheries co-management: lessons from Asia. *Marine Policy*, 25: 197 – 208.

Rees, W.E. (1991): Conserving natural capital: the key to sustainable landscapes. *International Journal of Canadian Studies* 4: 7 – 27.

Sen, S. and J. Raakjaer Nielsen (1996): Fisheries co-management: a comparative analysis. *Marine Policy*, 20: 405 – 418.

Stephenson, R.L. (1997): Successes and failures in the management of Atlantic herring fisheries: Do we know why some have collapsed and others survived. Hancock, D.A., Smith, D.C., Grant, A. and J.P. Beumer (Eds.). 2nd World Congress Developing and sustaining world fisheries resources. Part 11, 49-54.

Wilson, J.A., Acheson, J.M., Metcalfe, M. and P. Kleban (1994): Chaos, complexity and community management of fisheries. *Marine Policy*, 18: 291 – 305.

Glossary

AARSA	–	Automobile Association of South Africa
AMPL	–	A Modeling Language for Mathematical Programming
BCLME	–	Benguela Current Large Marine Ecosystem
BENEFIT	–	Benguela Environment Fisheries Interaction and Training Programme
BODC	–	British Oceanographic Data Centre
CBD	–	Convention on Biological Diversity
CBIF	–	Canadian Biodiversity Information Facility
CBS	–	Central Bureau of Statistics
CCAMLR	–	Commission for the Conservation of Antarctic Marine Living Resources
CITES	–	Convention on International Trade in Endangered Species of Wild Fauna and Flora
COADS	–	Comprehensive Ocean-Atmosphere Data Set
COMESA	–	Common Market for Eastern and Southern Africa
CORSA	–	Cloud and Ocean Remote Sensing around Africa data set
CPUE	–	Catch per unit effort
CSS	–	South African Centre for Statistical Services
CVM	–	Contingent Valuation Method
DFID	–	British Department for International Development
DVM	–	Diel vertical migration
DWF	–	Distant Water Fleets
DWFWG	–	Deep Water Fishery Working Group
EBC	–	Eastern Boundary Currents
EBCD	–	European Bureau for Conservation and Development
EEZ	–	Exclusive Economic Zone
EU	–	The European Union
FAO	–	Food and Agriculture Organization of the United Nations
FiB	–	Fishing-in-balance
FRS	–	Flow Relaxation Scheme
GDP	–	Gross Domestic Product
GEBCO	–	General Bathymetric Chart of the Oceans
GNI	–	Gross National Income
GSI	–	Gonadosomatic Indices
GSP	–	Generalized System of Preferences
HF	–	High frequency range

ICCAT	–	International Commission for the Conservation of Atlantic Tunas
ICES	–	International Council for the Exploration of the Sea
ICSEAF	–	International Commission for South East Atlantic Fisheries
IHO	–	International Hydrographic Organization
IIM	–	Instituto de Investigação Marinha
IMR	–	Institute of Marine Research in Norway
IOC	–	Intergovernmental Oceanographic Commission
ITIS	–	Integrated Taxonomic Information System
ITQ	–	Individual Transferable Quota
IUCN	–	International Union for Conservation of Nature and Natural Resources
IUU	–	Illegal, Unreported and Unregulated fishing
MCS	–	Monitoring, Control and Surveillance
MET	–	Ministry of Environment and Tourism
MFMR	–	Ministry of Fisheries and Marine Resources
MRA	–	Marine Resources Act
MRAC	–	Namibian Marine Resources Advisory Council
MSY	–	Maximum Sustainable Yield
NatMIRC	–	National Marine Information and Research Centre
NCAR	–	National Center for Atmospheric Research
NCEP	–	National Centers for Environmental Prediction
NDP	–	Net Domestic Product
NNI	–	Net National Income
NNP	–	Namib Naukluft Park
NOAA	–	National Oceanic and Atmospheric Administration
NORAD	–	Norwegian Agency for Development Cooperation
NORWECOM	–	Norwegian Ecological Model System
OECD	–	Organisation for Economic Co-operation and Development
OMP	–	Operational Management Procedure
POM	–	Princeton Ocean Model
PS	–	Permanent Secretary
QMA	–	Quota Management Area
SACU	–	Southern Africa Customs Union
SADC	–	Southern African Development Community
SAUP	–	Sea Around Us Project
SCOR	–	Scientific Committee on Oceanic Research
SCP	–	Skeleton Coast Park
SEAFO	–	South East Atlantic Fisheries Organisation
SeaWiFS	–	Sea-viewing Wide Field-of-view Sensor Project
SEEA	–	UN's System of Integrated Environmental and Economic Accounts
SFA	–	Sea Fisheries Act
SFAC	–	Sea Fishery Advisory Council
Sida	–	Swedish International Development Agency
SNA	–	System of National Accounts
SS	–	Sum of Squares
SSB	–	Spawning Stock Biomass
SST	–	Sea Surface Temperature
TAC	–	Total Allowable Catch
TCM	–	Travel Cost Method

TOR	–	Terms of Reference
UNCED	–	United Nations Conference on Environment and Development
UNDP	–	United Nations Development Programme
UNEP	–	United Nations Environment Programme
UNEP-WCMC	–	United Nations Environment Programme's World Conservation Monitoring Centre
USAID	–	The United States Agency for International Development
VMS	–	Vessel Monitoring System
VPA	–	Virtual Population Analysis
WCRA	–	West Coast Recreation Area
WSP	–	Wind Stability Parameter
WTA	–	Willingness to Accept
WTP	–	Willingness to Pay
WWF	–	World Wildlife Fund

Contributors

CLAIRE WINIFRED ARMSTRONG is an associate professor at the Department of Economics and Management, the Norwegian College of Fishery Science, University of Tromsø, Norway.

BJØRN ERIK AXELSEN is a scientist at the Institute of Marine Research in Bergen, Norway.

JONATHAN BARNES is currently with the Ministry of Environment and Tourism in Namibia.

GRAÇA BAULETH-D'ALMEIDA is a scientist at the National Marine Information and Research Centre in Swakopmund, Namibia.

PER ERIK BERGH is currently working for Nordenfjeldske Development Services – a Norwegian maritime and fisheries consultancy company.

GABRIELLA BIANCHI is a fishery resources officer of the Food and Agriculture Organization of the United Nations, Rome, Italy.

DAVID BOYER was a scientist with the Ministry of Fisheries and Marine Resources in Namibia for many years but is currently a free-lance consultant, primarily assisting research groups around the world to develop acoustic assessment surveys.

SANDY DAVIES is currently working for the Southern African Development Community in Botswana.

ANNA ERASTUS is currently Director of Policy, Planning and Economics, the Ministry of Fisheries and Marine Resources, Namibia.

ABRAHAM IYAMBO is Minister for Fisheries and Marine Resources in Namibia and a member of the Namibian Parliament.

HASHALI HAMUKUAYA is Director of the Marine Living Resources Activity Centre of the BCLME in Namibia.

JOHANNA J. HEYMANS is a post doctoral fellow in the Marine Mammal Unit of the Fisheries Centre (UBC), Vancouver, Canada.

JOHANNES ANDRIES HOLTZHAUSEN is a scientist at the National Marine Information and Research Centre in Swakopmund, Namibia.

CAROLA KIRCHNER is a scientist at the National Marine Information and Research Centre in Swakopmund, Namibia.

JENS-OTTO KRAKSTAD is a former scientist at the Institute of Marine Research in Bergen, and the National Marine Information and Research Centre in Swakopmund, Namibia.

GLENN-MARIE LANGE is a research scholar with the Center for Economy, Environment and Society of the Earth Institute at Columbia University in the USA.

ARNE-CHRISTIAN LUND is a doctoral candidate in mathematical economics at the Norwegian School of Economics and Business Administration, Bergen, Norway.

ELISABETH LUNDSØR is presently with the Norwegian Institute of Marine Research, Bergen, Norway

JAMES MACGREGOR currently works for the International Institute for Environment and Development, London, UK.

ORTEN MSISKA is a senior lecturer in the Department of Natural Resources and Conservation, University of Namibia, Windhoek, Namibia.

PAUL NICHOLS is currently employed as the FAO Special Adviser to the Minister for Fisheries and Marine Resources in Namibia.

BENDIGT MARIA OLSEN, has worked for Swedish International Development Agency (SIDA), UNESCO, World Food Programme (WFP) and affiliated with the Chr. Michelsens Institute, Bergen, Norway.

MARIA LOURDES D. PALOMARES is a research associate at the Fisheries Centre, University of British Columbia, Vancouver, Canada.

DANIEL PAULY is Director of the Fisheries Centre, University of British Columbia, Vancouver, Canada.

ALISON SAKKO has been involved in various projects for the Ministry of Environment and Tourism and the Ministry of Fisheries and Marine Resources in Namibia.

LEIF K. SANDAL is a professor at the Norwegian School of Economics and Business Administration (Department of Finance and Management Science), Bergen, Norway.

MORTEN D. SKOGEN is currently a scientist at the Institute of Marine Research in Bergen, Norway.

STEIN IVAR STEINSHAMN is at present a senior research fellow at the Institute for Research in Economics and Business Administration (Centre for Fisheries Economics) in Bergen, Norway.

USSIF RASHID SUMAILA is Director of the Fisheries Economics Research Unit, Fisheries Centre, University of British Columbia, Vancouver, Canada.

NICO WILLEMSE is currently with the Ministry of Environment and Tourism in Namibia.

FREDRIK ZEYBRANDT is currently a consultant in Sweden, working on the economic analysis of wildlife tourism and aspects of techniques for environmental evaluations.